CHEMICAL ENGINEERING MATHEMATICS

CHEMICAL ENGINEERING MATHEMATICS

Alapati Suryanarayana

B.Tech, M.S. Ph.D., L.M.I.I.Ch.E., L.F.I.E., L.M.I.S.T.E.
Retired Professor and HoD
Chemical Engineering Department
N.I.T. Rourkela, Odisha (India)
Ex Principal TRRECW &
Professor and HoD
Chemical Engineering Department
RRSCET, Hyderabad

TECHSAR™

NEW DELHI

Published and Printed by

TechSar LLP
4435-36/7, Ansari Road, Daryaganj,
New Delhi – 110 002 (India)
E-mail: tech_info@techsarworld.com
Website: www.techsarworld.com

ISBN: 978-93-90620-53-1

Reprint 2026

Dedicated to

Family Members

Vanajakshi, Madhavi, Nagalakshmi,

KSR, NMR, KLR, KRK, NIK and NSS

Preface

This book covers the applications of Mathematics in Chemical Engineering and other fields. Theoretical part is summarized in the beginning of the chapters, so that the reader does not have to refer other books of Mathematics. The applications are covered step-by-step starting from easy ones to tough problems. All background and requisite materials are included for readers' convenience.

The various topics included are: Elementary problems, Formulation of equations, Ordinary differential equations, Solutions by series methods, Partial differentiation, Partial differential equations, Numerical solutions, Statistical methods, Optimization problems, Laplace transforms and other mathematical methods.

The book is useful for B.Tech (Chemical Engineering) students, pharmacy students and for practising engineers. In transport phenomena subject, sometimes the equations are not fully solved. In those cases, this book will be of immense help in seeing the solutions. In subjects like fluid mechanics, heat transfer and mass transfer, the contents of this book will be useful to understand the solutions.

Alapati Suryanarayana

Contents

1

Introduction and Elementary Applications

1.1 DIFFERENTIATION FORMULAE

1. $d\left(\frac{u}{v}\right) = \frac{v\,du - u\,dv}{v^2}$
2. $d(u^n) = n\,u^{n-1}\,du$
3. $d(u^v) = v\,u^{v-1}\,du + u^v\,(\log u)\,dv$
4. $d(a^u) = a^u \log a\,du$
5. $d(\log u) = \frac{du}{u}$
6. $d(\log a^u) = \frac{\log a^e\,du}{u}$
7. $d(\sin u) = \cos u\,du$
8. $d\,(\cos u) = -\sin u\,du$
9. $d(\tan u) = \sec^2 u\,du$
10. $d(\cot u) = -\mathrm{cosec}^2 u\,du$
11. $d(\sec u) = \tan u \sec u\,du$
12. $d(\mathrm{cosec}\,u) = -\cot u\,\mathrm{cosec}\,u\,du$
13. $d(\tan^{-1} u) = \frac{du}{(1+u^2)}$
14. $d(\sin^{-1} u) = \frac{du}{(1-u^2)^{1/2}}$
15. $d(\sinh u) = \cosh u\,du$
16. $d(\cosh u) = \sinh u\,du$

17. $d(\tanh u) = \text{sec}^2\text{h}u\,du$

18. $d(\sinh^{-1}u) = \dfrac{du}{(u^2+1)^{1/2}}$

19. $d(\cosh^{-1}u) = \dfrac{du}{(u^2-1)^{1/2}}$

20. $d(\tanh^{-1}u) = \dfrac{du}{1-u^2}$

1.2 INTEGRATION FORMULAE

1. $\int \dfrac{dx}{x} = \ln x$

2. $\int e^x\,dx = e^x$

3. $\int a^x\,dx = \dfrac{a^x}{\ln a}$

4. $\int \dfrac{dx}{(a+bx)} = \dfrac{1}{b}\ln(a+bx)$

5. $\int \dfrac{dx}{(a+bx^2)} = \dfrac{1}{(ab)^{1/2}}\tan\dfrac{-x}{\left(\frac{b}{a}\right)^{\frac{1}{2}}}$ (a and b are positive).

6. $\int \dfrac{x\,dx}{(a+bx)} = \dfrac{x}{b} - \dfrac{a}{b^2}\ln(a+bx)$

7. $\int (a+bx)^{1/2}dx = \dfrac{2}{3b}(a+bx)^{3/2}$

8. $\int x\sqrt{(a+bx)}dx = \dfrac{2(3bx-2a)(a+bx)^{3/2}}{15b^3}$

9. $\int \dfrac{dx}{(a+bx)^{1/2}} = \dfrac{2(a+bx)^{1/2}}{b}$

10. $\int \dfrac{dx}{(a+bx^2)^2} = \dfrac{x}{2a(a+bx^2)} + \dfrac{1}{2a}\int \dfrac{dx}{(a+bx^2)}$

11. $\int \dfrac{dx}{a+bx^2} = \dfrac{1}{(ab)^{1/2}}\tan^{-1} x\sqrt{\dfrac{b}{a}}$ (a and b are positive)

12. $\int \dfrac{dx}{a+bx+cx^2} = \dfrac{2}{\sqrt{-q}}\tan^{-1}\dfrac{2cx+b}{\sqrt{(1-q)}}$ ($b^2 < 4ac$)

$$= \frac{1}{\sqrt{-q}} \log\left[\frac{2cx + b - \sqrt{-q}}{2cx + \sqrt{-q}}\right] \qquad (b^2 > 4ac)$$

$$= \frac{-2}{2cx + b} (b^2 = 4ac;\ q = 4acb^2)$$

13. $\int (x^2 \pm a^2)^{1/2}\, dx = \frac{1}{2}[x(x^2 \pm a^2)^{1/2} - a^2 \log x + (x^2 + a^2)^{1/2}]$

14. $\int \frac{dx}{(x^2 + a^2)^{1/2}} = \log(x + \sqrt{x^2 \pm a^2})$

15. $\int (a^2 - x^2) dx = \frac{1}{2}\left[x(x^2 - a^2) + a^2 \sin^{-1} \frac{x}{a}\right]$

16. $\int -\frac{dx}{(a^2 - x^2)} = \sin^{-1}\left(\frac{x}{a}\right)$

1.3 OTHER RELATIONS: π = 3.143(22/7)

Area of rectangle = $a \times b$ (a, b are the sides)

Area of square = $a \times a = a^2$(a is side)

Area of a parallelogram = $ab \sin \theta$ (a, b are the sides)

Area of triangle = $\frac{1}{2} \times b \times h$

Area of rhombus = $\frac{1}{2}$ × Product of diagonals

Area of trapezium = $\frac{1}{2} \times (a \times b)h$

Area of a polygon of equal sides of length $a = \frac{1}{4}\, na^3 \cot\left(\frac{180}{n}\right)$

Area of a circle = $3.143r^2 = \frac{3.143d^2}{4}$

Area of a circular sector = $\left(\frac{3.143}{360}\right)\theta\, r^2$

Area of an ellipse = $3.143 \times a \times b$

Area of a sphere = $4 \times 3.143\ r^2$

Area of a cylinder = $2 \times 3.143 \times r \times h$

Area of a prism = $\frac{1}{3} \times ph$ (cone or pyramid) = $\frac{1}{3} \times (p \times \text{slant height})$

Area of a spherical segment = $2 \times 3.143 \times r \times h$

Volume of rectangle parallelopiped = $a \times b \times c$

Volume of sphere = $\frac{4}{3} \times (3.143 \times r^3)$

Volume of cone or pyramid = $3.143\ r^2 \times \frac{h}{2}$

1.4 TRIGONOMETRIC RELATIONS

$\sin(90 - x) = \cos x$ $\quad\sin(90 + x) = \cos x$

$\cos(90 - x) = \sin x$ $\quad\text{cosec}(90 - x) = \sec x$

$\sec(90 - x) = \text{cosec}\, x$ $\quad\tan(90 - x) = \cot x$

$\cot(90 - x) = \tan x$ $\quad\sin(180 + x) = \sin x$

$\cos(180 - x) = -\cos x$ $\quad\tan(180 - x) = \tan x$

$\cot(180 + x) = \cot x$ $\quad\sin(-x) = -\sin x$

$\cos(-x) = \cos x$ $\quad\tan(-x) = -\tan x$

$\sin(x \pm y) = \sin x \cos y \pm \cos x \sin y$

$\cos(x \pm y) = \cos x \cos y \mp \sin x \sin y$

$\sin 2x = 2 \sin x \cos x$

$\cos 2x = 2 \cos^2 x - 1$

$\sinh(-x) = -\sinh x$ $\quad\cosh(-x) = \cosh x$

$\tanh(-x) = -\tanh x$

Series:

$$e^{-x} = 1 - x + \frac{x^2}{2} - \frac{x^3}{2 \times 3} + \frac{x^4}{2 \times 3 \times 4}$$

$$\sin x = x - \frac{x^3}{3 \times 2} + \frac{x^5}{5!} - \frac{x^7}{7!}$$

$$\cos x = 1 - \frac{x^2}{2!} + \frac{x^4}{4!} - \frac{x^6}{6!}$$

$$\tan x = x + \frac{x^3}{(3!)} + \frac{2x^5}{25} + \frac{17x^7}{315} + \ldots$$

1.5 TYPES OF MATHEMATICAL EQUATIONS

The relation of variables in a system or a process are represented mathematically in terms of algebraic equations (final solutions) or differential equations (not final solutions).

Examples of algebraic equations are: Overall mass balance in a continuous distillation column at steady state and component balance at steady state.

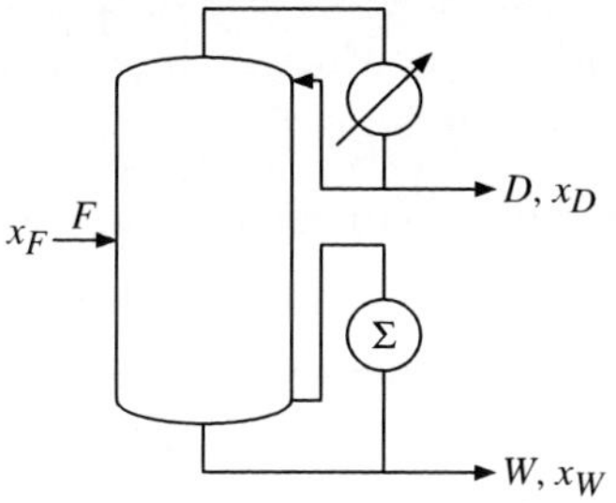

$$F = D + W$$

$$F \times x_F = D \times x_D + W \times x_W$$

Fig. 1.1

A differential equation is an equation which involves differential coefficients or differentials.

e.g., $$\frac{dy}{dx} + Py = Q; \quad \frac{d^2y}{dx^2} + n^2 y = 0$$

The differential equations are divided into two types. (1) Ordinary differential equations and (2) Partial differential equations.

Examples of PDE are:

$$\frac{d^2y}{dt^2} = c^2 \frac{d^2y}{dx^2}; \quad \frac{dT}{dt} = K\frac{d^2T}{dx^2} \quad \text{(unsteady state heat conduction equation)}$$

Ordinary differential equations are further classified as : linear and non-linear type. Examples of differential equations are:

$$q = -\frac{KAdT}{dx} \text{ and } N_{AB} = -D_{AB}\frac{dC}{dx}$$

Formulating Mathematical Equations in Chemical Engineering Problems

The various methods used are:

1. *Empirical equations*
2. *Dimensional analysis*
3. *Fundamental equations approach.*

1.5.1 Empirical Equations

In these, experimental data or, observations are plotted in such a way to give straight line form of relation ($y = mx + c$). From straight line, slope and intercept values, are obtained. From slope (m) and intercept (c) the final form of equation is obtained. This is a trial and error type method. From the form of equation obtained based on data, theoretical explanation may be possible.

(a) $y = a + bx$. Plot y values on vertical axis and plot x values on horizontal axis. a is intercept (*at* $y = 0$) and b is slope. This plot is on ordinary graph paper.

(b) $y = ax^n$. Plot log y versus log x or y versus x on log-log coordinates. $\ln y = \ln a + n \ln x$(taking logarithms on both sides). $\ln a$ is intercept *at* $\ln x = 0$ and n is slope.

(c) $y = c + ax^n$; $(y - c) = ax^n$. Plot $\ln (y - c)$ versus $\ln x$ or $(y - c)$ versus x on log-log coordinates. n is slope and $\ln a$ is intercept.

(d) $y = ae^{bx}$; Plot $\ln y$ versus x or y versus x on semilog coordinates. $\ln y = \ln a + bx \ln e$. $\ln y = \ln a + bx$. $\ln a$ is intercept and b is slope.

(e) $y = ab^x$. Plot $\ln y$ versus x or y versus x on semilog coordinates. $\ln y = \ln a + x \ln b$; $\ln a$ is intercept and $\ln b$ is slope.

(f) $y = a + \dfrac{b}{x}$. Plot y versus $\dfrac{1}{x}$. a is intercept and b is slope.

(g) $y = \dfrac{x}{a + bx}$. Plot $\dfrac{x}{y}$ versus x or $\dfrac{1}{y}$ versus $\dfrac{1}{x}$. a is slope and b is intercept.

(h) $y = a + bx + c\,x^2$. Plot $\dfrac{y - y_n}{x - x_n}$ versus x, x_n, u_n are the coordinates on smooth curve.

(i) $y = \dfrac{x}{a + bx} + c \cdot y - c = \dfrac{x}{a + bx}$. Plot $\dfrac{1}{(y - c)}$ versus $\dfrac{1}{x}$. a is slope and b is intercept. First plot y versus x. This gives c value (when $x = 0$).

Example 1.1 $y = ae^{bx}$. Below is the experimental data of distribution of acetic acid – water – petroleum ether with sodium sulphate (saturated). X = acetic acid composition in water phase, Y = acetic acid composition in extract phase, X_0 = composition of acetic acid in water phase without salt. X_S is salt weight per cent.

S. No.	Y	X	X_S	X_0	$\ln\left(\dfrac{X_0}{X}\right)$
1	0.230	3.101	23.50	4.605	0.2813
2	0.423	7.990	21.05	10.766	0.2982
3	0.716	15.881	15.44	19.526	0.2085
4	0.917	23.116	11.83	26.753	0.1461
5	1.195	27.785	9.60	31.680	0.1312
6	1.311	35.55	7.64	37.629	0.0582
7	2.055	45.199	4.55	47.020	0.0397
8	2.299	50.275	3.73	52.793	0.0489
9	2.871	52.457	2.85	53.611	0.0218
10	3.298	55.800	2.955	57.319	0.0269

11	4.273	61.665	0.80	62.151	0.00785
12	4.380	59.257	0.941	62.441	0.0523
13	3.157	59.247	2.424	62.465	0.0527
14	3.474	59.334	1.5389	64.175	0.0784
15	5.737	71.205	0.0153	72.302	0.0153

Solution: Plotting $\ln\left(\frac{X_0}{X}\right)$ on vertical axis and X_S on horizontal axis gives a straight line relationship. Plot is shown in Fig. 1.2. $\ln\left(\frac{X_0}{X}\right) = K_S X_S$; K_S for sodium sulphate is: $\frac{1.60}{100} = 0.0160$ (slope). The final equation is: $\frac{X_0}{X} = e^{0.0160\, X_S}$.

Example 1.2 $y = ae^{bx}$. Distribution data of acetic acid–water–carbon tetrachloride are given below, for sodium chloride salt at saturation. Fit an empirical equation.

S. No.	***X***	***Y***	Y_S	X_0	$\ln\left(\frac{X_0}{X}\right)$
1	19.98	1.28	13.77	33.00	0.5018
2	22.71	1.86	13.05	36.50	0.4745
3	23.30	2.77	11.33	45.00	0.6582
4	32.03	2.93	7.40	47.50	0.3941
5	33.59	2.40	6.66	42.00	0.2234
6	41.00	2.63	6.26	57.00	0.3295
7	50.64	2.88	5.14	58.50	0.1443
8	53.13	5.00	4.21	59.00	0.1048
9	55.74	6.56	4.11	63.00	0.1224
10	60.85	11.75	4.00	72.50	0.1752
11	66.92	12.52	3.79	74.00	0.1006
12	69.03	14.91	2.77	78.50	0.1286
13	72.50	15.76	2.54	81.00	0.1109
14	72.64	17.23	2.15	83.50	0.1393

Solution: Plotting $\ln\left(\frac{X_0}{X}\right)$ on vertical axis and X_S on horizontal axis gives a straight line relation shown in Fig. 1.3. The final equation is:

$$\ln\left(\frac{X_0}{X}\right) = 0.464\, X_S;\ K_S\ (\text{slope}) = \frac{4.64}{100} = 0.\ 464.$$

$$\frac{X_0}{X} = e^{0.0464\, X_S}$$

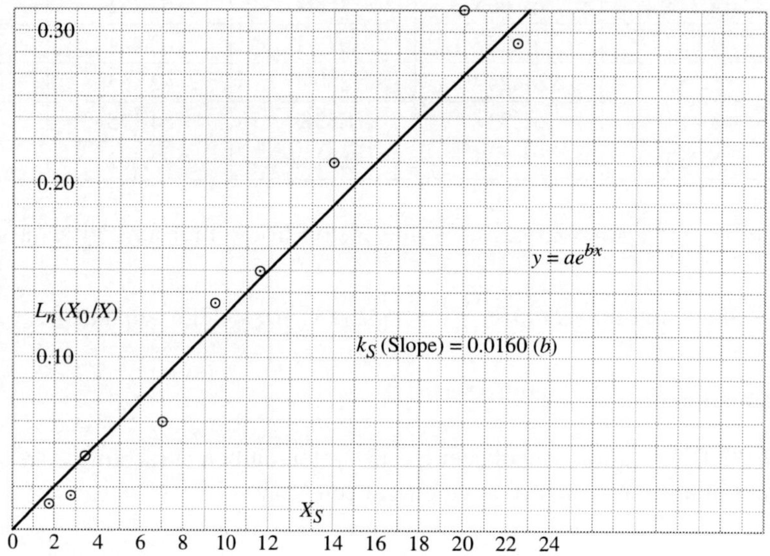
0.30
0.20
0.10
$L_n(X_0/X)$
$y = ae^{bx}$
k_S (Slope) = 0.0160 (b)
X_S
0
2
4
6
8
10
12
14
16
18
20
22
24

Fig. 1.2

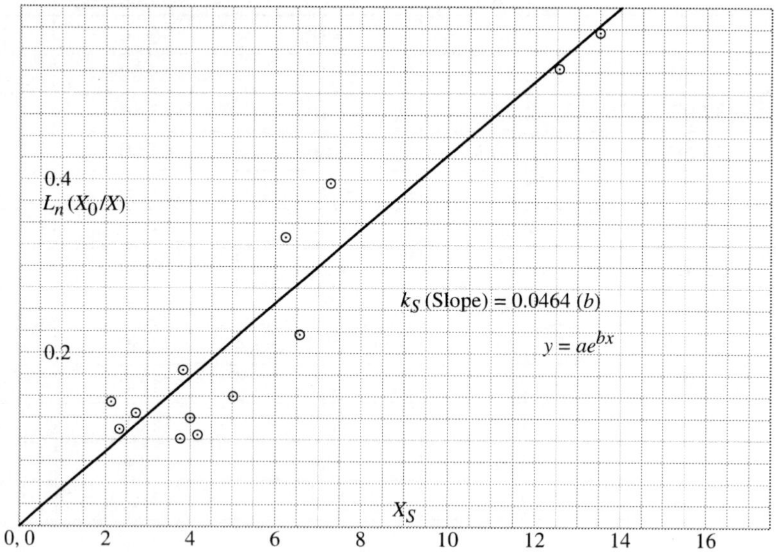
0.4
$L_n(X_0/X)$
0.2
k_S (Slope) = 0.0464 (b)
$y = ae^{bx}$
X_S
0, 0
2
4
6
8
10
12
14
16

Fig. 1.3

Example 1.3 $y = \dfrac{x}{a+bx}$. The following data are obtained in salt effect of potassium sulphate in propionic acid–water–benzene system.

S. No.	Y	X	X_0	$(x)X_S$	$\ln\left(\frac{X_0}{X}\right)(y)$	$\frac{1}{x}$	$\frac{1}{y}$
1	0.0	0.0	0.0	–	–	–	–
2	10.5	8.70	14.5	5.89	0.511	16.98	1.96
3	16.0	13.50	21.50	4.68	0.465	21.37	2.15
4	23.0	18.67	28.50	3.52	0.423	28.41	2.36
5	31.50	22.39	37.0	2.72	0.480	36.77	2.083
6	36.5	24.12	41.5	2.27	0.543	44.05	1.89
7	45.5	26.72	49.0	1.96	0.606	51.0	1.65
8	49.59	28.91	53.00	1.88	0.606	53.20	1.65
9	55.65	30.65	58.00	1.59	0.638	62.89	1.57
10	61.85	33.00	64.00	1.36	0.662	75.53	1.52

X = Composition of propionic acid in water phase with salt.

Y = Composition of propionic acid in extract phase with salt.

X_0 = Composition of propionic acid without salt; X_S = salt weight percent in the solution (both layers combined).

Solution: Taking $\ln\left(\dfrac{X_0}{X}\right)$ as y and X_S as x and plotting $\dfrac{1}{y}$ on vertical axis and $\dfrac{1}{x}$ on horizontal axis gives a straight line relation. Plotting $\dfrac{1}{\ln}\left(\dfrac{X_0}{X}\right)$ on vertical axis and $\dfrac{1}{X_S}$ on horizontal axis, gives a (Fig. 1.4) straight line relation. The equation is : $\ln\left(\dfrac{X_0}{X}\right) = \dfrac{X_S}{a+b\,X_S}$ where a is slope and b is intercept.

Here a is –0.018 and 2.68 = b. Final equation is: $\ln\left(\dfrac{X_0}{X}\right) = \dfrac{X_S}{2.68X_S - 0.018}$.

Example 1.4 $y = ax^n$. The following data are obtained in a fluidization column for separation of a binary solid mixture. Fit an equation for $\frac{h_S}{D_C}$ with $\frac{h_t}{D_C}$.

S. No.	h_t	h_S	$\frac{h_S}{D_C}$	$\frac{h_t}{D_C}$	$\ln\left(\frac{h_S}{D_C}\right)$	$\ln\left(\frac{h_t}{D_C}\right)$
1	14.0	6.9	3.63	7.36	1.289	1.996
2	18.0	7.6	4.00	9.47	1.386	2.248
3	22.0	8.6	4.47	11.5	1.497	2.442
4	26.0	9.50	5.00	13.6	1.609	2.610
5	30.0	10.4	6.47	15.78	1.699	2.259
6	34.0	12.6	6.63	17.80	1.892	2.879

Solution: Plotting $\ln\frac{h_S}{D_c}$ on vertical axis and $\ln\frac{h_t}{D_c}$ on horizontal axis gives a straight line relationship with a slope of 0.567 and intercept as –1.4. Antilog of –1.4 is 0.2466.

The equation is: $\frac{h_S}{D_c} = 0.2466\left(\frac{h_t}{D_c}\right)^{0.567}$, (Fig. 1.5).

Example 1.5 $y = mx + c$ (straight line). The following data are given, fit an equation for the given data.

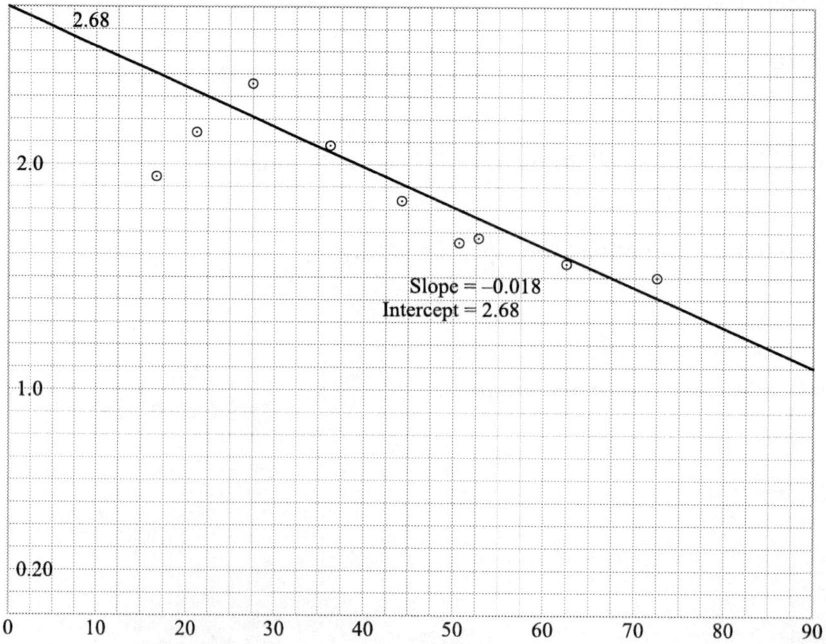

Fig. 1.4

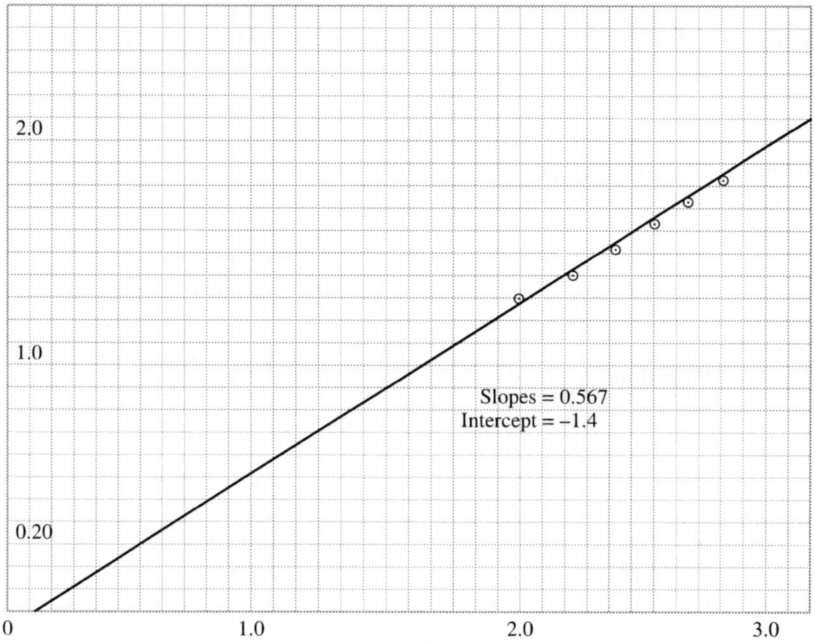

Fig. 1.5

w	50	70	100	120
p	12	15	21	25

Solution: Plotting *p* values on vertical axis and *w* values on horizontal axis gives a straight line relation (Fig. 1.6).

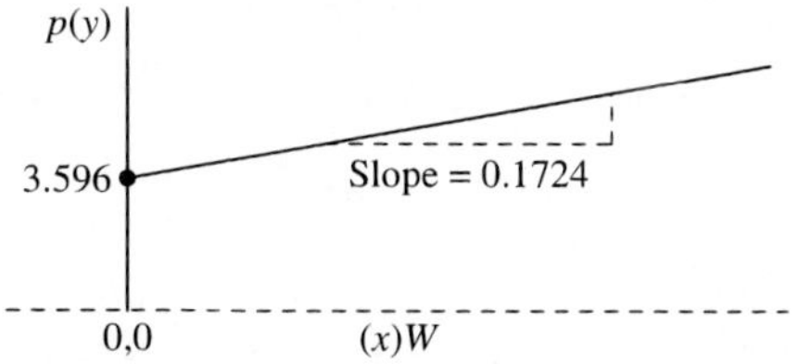

Fig. 1.6

The equation is: $P = 0.1724 \times W + 3.596$

1.5.2 Dimensional Analysis Approach

Right now two methods are available to obtain mathematical equation. (1) Rayleigh's method and (2) Buckingham's method.

Rayleigh's method. This method is used for determining the expressions for a variable which depends upon maximum three or four variables only. If the number of independent variables is more than four, then it is very difficult to find expression for the dependent variable. If *X* is a variable which depends on X_1, X_2, X_3 variables, then by Rayleigh's method, X is a function of X_1, X_2, X_3. This is written as:

$$X = KX_1^{a}X_2^{b}X_3^{c}$$

where K is a constant and a, b, c are also constants (powers). The values of a, b, c, are obtained by equating the powers of the fundamental dimensions (L, M, T) on both sides. K is evaluated by experimental observations.

Example 1.6 Find the expression for power P, developed by the pump when P depends upon the head, H, the discharge Q, and specific weight of the fluid.

Solution: Power P is a function of:

(1) Head, H; (2) Discharge Q; (3) Specific weight of fluid, W.

$$P = KH^{a}\, Q^{b}\, W^{c} \qquad \ldots(1)$$

where K is a constant (dimensionless)

The dimensions of P are: $\dfrac{ML^2}{T^3}$; Dimensions of H are: L; Dimensions of Q are: $\dfrac{L^3}{T}$; Dimensions of W are $\dfrac{M}{(L^2\, T^2)}$

Substituting dimensions in Eq. (1) gives

$$ML^2T^{-3} = KL^{a}(L^3T^{-1})^{b}(ML^{-2}T^{-2})^{c}$$

$$ML^2T^{-3} = K\, L^{a+3b-2c}\, M^{c}\, T^{-b-2c}$$

Equating the dimensions on both sides: power of M : $1 = c$

Powers of L: $2 = a + 3b - 2c$; Power of T: $-3 = -b - 2c$; Giving $a = 1$; $b = 1$; $c = 1$.

The equation is: $P = KHQW$ **Ans.**

Example 1.7 Find an expression for the drag force on a smooth sphere of diameter D moving with an uniform velocity of V in a fluid of density ρ and dynamic viscosity μ.

Solution: Drag force F is a function of : (1) Diameter, D (2) Velocity, V (3) Density, ρ and (4) Viscosity, μ.

$$F = KD^{a}\, V^{b}\, \rho^{c}\, \mu^{d}$$

where K is a dimensionless factor.

Substituting dimensions on both sides,

$$MLT^{-2} = KL^{a}(LT^{-1})^{b}(ML^{-3})^{c}(ML^{-1}T^{-1})^{d}$$

$$MLT^{-2} = KL^{(a+b-3c-d)}\, M^{(c+d)}\, T^{(-b-d)}$$

Equating the powers of *M*, *L*, *T*:

For *M*: $1 = c + d$; For *L*: $1 = a + b - 3c - d$; For *T*: $-2 = -b - d$

Here viscosity has a vital role, hence, *a*, *b*, *c* are expressed in terms of *d* (which is the power of viscosity).

$c = 1 - d$; $b = 2 - d$; $a = 1 - b + 3c + d$; $a = 2 - d$; substituting *a*, *b*, *c* values in equation of *F*,

$$F = KD^{2-d}\, V^{2-d}\, \rho^{1-d}\, \mu^{d} = KD^2\rho V^2 \left(\frac{\mu}{\rho VD}\right)^d$$

$$F = K\,\rho D^2\, V^2\, \phi\left(\frac{\mu}{DV\rho}\right)$$ **Ans.** *d* and *K* are obtained by the experiments.

Example 1.8 Pressure drop in conical packed beds. The pressure drop in the bed is a function of the bed, solid and fluid properties. The dependence is expressed as

Solution:
$$\Delta P = f\left(\rho_p,\, d_p,\, h_s,\, \mu,\, D_c,\, g_c,\, \tan\frac{\alpha}{2}, V_0\right)$$

Δp = pressure drop, ρ_p = particle density, d_p = particle diameter, h_S = height of static bed, μ = viscosity of fluid, D_C = column diameter, g_c = gravitational constant; α = angle of cone, V_0 = velocity of fluid. By equating the dimensions (*L*, *M*, *T*) on both sides,

$$f = \frac{\Delta p\, g_c\, d_p}{2\rho V_0^2\, 2h_S} = K(\text{Re})^a \left(\frac{D_c}{d_p}\right)^b \left(\frac{h_s}{D_c}\right)^c \left(\tan\frac{\alpha}{2}\right)^d$$

where *a*, *b*, *c*, *d* and *K* are determined by experiments.

Example 1.9 Onset of semifluidization velocity in conical vessels.

$$G_{mf} = f\left(d_p, \mu, h_S, D_c, \rho_S, \rho_f, \tan\frac{\alpha}{2}\right)$$

where p = particle; S = solid; μ = viscosity of fluid; f = fluid; ρ = density

Solution: Using dimensions (*L*, *M*, *T*) on both sides:

$$\frac{d_p\, G_{mf}}{\mu} = 2.32\times 10^4 \left(\frac{h_S}{D_c}\right)^{0.73} \left(\frac{D_c}{d_p}\right)^{-1.68} \left(\frac{\rho_S}{\rho_f}\right)^{1.70} \left(\tan\frac{\alpha}{2}\right)$$

For onest of fluidization:

$$\frac{d_p G_{of}}{\mu} = 1.65\times 10^6 \left(\frac{h_S}{D_c}\right)^{0.89} \left(\frac{D_c}{d_p}\right)^{-2.70} \left(\frac{\rho_S}{\rho_f}\right)^{1.65} \left(\tan\frac{\alpha}{2}\right)^{2.13}$$

Example 1.10 Fluctuation Ratio (*r*).

$$r = f(G_f, G_{of}, D_o, d_{pm}, \tan\theta)$$

where G_f = fluidization mass velocity; G_{of} = minimum fluidization velocity; D_o = inlet diameter of cone; d_{pm} = mean particle diameter; tan θ = tan of angle of cone.

Solution: $$r = 9.8 \times 10^2 \left(\frac{G_f}{G_{of}}\right)^{1.06} \left(\frac{D_o}{d_{pm}}\right)^{-1.97} \left(\frac{h_S}{D_o}\right)^{-0.25} (\tan\theta)^{-0.25}$$

for homogenous mixture.

Example 1.11 Separation of binary solid mixtures by fluidization.

The parameters involved are: (1) shape of the particles; (2) size of the particles; (3) density of the particles; (4) minimum fluidization velocity; (5) operating fluidization velocity; (6) column diameter.

$$h_S \text{ (height of small size aprticles)} = f\,(h_t, d_S, d_b, G, G_{mf}, \rho_{SS}, \rho_{sb}, D_C)$$

where h_t = total bed height;

d_S = diameter of small size particles;

d_b = diameter of big size particles;

G = operating fluidization velocity;

G_{mf} = minimum fluidization velocity;

ρ_{SS} = density of small size particles;

ρ_{sb} = density of big size particles,

D_c = column diameter; shape factors are neglected.

Solution: Using Rayleigh's method of dimensional analysis

$$h_S = (h_t^a, d_S^b, d_b^c, G_{mf}^d, G^e, \rho_{SS}^f, \rho_{sb}^g, D_c^h)$$

Putting dimensions (L, M, T) on both sides,

$$L^1 = (L^a L^b L^c L^d)\left(\frac{M}{L^2T}\right)^d \left(\frac{M}{LT^2}\right)^e \left(\frac{M}{L^3}\right)^f \left(\frac{M}{L^3}\right)^g$$

Equating the dimensions on both sides:

For L: $a + b + c + h - 2h - 2e - 3f - 3g = 1$

For M: $d + c + f + g = 0$

For T: $0 = -d - e$; $d = -e$; $f = -g$; $h = 1 - a = b - c$

$$\frac{h_S}{h_t} = K\left(\frac{d_S}{h_t}\right)^b \left(\frac{d_b}{h_t}\right)^c \left(\frac{D_c}{h_t}\right)^h \left(\frac{G_{mf}}{G}\right)^e \left(\frac{\rho_{Sb}}{\rho_{SS}}\right)^g$$

K, b, c, h, e, g values are obtained by experiments.

Final equation: $$\frac{h_S}{h_t} = 0.205\left(\frac{d_S}{h_t}\right)^{0.810} \left(\frac{d_b}{h_t}\right)^{0.346} \left(\frac{D_C}{h_t}\right)^{0.832} \left(\frac{G_{mf}}{G}\right)^{0.125} \times \left(\frac{\rho_{Sb}}{\rho_{SS}}\right)^{0.151}$$

Example 1.12 Separation of binary solid mixtures by fluidization.

$$h_S = f(h_t^a, d_S^b, d_b^c, G^d, G_{mf}^e, \rho_{SS}^f, D_c^h, \rho_{Sb}^g)$$

Solution: Applying Rayleigh's method,

$$h_S = (K h_t^a\, d_S^b\, d_b^c\, G^d\, G_{mf}^e\, \rho_{SS}^f\, \rho_{Sb}^g\, D_c^h)$$

Putting the dimensions (L, M, T) on both sides

$$L_1 = K\left(L^a L^b L^c \left(\frac{M}{TL^2}\right)^d, \left(\frac{M}{TL^2}\right)^e \left(\frac{M}{L^3}\right)^f \left(\frac{M}{L^3}\right)^g L^h\right)$$

Collecting powers of L, M, T:

$$L: 1 = a + b + h - 2d - 2e - 3f - 3g;\; M: 0 = d + e + f + g;\; T: 0 = -d - e.$$

$$\frac{h_S}{D_c} = K\left(\frac{h_t}{D_c}\right)^a \left(\frac{d_S}{D_c}\right)^b \left(\frac{d_b}{D_c}\right)^c \left(\frac{G_{mf}}{G}\right)^e \left(\frac{\rho_{Sb}}{\rho_{SS}}\right)^g$$

By experiments k, a, b, c, e, f, g are calculated.

The final equation is

$$\frac{h_S}{D_c} = 0.776\left(\frac{h_t}{D_c}\right)^{0.467} \left(\frac{d_S}{D_c}\right)^{-0.1387} \left(\frac{d_b}{D_c}\right)^{-0.2027} \left(\frac{G_{mf}}{G}\right)^{0.83} \left(\frac{\rho_{Sb}}{\rho_{SS}}\right)^{0.074}$$

Example 1.13 Pressure drop in conical packed beds.

$$P \text{ (pressure drop)} = f\left(\rho_p, d_p, h_s, \mu, D_c, g_c, \tan\frac{\alpha}{2}, V_0\right)$$

where f = is a function of;

ρ_p = density of particle;

d_p = diameter of particle;

h_S = static bed height;

μ = viscosity;

D_c = column diameter;

g_c = gravitation constant;

α = angle of cone;

V_o = velocity of fluid at the entrance of the cone.

Solution: Putting dimensions (L, M, T) on both sides, equating and grouping them gives

$$f = \frac{\rho g_c d_p}{(2\rho_p V_o^2 h_S)} = K(\text{Re})^a \left(\frac{h_S}{D_c}\right)^b \left(\frac{D_c}{d_p}\right)^c \left(\tan\frac{\alpha}{2}\right)^d$$

by experiments,

$$f = 2.33 \times 10^3 \times \text{Re}^{-0.86}\left(\frac{h_S}{D_c}\right)^{-0.60}\left(\frac{D_c}{d_p}\right)^{-1.63}\left(\tan\frac{\alpha}{2}\right)^{-2.71}$$

Buckingham's theorem. The Rayleigh's method of dimensional analysis becomes more laborious if the variables are more than the number of fundamental dimensions (L, M, T). The drawback is avoided by using Buckingham's theorem.

This states if there are n variables (independent) and dependent in a physical phenomena. If these variables contain m fundamental dimensions (L, M, T), then the variables are arranged into (n–m) dimensionless terms. Each term is a π term.

If $x_1, x_2, x_3, \ldots x_n$ are the variables involved in a physical problem, let x_1 be the dependent variable and others ($x_2, x_3, \ldots x_n$ are the variables involved (independent) on which x_1 depends,

$$x_1 = f(x_2, x_3, \ldots, x_n) \qquad \ldots(1)$$

Equation (1) can also be written as

$$f(x_1, x_2, \ldots, x_n) = 0 \qquad \ldots(2)$$

Equation (2) becomes as

$$f(\pi_1, \pi_2, \pi_3, \ldots, \pi_{n-m}) = 0 \qquad \ldots(3)$$

Each term contains $m + 1$ variables, m is the number of fundamental dimensions and are also called repeating variables. If x_2, x_3, x_4 are repeating variables and the fundamental dimensions (M, L, T) $m = 3$. Then each term is

$$\begin{aligned} \pi_1 &= x_2^{a1} x_3^{b1} x_4^{c1} \\ \pi_2 &= x_2^{a2} x_3^{b2} x_4^{c2} \\ \pi_{n-m} &= x_2^{a_{n-m}} x_3^{b_{n-m}} x_4^{c_{n-m}} \end{aligned} \qquad \ldots(4)$$

Each equation is solved by the principle of dimensional homogeneity and values of a_1, b_1, c_1, etc., are obtained. These values of π are substituted in Eq. (3). The final equation is obtained by expressing any one of the terms as a function of others as

$$\left.\begin{aligned} \pi_1 &= \phi(\pi_2, \pi_3, \pi_4, \pi_{n-m}) \\ \pi_2 &= \phi(\pi_1, \pi_3, \pi_4, \pi_{n-m}) \end{aligned}\right\} \qquad \ldots(5)$$

Method of selecting repeating variables. The number of repeating variables is equal to the number of fundamental dimensions of the problem. Repeating variable is governed by the following considerations.

1. The dependent variable should not be selected as repeating variable.
2. The repeating variables should be chosen in such a way, that one variable contains geometry property, other variable contains flow variable property and third variable contains fluid property.

Variables with geometry property are: (1) length, l; (2) diameter, d; (3) height, H; (4) angle, etc.

Variables with fluid property are: (1) visosity, μ; (2) density, ρ; (3) specific weight, w etc.

Variables with flow property are: velocity, acceleration, etc.

3. The repeating variables together must have the same number of fundamental dimensions.

4. The repeating variables selected should not form a dimensionless group.

5. No two repeating variables should have the same dimensions. In fluid mechanics problems the choice of repeating variables: (1) d, ρ; (2) 1, V, ρ, (3) 1, v, μ; (4) d, v, μ.

π Theorem

Example 1.14 The pressure difference dP in a pipe of inside diameter D and length l due to turbulent flow depends on the velocity v of the fluid in the pipe, viscosity μ of fluid and density ρ of fluid and roughness k of the pipe. Using Buckingham's π theorem, obtain an expression for dP.

Solution: Given: dP is a function of : D, l, v, μ, ρ, k

$$dP = f(D, l, v, m, \rho, k) \quad \text{or} \quad f_1(dP, D, l, v, \mu, \rho, k) = 0 \qquad ...(1)$$

Total number of variables, $n = 7$

Writing dimensions of each variable,

$dP = ML^{-1}T^{-2}$ (dimensions of pressure); $D = L$; $1 = L$; $V = LT^{-1}$; $\mu = ML^{-1}T^{-1}$; $\rho = ML^{-3}$; $k = L$

Number of fundamental dimensions, $m = 3$

Number of terms $= n - m = 7 - 3 = 4$

Equation (1) is grouped in 4 terms as:

$$f_1(\pi_1, \pi_2, \pi_3, \pi_4) = 0 \qquad ...(2)$$

Each term contains $m + 1 = 3 + 1 = 4$ variables. The four π terms are:

$$\pi_1 = D^{a1}\, v^{b1}\, \rho^{c1}\, dP; \;\; \pi_2 = D^{a2}\, v^{b2}\, \rho^{c2}\, l$$

$$\pi_3 = D^{a3}\, v^{b3}\, \rho^{c3}\, 1; \;\; \pi_4 = D^{a4}\, v^{b4}\, \rho^{c4}\, k$$

First term: $\pi_1 = D^{a1}\, v^{b1}\, \rho^{c1}\, dP$; Putting dimensions on both sides, $M^0 L^0 T^0 = L^{a1}(LT^{-1})^{b1} (ML^{-3})^{c1} (ML^{-1}T^{-2})$

Equating the powers of L, M, T on both sides:

Power of $M: 0 = c_1 + 1 : c_1 = -1$; Powers of L: $0 = a_1 + b_1 + -3\, c_1 - 1$

Powers of T: $0 = -b_1 - 2$; $b_1 = -2$; $a_1 = 0$

Substituting the values of a_1, b_1, c_1 in 1 gives: $\pi_1 = D^{C} v^{-2} \rho^{-1}\, dP = \dfrac{dP}{\rho V^2}$

Second term is: $\pi_2 = D^{a2}\, v^{b2}\, \rho^{c2}\, 1$; putting dimensions ($L, M, T$) on both sides, $M^0 L^0 T^0 = L^{a2} (LT^{-1})^{b2} (ML^{-3})^{c2} L = L^{(a_2 + b_2) - (3c_2 + 1)} M^{c2} T^{-b2}$.

Equating the powers on both sides : $0 = c_2$; $c_2 = 0$

$$0 = a_2 + b_2 - 3c_2 + 1;\ a_2 = -1;\ 0 = -b_2;\ b_3 = 0$$

$$\pi_2 = D^{-1} v^0\, p^0\, 1;\ \pi_2 = \frac{1}{D}$$

Third term is: $\pi_3 = D^{a_3}\, V^{b_3}\, \rho^{c_3}\, \mu$;

Putting dimensions on both sides, $L^0\, M^0\, T^0 = L^{a_3 + b_3 - 3c_3 - 1}\, M^{c_3 + 1}\, T^{-b_3 - 1}$;

Equating the dimensions on both sides; $0 = a_3 + b_3 - 3c_3 - 1$; $0 = c_3 + 1$; $0 = -b_3 - 1$; gives:

$$a_3 = -1;\ c_3 = 1\ \pi_3 = D^{-1}\, v^{-1}\, f^{-1}\, \rho^{-1}\, \mu = \frac{\mu}{DV\rho}$$

Fourth term is: $D^{a4}\, v^{b4x_4}\, \rho^{c4}\, K$,

Putting dimensions (L, M, T) on both sides gives:

$$M^0\, L^0\, T^0 = L^{a_4}(LT^{-1})^{b_4}\, (ML^{-3})^{c_4}\, L$$

$$0 = a_4 + b_4 + c_4 + 1;\ 0 = b_4;\ 0 = c_4;\ a_4 = -1$$

$$\pi_4 = D^{-1} V^0 \rho^0 k;\ \pi_4 = \frac{k}{D}$$

Substituting π_1, π_2, π_3, π_4 values in Eq. (2) gives

$$f_1 = \left(\frac{dP}{\rho V^2}, \frac{1}{D}, \frac{\mu}{DV\rho}, \frac{k}{D}\right) = 0 \text{ or } \frac{dP}{\rho V^2} = \phi\left(\frac{1}{D}, \frac{\mu}{DV\rho}, \frac{k}{D}\right)$$

Expression for h_f. It is observed that pressure drop dP is a linear function of $\frac{1}{D}$, so it is taken out of the function.

$$\frac{dP}{\rho V^2} = \frac{1}{D}\phi\left(\frac{\mu}{DV\rho}, \frac{k}{D}\right);\ \frac{dP}{\rho} = \frac{V_1^2}{D}\phi\left(\frac{\mu}{DV\rho}, \frac{k}{D}\right)$$

$\frac{DV\rho}{u}$ is Reynolds number, $\frac{k}{D}$ is roughness factor.

$\phi\left(\frac{\mu}{DV\rho}, \frac{k}{D}\right)$ is taken as $2f$ where f is the coefficient of friction.

$$\frac{dP}{(\rho g)} = V_1^2 \frac{4f}{(gD^2)};\ \frac{dP}{(\rho g)} = 4f\frac{LV^2}{(D2g)} \quad \textbf{Ans.}$$

Example 1.15 Using π theorem, show that the discharge Q consumed by an oil ring is given by:

$$Q = Nd^3\phi\left(\frac{\mu}{\rho^2 ND}, \frac{\sigma}{\rho^3 N^2 D^3}, \frac{w}{pN^2 d}\right)$$

where d is the internal diameter of the ring, N is rotational speed, ρ is density, μ is viscosity, σ is surface tension, and w is specific weight of the oil.

Solution: $Q = f(d, N, \rho, \mu, \sigma, w)$ or $f_1(Q, d, N, \rho, u, \sigma, w) = 0$ ...(1)

Total number of variables are : 7

Dimensions of the variables are: $Q = L^3 T^{-1}$; $d = L$; $N = T^{-1}$; $\rho = ML^{-3}$; $u = ML^{-1} T^{-1}$; $\sigma = MT^{-2}$; $w = ML^{-2} T^{-2}$

Total number of fundamental dimensions are: $m = 3$

Total number of terms = $n - m = 7 - 3 = 4$

Equation (1) becomes: $f(\pi_1, \pi_2, \pi_3, \pi_4) = 0$ ...(2)

Choosing d, N, ρ as repeating variables, the π terms are:

$$\pi_1 = d^{a1} N^{b1} \rho^{c1} Q; \pi_2 = d^{a2} N^{b2} \rho^{c2} \mu$$

$$\pi_3 = d^{a3} N^{b3} \rho^{c3} \sigma; \pi_4 = d^{a4} N^{b4} \rho^{c4} w$$

Putting dimensions (L, M, T) on both sides for ρ_1 and equating them,

$$\pi_1 = L^0 M^0 T^0 = L^{a1} (T^{-1})^{b1} (ML^{-3})^{c1} (L_3 T^{-1})$$

$$= L^{a1 + 3 - 3c1} M^{c1} T^{-b1-1}$$

$$0 = -a_1 - 3c_1 + 3; 0 = c_1; 0 = -c_1;$$

$$0 = -b_1 - 1; \text{gives: } a_1 = -3; b_1 = -1; c_1 = 0$$

$$\pi_1 = d^{-3} N^{-1} \rho^0 Q = \frac{Q}{d^3 N}$$

Second term is : $\pi_2 = d^{a2} N^{b2} \rho^{c2} \mu$; substituting the dimensions on both sides gives:

$$\pi_2 = \frac{\mu}{(\rho N d)}$$

(π_3) Third term : $M^0 L^0 T^0 = L^{a3} (T^{-1})^{b3} (ML^{-3}) (MT^{-2})^{c3} = L^{a3-3c3} M^{c3+1} T^{-2-b3}$

For M: $0 = c_3 + 1$; For L: $0 = a_3 - 3c_3$; For T : $0 = -2 - b_3$; Gives: $a_3 = -3$; $b_3 = -2$; $c_3 = -1$;

$$\pi_3 = d^{-3} N^{-2} \rho^{-1} \sigma = \frac{\sigma}{d^3 N^2 \rho}$$

Fourth term is: $\pi_4 = d^{a4} N^{b4} \rho^{c4} w$; putting dimensions ($L$, M, T) on both sides:

$$M^0 L^0 T^0 = L^{a4} (T^{-1})^{b4} (MT^{-3})^{c4} (ML^{-2}T^{-2}) = M^{c4+1} L^{a4-2-3c4} T^{-b4-2}$$

Equating the powers of L, M, T gives : $0 = c_4 + 1$; $0 = a_4 - 3c_4$; $0 = -b_4 - 2$

$$c_4 = -1; b_4 = -2; a_4 = -1$$

$$\pi_4 = d^{-2} N^{-2} p^{-1} w = \frac{w}{\rho d^2 N^2}$$

Substituting $\pi_1, \pi_2, \pi_3, \pi_4$ in $f(\pi_1, \pi_2, \pi_3, \pi_4) = 0$ gives:

$$f\left(\frac{Q}{d^3N}, \frac{\mu}{\rho N d^2}, \frac{\sigma}{d^3N^2\rho}, \frac{w}{d^2N^2\rho}\right) = 0$$

or $$\frac{Q}{d^3N} = f_1\left(\frac{\mu}{\rho N d^2}, \frac{\sigma}{d^3N^2\rho}, \frac{w}{d^2N^2\rho}\right) \textbf{ Ans.}$$

Example 1.16 On fluidization (using for separation of solids) h_S depends on h_t, d_S, d_b, G_{mf}, G, ρ_{ss}, ρ_{sb}, D_c) or it can be written as $f_1(h_s, h_t, d_s, d_b, G_{mf}, G, \rho_{ss}, \rho_{sb}, D_c) = 0$

Solution: Total number of variables: $9(n)$

Number of fundamental dimensions = $3(m)$

Number of dimensionless terms : $9 - 3 = 6$

The six terms are : $\pi_1, \pi_2, \pi_3, \pi_4, \pi_5, \pi_6$.

Hence, $f(\pi_1, \pi_2, \pi_3, \pi_4, \pi_5, \pi_6) = 0$

Each term contains $m + 1$ variables, where $m = 3$ and is also repeating variables. Choosing D_c, G_{mf}, ρ_{ss} as repeating variables, we have

$$\pi_1 = D_c^{a_1} G_{mf}^{b_1} \rho_{ss}^{c_1} h_t; \quad \pi_2 = D_c^{a_2} G_{mf}^{b_2} \rho_{ss}^{c_3} d_s$$

$$\pi_3 = D_c^{a_3} G_{mf}^{b_3} \rho_{ss}^{c_3} d_b; \quad \pi_4 = D_c^{a_4} G_{mf}^{b_4} \rho_{ss}^{c_4} G$$

$$\pi_5 = D_c^{a_5} G_{mf}^{b_5} \rho_{ss}^{c_5} \rho_{sb}; \quad \pi_6 = D_c^{a_6} G_{mf}^{b_6} \rho_{ss}^{c_6} h_s$$

First term : $D_c^{a_1} G_{mf}^{b_1} \rho_{ss}^{c_1} h_t$; putting dimensions on both sides:

$$M^0 L^0 T = L^{a_1}\left(\frac{M}{L^2T}\right)^{b_1}\left(\frac{M}{L^3}\right)^{c_1} L$$

$$= L^{(a_1 - 2b_1 - 3c_1 + 1)} M^{b_1 + c_1} T^{-b_1}$$

Equating the power of L, M, T on both sides:

Powers of M: $0 = b_1 + c_1$; Powers of L: $0 = a_1 - 2b_1 - 3c_1 + 1$; Power of T : $0 = -b_1$ giving: $c_1 = 0$, $b_1 = 0$, $a_1 = -1$; $\pi_1 = D^{-1}h_t$

$\pi_2 = D_c^{a_2} G_{mf}^{b_2} \rho_{ss}^{c_2} d_s$; giving dimensions on both sides,

$$M^0 L^0 T^0 = L^{a_2}\left(\frac{M}{L^2T}\right)^{b_2}\left(\frac{M}{L^3}\right)^{c_2} L^1$$

$$= M^{b_2 + c_2} L^{(a_2 - 2b_2 - 3c_2 + 1)} T^{-b_2}$$

Equating the powers : For M : $0 = b_2 + c_2$; For L : $0 = a_2 - 2b_2 - 3c_2 + 1$; For T : $0 = -b_2$; Giving : $b_2 = 0$, $a_2 = -1$, $c_2 = 0$

$$\pi_2 = D_c^{-1} d_s = \frac{d_s}{D_c}$$

$$\pi_3 = D_c^{-1}\, d_b = \frac{d_b}{D_c}$$

Putting dimensions (*L*, *M*, *T*) on both sides,

$$M^0 L^0 T^0 = M^{b_3+c_3}\, L^{(a_3-2b_3-3c_3+1)}\, T^{-b_3}$$

Equating the powers of *L*, *M*, *T*

$$a_3 = -1,\ b_3 = c_3 = 0;\ \text{giving: } \pi_3 = D_c^d b = \frac{d_b}{D_c}$$

$$\pi_4 = D_c^{a_4}\, G_{mf}^{b_4}\, \rho_{ss}^{c_4}\, G$$

Putting dimensions (*L*, *M*, *T*) on both sides,

$$M^0 L^0 T^0 = L^{a_4}\left(\frac{M}{L^2 T}\right)^{b_4}\left(\frac{M}{L^3}\right)^{c_4}\frac{M}{L^2 T}$$

$$= M^{b_4+c_4+1}\, L^{(a_4-2b_4-3c_4-2)} \times T^{-b_4-1}$$

$$0 = b_4 + c_4 + 1;\ 0 = -1 + c_4 + 1,\ c_4 = 0;\ a_4 - 2b_4 - 3c_4 - 2;$$

$$0 = -b_4 - 1;\ c_4 = 0,\ a_4 = 0,\ b_4 = -1$$

$$\pi_4 = G_{mf}^{-1}\, G = \frac{G}{G_{mf}}$$

$$\pi_5 = D_c^{a_5}\, G_{mf}^{b_5}\, \rho_{ss}^{c_5}\, \rho_{sb}$$

Putting dimensions on both sides,

$$M^0 L^0 T^0 = L^{a_5-2b_5-3c_5-3}\, M^{b_5+c_5+1}\, T^{-b_5}$$

Powers of M : $0 = b_5 + c_5 + 1$

Powers of L : $0 = a_5 - 2b_5 - 3c_5 - 3$

Powers of T : $0 = -b_5$ giving $b_5 = 0,\ a_5 = -1$

$$\pi_5 = \rho_{ss}^{-1}\, \rho_{sb} = \frac{\rho_{sb}}{\rho_{ss}}$$

$\pi_6 = D_c^{a_6}\, G_{mf}^{b_6}\, \rho_{ss}^{c_6}\, h_s$; putting dimensions on both sides,

$$M^0 L^0 T^0 = L^{a_6-3c_6+1}\, M^{b_6+c_6}\, T^{-b_6}$$

Equating the powers $\quad 0 = a_6 + 1 - 3c_6;\ 0 = b_6 + c_6;$

$$0 = -b_6;\ a_6 = -1,\ b_6 = 0,\ c_6 = 0$$

$$\pi_6 = D_c^{-1}\, h_s = \frac{h_s}{D_c}$$

$$f(\pi_1, \pi_2, \pi_3, \pi_4, \pi_5, \pi_6) = 0;\ f\left(\frac{h_t}{D_c}, \frac{d_s}{D_c}, \frac{d_b}{D_c}, \frac{G}{G_{mf}}, \frac{\rho_{sb}}{\rho_{ss}}, \frac{h_s}{D_c}\right)$$

or
$$\frac{h_s}{D_c} = f\left(\frac{h_t}{D_c}, \frac{d_s}{D_c}, \frac{d_b}{D_c}, \frac{G}{G_{mf}}, \frac{\rho_{sb}}{\rho_{ss}}\right)$$

or
$$\frac{h_s}{D_c} = K\left(\frac{h_t}{D_c}\right)^a \left(\frac{d_s}{D_c}\right)^b \left(\frac{d_b}{D_c}\right)^c \left(\frac{G}{G_{mf}}\right)^d \left(\frac{\rho_{sb}}{\rho_{ss}}\right)^e$$

K, a, b, c, d, e are obtained by the experiments.

If we take h_t, G_{mf}, ρ_{ss} as repeating variables, then we get a different equation as

$$\frac{h_s}{h_t} = f\left(\frac{d_s}{h_t}, \frac{d_b}{h_t}, \frac{D_c}{h_t}, \frac{G_{mf}}{G}, \frac{\rho_{sb}}{\rho_{ss}}\right)$$

or
$$\frac{h_s}{h_t} = K\left(\frac{d_s}{h_t}\right)^a \left(\frac{d_b}{h_t}\right)^b \left(\frac{D_c}{h_t}\right)^c \left(\frac{G_{mf}}{G}\right)^d \left(\frac{\rho_{sb}}{\rho_{ss}}\right)^e$$

where a, b, c, d, e are constants, obtained by the experiments.

1.5.3 Fundamental Equations

Example 1.17 *Incipient fluidization velocity in conical vessels*. Equating force of particles (mass) with derived equation.

$$\Delta P_{mf} = \frac{M(\rho_p - \rho_f)}{\rho_p A_m} = \cos\frac{\alpha}{2}\left(C_1 A \frac{R_0}{R}\right)(R - R_0)V_{0mf} + C_2 B\left(\frac{R_0}{3R^3}\right)(R^3 - R_0^3)V_{0mf}^2$$

where ΔP_{mf} = Pressure drop;

M = Mass of particles (in conical vessel),

α = Angle of cone, C_1A and

C_2B = Ergun constants;

R, R_0 = Radial distance from apex of cone to terminal and bottom points of bed, respectively.

V_{0mf} = Minimum fluidization velocity.

Example 1.18 $q = -KA\dfrac{dT}{dx}$; q = rate of conduction heat transfer, k = thermal conductivity of the materials, A = area of heat transfer, d = difference, T = temperature, dx = thickness.

Example 1.19 $q = hA\,dT$ (convection heat transfer), where q is the rate of heat flow, h = convection heat transfer coefficient, A = area, dT is temperature difference.

Example 1.20 $\tau = \mu \frac{du}{dy}$; τ = Shear stress, μ = Viscosity, $\frac{du}{dy}$ is velocity gradient.

Example 1.21 In process control for PID control the manipulated variable equation is

$$m(t) = K_c\, e(t) + T_d \frac{de}{dt} + \frac{1}{T_i} \int e\, dt$$

where $m(t)$ is a manipulated variable, K_c is proportional sensitivity, $e(t)$ is error, T_d is derivative time, t = time, T_i is integral time, d is differential.

Example 1.22 $P = h\rho g$ where P is pressure, h = height of liquid, ρ is density, g is acceleration due to gravity.

Example 1.23 Settling velocity equation (Stokes law) :

$$d_p = \sqrt{\frac{18\eta u}{(\rho_p - \rho_f)g}}$$

where d_p is particle diameter, η is viscosity, u = settling velocity, ρ_p is density of particle, ρ_f is density of fluid, g is acceleration due to gravity.

Example 1.24 $H = U + PV$ where U is internal energy, P is pressure and V is volume, and H is enthalpy.

Example 1.25 Law of conservation of mass. At steady state, input amounts = output amounts.

Example 1.26 Pressure diffusion : $N_A = \frac{4gd(P - P')}{2MRT1}$ where N_A is molar flux, $\frac{\text{moles}}{\text{unit}}$, g = acceleration due to gravity, M is molecular weight, R is gas constant, d = Pore diameter, l is pore length, P is upstream pressure, P' is downstream pressure,

Example 1.27 Gilliland equation for diffusion coefficient of gases

$$D = \frac{0.0043T^{1.5}\left(\frac{1}{M_A} + \frac{1}{M_B}\right)}{P(V_A^{1/3} + V_B^{1/3})}$$

where D is diffusion coefficient, T is absolute temperature, M_A is molecular weight of A and M_B is molecular weight of B, V_A is specific volume of A, V_B is specific volume of B.

Example 1.28 Diffusion coefficient of liquids

$$D_{AB} = 0.000929\, T^{1.5}\left(\frac{1}{M_A} + \frac{1}{M_B}\right)^{0.50}$$

where D_{AB} = diffusion coefficient of liquids, T = absolute temperature, in °K, M_A = molecular weight of A, M_B = molecular weight of B.

Example 1.29 Diffusion coefficients of liquids

$$D_{AB} = 9.4 \times 10^{-8} \frac{(\psi_B M_B)^{1/2} T}{(\mu V_A^{0.6})}$$

where V_A = molar volume of solute A; T = absolute temperature; ψ_B = a factor varying with liquid (2.5 for water); μ = viscosity of liquid.

Example 1.30 Diffusion coefficients of gases

$$\frac{PD_{AB}}{(P_{CA} P_{CB})^{1/3} (T_{CA} T_{CB})^{5/2} \left(\frac{1}{M_A} + \frac{1}{M_B}\right)^{1/2}} = \left[\frac{T}{(T_{CA} T_{CB})^{1/2}}\right]^b$$

where D_{AB} = Diffusion coefficient; P = Pressure; M_A = Molecular weight of A; M_B = Molecular weight of B; C = Critical value; T = Absolute temperature;

Example 1.31 Chapman–Enskog formula for estimating diffusion coefficients of gases

$$D_{AB} = \frac{0.00018583 T^3 \left(\frac{1}{M_A} + \frac{1}{M_B}\right)^{1/2}}{P\sigma\, AB^2\, \Omega D_{AB}}$$

where D_{AB} = diffusion coefficient; T = absolute temperature; M_A = molecular weight of A; M_B = molecular weight of B; P = pressure; σ_{AB} = collision diameter; ΩD_{AB} = function.

1.6 ELEMENTARY METHODS/PROBLEMS

(a) Gradients/Slopes

Generate expressions for benzene concentration gradients $\frac{dc}{dt}$ at $t = 4.0$. Use several differentiation methods. The data is given below.

Time	0	1	2	3	4	5	6	7
Conc. of benzene, mg/L	7.46	6.413	3.80	2.70	2.01	1.63	1.34	1.17

***Solution: Method 1*:** Using data points above and below slope is, i.e., 3, 5. Slope = $\frac{(c_5 - c_3)}{(t_5 - t_3)}$ = $\frac{(1.63 - 2.70)}{(5.0 - 3.0)}$ = –0.555 **Ans.**

***Method 2*:** Using table of values:

$$\text{Slope 1} = \frac{(c_4 - c_3)}{(t_4 - t_3)} = \frac{(2.01 - 2.70)}{(4 - 3)} = -0.69$$

$$\text{Slope 2} = \frac{(c_5 - c_4)}{(t_5 - t_4)} = \frac{(1.63 - 2.01)}{(5 - 4)} = -0.38$$

$$\text{Slope}_{\text{avg.}} = \left(\frac{(-0.69) + (-0.38)}{2}\right) = -0.535 \textbf{ Ans.}$$

***Method 3*:** This method consists of using any three data points and fitting a curve to it. The equation of the curve is a second order equation and solving with the three data points. The derivative is calculated at the desired point (here at 4). Here the equation obtained is :

$$y = ax^2 + bx + c = 0.155t^2 - 1.775\ t + 6.63 = c$$

$$\frac{dc}{dt} = 0.311t - 1.775;\ \text{at } t = 4.0;\ \frac{dc}{dt} = -0.535 \textbf{ Ans.}$$

***Method 4*:** Uses least squares. All data points are used to generate a second order polynomial equation. This equation is differentiated and evaluated at the point where the value of the derivative is required. Microsoft Excel can be used to generate regression equation. Once all coefficients are known, the equation is analytically differentiated.

$$c = 0.1626t^2 - 1.9905t + 7.318;\ \frac{dc}{dt} = 0.3252t - 1.9905$$

$$\text{at } t = 4.00;\ \frac{dc}{dt} = -0.6897 \textbf{ Ans.}$$

***Method 5*:** This method uses 5 data points to generate a five-coefficient (fourth order) model using an equation of the form, $f = A + Bx + Cx^2 + Dx^3 + Ex^4$ equation. This method is known as interpolating. A set of equations is used to evaluate numerical derivatives from the interpolating polynomials. The equations are listed below.

$$f'(x_0) = \frac{(-25f_0 + 48f_1 - 36f_2 + 16f_3 - 3f_4)}{(12h)}$$

$$f'(x_1) = \frac{(-3f_0 - 10f_1 + 18f_2 - 6f_3 + f_4)}{(12h)}$$

$$f'(x_i) = \frac{(f_{i-2} - 8f_{i-1} + 8f_{i+1} - f_{i+2})}{(12h)}$$

$$f'(x_{n-1}) = \frac{(-f_{n-4} + 6f_{n-3} - 18f_{n-2} + 10f_{n-1} + 3f_n)}{(12h)}$$

$$f'(x_n) = \frac{(3f_{n-4} - 16f_n + 36f_{n-2} - 48f_{n-1} + 25f_n)}{(12h)}$$

where $h = x_{i+1} - x_i$; f = function evaluated at i.

For example, the equation obtained from dataset from 1.0 to 5.0

(t = 1.0, 0, 2.0, 3.0, 4.0, 05) using the equation given above is:

$$f(x) = 0.0012x^4 + 0.002x^3 + 0.2616x^2 - 2.34x + 7.467$$

All these equations are evaluated for each value of x and $f(x)$. The value obtained for point 4 is 0.5448. **Ans.**

(b) Differentiation

Example 1.32 At a given instant the sides of a rectangle are 4′ and 3′ respectively. They are increasing at a rate of 1.5′/sec and 0.5′/sec. Find the rate at which the area is increasing at that instant?

Solution: See Fig. 1.7:

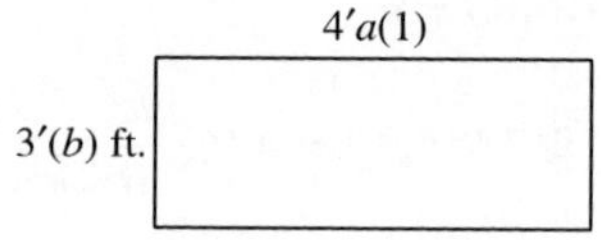

Fig. 1.7

Given: $$\frac{dl}{dt} = 1.5; \ \frac{db}{dt} = 0.5$$

$$A = 1 \times b; \ \frac{dA}{dt} = 1\frac{db}{dt} + b\frac{dl}{dt}$$

$$= 4.0 \times 0.5 + 3 \times 1.5 = 6.5 \text{ ft}^2/\text{sec}.$$

Example 1.33 The temperature at any point (x, y, z) in space is $t = 400\,xyz^2$. Find the highest temperature on the surface of a unit sphere.

Solution: $$t = 400\,xyz^2 + \lambda(x^2 + y^2 + z^2 - 1) \quad \ldots(1)$$

$$\frac{dt}{dx} = 400\,yz^2 + 2\lambda x = 0 \quad \ldots(2)$$

$$\frac{dt}{dy} = 400\,xz^2 + 2\lambda y = 0 \quad \ldots(3)$$

$$\frac{dt}{dz} = 800\,xyz + 2\lambda z = 0 \quad \ldots(4)$$

Multiplying Eq. (2) by y and Eq. (3) by x and subtracting gives

$t = x$; Eq. (3) $\times\, z$ – Eq. (4) $\times\, y$ gives : $z = y$; so $x = y = z$

$t = 400\,x^4$ **Ans.**

Example 1.34 If $z = 2xy^2 - 3x^2y$ and if x increases at the rate of 2 cm/second and it passes through the value of $x = 3$ cm, show that if y is passing through the value $y = 1$ cm, y must be decreasing at the rate of $\frac{32}{15}$ cm/sec in order that z remains constant.

Solution: $z = f(x, y) = 2\,xy^2 - 3\,x^2y$; $\frac{dz}{dt} = \frac{dz}{dx} \times \frac{dx}{dt} + \frac{dz}{dy} \times \frac{dy}{dt}$. Here $\frac{dz}{dt} = 0$, since z is constant,

$$\frac{dx}{dt} = 2; \frac{dy}{dt} = \frac{32}{15}.$$

$$0 = \frac{dz}{dx} \times 2 + \frac{dz}{dy} \times \frac{dy}{dt}; \; -2\frac{dz}{dx} = \frac{dz}{dy} \times \frac{dy}{dt}; \; \frac{dy}{dt} = -2\frac{dz}{dx} - 2\frac{dz/dx}{dz/dy}; \; \frac{dz}{dx} \text{ at constant } y = 2y^2 - 6xy;$$

$\frac{dz}{dy}$ at constant x is $4xy - 3x^2$. So $\frac{dy}{dt} = \frac{2(2y^2 - 6xy)}{(4xy - 3x^2)}$

$$= -\frac{2(2 - 6 \times 3 \times 1)}{4 \times 3 \times 1 - 3 \times 9} = \frac{-2(2 - 18)}{(12 - 27)} = \frac{32}{15} \text{ cm/sec.} \quad \textbf{Ans.}$$

Example 1.35 Find the minimum value of $x^2 + y^2 + z^2$, given $xyz = a^3$.

Solution: $F = x^2 + y^2 + z^2$; $xyz = a^3$; $F = x^2 + y^2 + z^2 + \lambda(xyz)$

$$\frac{dF}{dy} = 2y + \lambda zx = 0; \; \frac{dF}{dz} = 2z + \lambda xy = 0; \; \frac{dF}{dx} = 2x + \lambda yz = 0 \qquad \ldots(1)$$

Multiplying $\frac{dF}{dx}$ by x and $\frac{dF}{dy}$ by y and subtracting gives: $x = y$;

Similarly, multiplying $\frac{dF}{dz}$ by z and $\frac{dF}{dy}$ by y and subtracting gives $y = z$.

So $x = y = z$; $xyz = x^3 = a^3$; $x = a$, $y = a$, $z = a$. **Ans.**

Example 1.36 If the thermodynamic variables P, V, T are connected by relation $(P, V, T) = 0$ show that: $\left(\frac{dp}{dt}\right)_V \left(\frac{dT}{dV}\right)_P \left(\frac{dV}{dP}\right)_T = 1$.

Solution:

$$dF = \frac{dF}{dP}dP + \frac{dF}{dV}dV + \frac{dF}{dT}dT \qquad \ldots(1)$$

$$dT = -\frac{dT}{dP}dP + \frac{dT}{dV}dV + \frac{dT}{dT}dT \qquad \ldots(2)$$

Substituting Eq. (2) in Eq. (1) gives

$$\left(\frac{dF}{dP} + \frac{dF}{dT}\frac{dT}{dP}\right)dP + \left(\frac{dF}{dV} + \frac{dF}{dT}\frac{dT}{dV}\right)dV = 0 \qquad \ldots(3)$$

$$\frac{dT}{dP} = \frac{-\frac{dF}{dP}}{\frac{dF}{dT}}; \; \frac{dP}{dT} = \frac{-\frac{dF}{dT}}{\frac{dF}{dP}} \qquad \ldots(4)$$

$$\frac{dV}{dP} = \frac{-\frac{dF}{dP}}{\frac{dF}{dV}}; \; \frac{dT}{dV} = \frac{-\frac{dF}{dV}}{\frac{dF}{dT}}$$

$$\left(\frac{dP}{dT}\right)_V\left(\frac{dT}{dV}\right)_P\left(\frac{dV}{dP}\right)_T = -\left(\frac{\frac{dF}{dV}}{\frac{dF}{dT}}\right)\times\left(\frac{-\frac{dF}{dP}}{\frac{dF}{dV}}\right)\times\frac{-\left(\frac{dF}{dT}\right)}{\left(\frac{dF}{dP}\right)} = -1 \quad \textbf{Ans.}$$

Example 1.37 (a) Given $P = \dfrac{4flV^2}{d2g}$. The errors are: $P = 2\%$ error; $l = \pm 2\%$ error, $V = \pm 1\%$ error, $d = \pm 1\%$ error, what is the % error in f?

Solution: $df = \dfrac{gddP}{21v^2} + \dfrac{ddgp}{21v^2} - \dfrac{2V_l dVgPd}{21^2 v^4} \quad \dfrac{-V^2 dl\, gPd}{21^2 v^4}$; $f = \dfrac{pd2g}{4v^2 l}$

Dividing by *f* Throughout

$$\frac{df}{f} = \frac{dP}{P} + \frac{dd}{d} - 2\frac{dV}{V} - \frac{dl}{1} = 2 + 1 - 2 - 2 = 7\% \text{ absolute.} \quad \textbf{Ans.}$$

So, maximum error is 7%.

(b) Find error in r : $\dfrac{1}{r} = \dfrac{1}{r_1} + \dfrac{1}{r_2}$; r_1 and r_2 by 2% error.

Solution:

$$-\frac{dr}{r^2} = -\frac{dr_1}{r_1^2} - \frac{dr_2}{r_2^2};\ r = \frac{r_1 r_2}{(r_1 + r_2)}$$

$$\frac{(-dr)}{r}\frac{1}{r} = -\frac{dr_1}{r_1}\times\frac{1}{r_1}\frac{-dr_2}{r_2}\times\frac{1}{r_2}$$

$$\frac{dr}{r} = r\left[\frac{dr_1}{r_1}\frac{1}{r_1} + \frac{dr_2}{r_2}\frac{1}{r_2}\right] = \frac{r_1 r_2}{(r_1 + r_2)}\left[\frac{0.02}{r_1} + \frac{0.02}{r_2}\right]$$

$$\frac{dr}{r} = \frac{r_1 r_2}{(r_1 + r_2)}\frac{(r_1 + r_2)}{r_1 r_2} \times 0.02 = 0.02 \text{ or } 2\% \text{ error.} \quad \textbf{Ans.}$$

Example 1.38 Find the total differential coefficient of x^2y with respect to x when x and y are connected *by* : $x^2 + xy + y^2 = 1$.

Solution: $\dfrac{dy}{dx} = -\dfrac{\frac{df}{dx}}{\frac{df}{dy}}$; $\dfrac{df}{dx} = 2x + y$; $\dfrac{df}{dy} = x + 2y$; $\dfrac{dy}{dx} = -\dfrac{(2x + y)}{x + 2y}$

$$u = x^2y;\ \frac{du}{dx} = \frac{du}{dx} + \frac{du}{dy}\frac{dy}{dx} = 2xy + x^2 - \left[\frac{(2x + y)}{x + 2y}\right] = \frac{x^2y + 4xy - 2x^3}{(x + 2y)} \quad \textbf{Ans.}$$

Example 1.39 If the kinetic energy $(K) = \dfrac{wV^2}{(2g)}$, find approximately the change in the kinetic energy as w changes from 49 to 49.5 and V changes from 1600 to 1590.

Solution: $K = \dfrac{wV^2}{2g}$; differentiating, $\dfrac{dK}{K} = \dfrac{dw}{w} + \dfrac{2dV}{V}$; 2 and g are constants.

Here $dw = 0.5,\ \dfrac{dw}{w} = \dfrac{0.5}{49}$; $dv = -10$; $\dfrac{dV}{V} = -\dfrac{10}{1600}$;

$$\frac{dK}{K} = \frac{0.5}{49} + \frac{2 \times 10}{1600} = 0.0225;\ \frac{dK}{K} = 2.25\%\ \textbf{Ans.}$$

(c) Integration

1. General

If $\dfrac{dp}{dv} = -\rho v$; v is velocity, $p = K\rho^r$; $a^2 = \dfrac{rp}{\rho}$. Prove : $\dfrac{a^2}{r-1} + \dfrac{v^2}{2r} = \text{constant}$.

Solution:

$$p = K\rho^r;\ \rho^r = \frac{p}{K};\ \rho = \left(\frac{p}{K}\right)^{1/r} = p^{1/r} K^{-1/r}$$

$$\frac{dp}{dv} = -\rho v = -p^{1/r} K^{-1/r} v;\ \int p^{-1/r} dp = -\int K^{-1/r} v\, dv$$

Integrating,

$$\frac{p^{1-1/r}}{\left(1 - \dfrac{1}{r}\right)} = -K^{-1/r} \frac{v^2}{2} + \text{constant}$$

$$\frac{rp}{r-1} = -K^{-1/r} \frac{v^2}{2} + \text{constant}$$

Using

$$a^2 = \frac{rp}{\rho} = \frac{rp}{p^{1/r}} K^{1/r}$$

$$a^2 = rp^{1-\frac{1}{r}} K^{\frac{1}{r}} = K^{\frac{1}{r}} p^{\frac{(r-1)}{r}} r$$

$$\frac{a^2}{(r-1)K^{1/r}} + K^{-1/r} \frac{v^2}{2} = \text{constant}$$

Multiplying with $K^{1/r}$ throughout, $\dfrac{a^2}{(r-1)} + \dfrac{v^2}{2} = \text{constant}$. **Ans.**

2. Fluid Flow

A viscous fluid flows through a long straight tube of radius R under constant pressure gradient, p per unit length. The velocity V of the fluid at a point which is at a distance r from the axis of the tube, where $0 \le r \le R$ is given by,

$$\frac{d}{dr}\left(\frac{r\,dV}{dr}\right) = -\frac{pr}{\mu}$$

Integrating, $\frac{r\,dV}{dr} = -\frac{pr^2}{2\mu} + c_1$ where c_1 is a constant.

$$\frac{dV}{dr} = \frac{-pr}{2\mu} + \frac{c_1}{r};\ dV = -\frac{pr\,dr}{(2\mu)} + \left(\frac{c_1}{r}\right)dr$$

Integrating, $V = -\frac{pr^2}{4\mu} + c_1 \ln r + c_2$ where c_2 is a constant.

$V = 0$ at $r = R$. Using this B.C., $0 = -\frac{pR^2}{4\mu} + c_1 \ln R + c_2$

Giving $c_2 = \frac{pR^2}{4\mu} - c_1 \ln R;\ \frac{dV}{dr} = 0$ at $r = 0$

$$\frac{r\,dV}{dr} = -\frac{pr^2}{2\mu} + c_1 \text{ gives: } 0 = 0 + c_1, \text{ giving } c_1 = 0$$

$$V = -\frac{pr^2}{4\mu} + \frac{pR^2}{4\mu} = \frac{p}{4\mu}(R^2 - r^2)$$

$$V = \frac{pR^2}{4\mu}\left(1 - \left(\frac{r}{R}\right)^2\right)$$

3. Reaction Kinetics

A chemical reaction takes place according to the law $\frac{dx}{dt} = k(a - x)$ where a is the initial concentration, k is a constant and x is the amount transformed in time. If $a = 9.5$, $x = 3.2$ at two minutes, find (a) the value of k and (b) x after 5 minutes.

Solution: $\frac{dx}{(a-x)} = k\,dt$; integrating, $-\ln(a - x) = kt + \text{constant}$

at $t = 0, x = 0$; gives $-\ln a = \text{cont}$; $-\ln(a - x) = kt - \ln a$

$$\ln\frac{(a)}{a-x} = kt$$

(a) $\ln\left(\dfrac{9.5}{9.5-3.2}\right) = k \times 2;\ k = 0.1925$ **Ans.**

(b) $\ln\left(\dfrac{9.5}{9.5-x}\right) = 5 \times 0.1925 = 0.9625;$ **Ans.** $x = 8.4$ **Ans.**

4. General

A substance evaporates at a rate proportional to the exposed surface. If the spherical naphthalene ball of radius 0.5 cm has a radius of 0.4 cm; after 6 months, how long will it take?

(a) For radius to be 0.25 cm.

(b) For volume of naphthalene ball to be half of what it was originally?

Solution: $\dfrac{dw}{dt} = K4\pi r^2$ where w = weight of substance, gm, t = time, months; ρ = density of ball, gm/cm^3.

$$w = \frac{4}{3}\pi r^3 \rho\ ;\ \frac{dw}{dt} = 4\pi r^2 \times \frac{\rho dr}{dt} = K4\pi r^2$$

$$\rho \frac{dr}{dt} = k;\ dr = k\frac{dt}{\rho}$$

$$\text{Integrating, } r = \frac{kt}{at} + c;\ \text{at } t = 0,\ r = 0.5;\ \text{giving } c = 0.5$$

$$r = 0.50 + \frac{kt}{\rho}\ ;\ t = 6 \text{ months,}$$

$$r = 0.40, \text{ giving } k = -0.0167\rho$$

$r = 0.5 - 0.0167t$ **Ans.**

(a) For $r = 0.25$ cm; $0.25 = 0.5 - 0.0167t$; $t = 15$ months.

(b) $\text{Volume} = \dfrac{4\pi r_i^3}{3};\ \dfrac{4\pi r_i^3}{6} = \dfrac{4\pi r_f^3}{3}$

$$r_i^3 = 2r_f^3;\ \left(\frac{r_f}{r_i}\right)^3 = 0.50;\ \frac{r_f}{r_i} = 0.7937$$

$r_f = 0.7937 r_i$; $r = 0.50 - 0.0167t$; $t = 6.18$ months. **Ans.**

5. Reaction Kinetics: Dissolving

The rate at which a solid substance dissolves directly as the product of difference between the instantaneous concentration and the undissolved amount. 20 kilograms of solute is dumped into a tank containing 120 kg of solvent and at the end of 12 minutes, the concentration is observed to be one part in thirty. Find the amount of solute in solution at any time, t if the saturation concentration is 1 part of solute to 3 parts of solvent.

Solution: Let q be the amount of material dissolved in time t.

Undissolved substance = $20 - q$

Concentration at time $t = \frac{q}{120}$; Saturation concentration = $\frac{1}{3}$

$$\frac{dq}{dt} = k(20-q)\left(\frac{1}{3}-\frac{q}{120}\right) = \frac{k}{120}(20-q)(40-q)$$

Using separation variable method,

$$\frac{dq}{(20-q)(40-q)} = \left(\frac{k}{120}\right)dt$$

By partial fractions,

$$\frac{A}{(20-q)} + \frac{B}{(40-q)} = \frac{1}{(20-q)(40-q)}\text{; } A(40-q) + B(20-q) = 1$$

Putting $q = 20$, $A = \frac{1}{20}$; Putting $q = 40$, $B = -\frac{1}{20}$

$$-\frac{1}{20}\left(\frac{1}{20-q} - \frac{1}{40-q}\right)dq = \frac{k}{120}dt$$

Integrating,

$$-\ln(20-q) + \ln(40-q) = \frac{kt}{6} + c$$

where c is a constant term. When $t = 0$, $q = 0$ gives: $-\ln 20 + \ln 40 = c$; $c = \ln 2$

$$\ln\frac{(40-q)}{20-q} = \frac{kt}{6} + 2;\ \ln\frac{(40-q)}{2(20-Q)} = 0.05889\frac{t}{6} = 0.0098t$$

$$\frac{40-g}{2(20-q)} = e^{0.0098\,t};\ \text{or } q = \frac{40(1-e^{-0.0098t})}{(1-2e^{-0.0098t})}\ \textbf{Ans.}$$

6. Heat Conduction

A pipe of 20 cm diameter contains steam at 150°C and is provided with a covering of 5 cm thickness, for which $k = 0.0025$. Temperature of the outer surface of the covering is 40°C. Find the temperature halfway through the covering at steady state.

Solution: See Fig. 1.8:

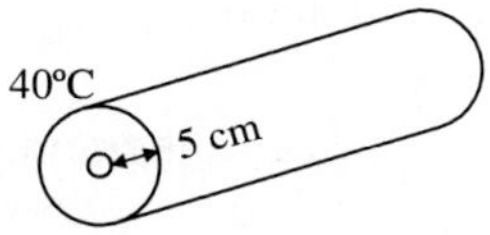

Fig. 1.8

$$q = -KA\frac{dT}{dx} = -K(2\pi x)\frac{dT}{dx}$$

$$dT = \frac{-q}{2\pi K}\ \frac{dx}{x}$$

Area of belt surface/unit length is one.

Integrating, $T = -\dfrac{q}{2\pi x}\ln x + c$ where c is a constant.

At $x = 10$, $T = 150$ºC; $$150 = \frac{-q}{2\pi K}\ln 10 + c \quad \text{...(1)}$$

At $x = 15$, $T = 40$ºC; $$40 = \frac{-q}{2\pi K}\ln 15 + c \quad \text{...(2)}$$

Subtracting the two equations:

$$110 = \frac{q}{2\pi K}\ln 1.5 \quad \text{...(3)}$$

Subtracting Eq. (4) from Eq. (1):

$$T - 150 = \frac{-q}{2\pi K}\ln 1.25 \quad \text{...(4)}$$

$$T \text{ At } x = 12.5,\ T = \frac{-q}{2\pi K}\ln 1.25 + c \quad \text{...(5)}$$

$$\frac{\text{Eq. 5}}{\text{Eq. 3}} = \frac{T - 150}{110} = \frac{-\ln 1.25}{1.5};\ T = 89.5\text{ºC}\ \textbf{Ans.}$$

7. Mass Balance

A tank contains 100 m^3 of fresh water. 2 m^3 of brine having a concentration of 1 kg of salt per m^3 of solution is run into the tank per minute and the mixture is kept uniform by mixing, runs out at the rate of 1 m^3/minute. What will be the exit brine concentration when the tank contains 150 m^3 of brine?

Solution: See Fig. 1.9:

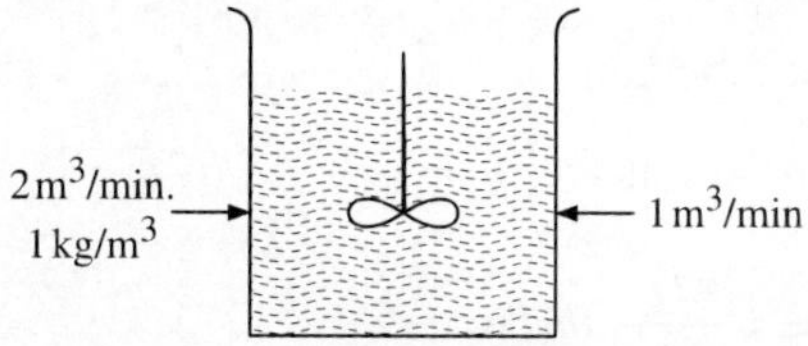

Fig. 1.9

By overall balance: Input – Output = Accumulation

$$2 - 1 = \frac{dV}{dt}; \quad \frac{dV}{dt} = 1; \text{ integrating,}$$

$$V = t + \text{constant, at } t = 0,\ V = 100$$

$$\text{So constant} = 100;\ V = 100 + t$$

By component balance : Input rate – Output rate = Rate of accumulation

$$2 \times 1 - 1 \times c = \frac{d}{dt}((100+t) \times c);\ 2 - c = 100\frac{dc}{dt} + c + \frac{t\,dc}{dt}$$

$$2 - 2c = \frac{dc}{dt}(100 + t);\ \frac{dc}{(2-2c)} = \frac{dt}{(100+t)}$$

$$\text{Integrating, } \ln\left(\frac{2}{2-2c}\right) = 2\ln\left(\frac{100+t}{100}\right)$$

$$\frac{2}{2-2c} = \left[\frac{100+t}{100}\right]^2;\ \text{at } V = 150 \text{ m}^3,\ t = 50 \text{ min}$$

$$c = 1 - \left[\frac{100}{100+50}\right]^2 = 0.555 \textbf{ Ans.}$$

8. General

Draining of a water tank (Fig. 1.10):

The volume of water leaving the tap exit is

$$dV_{\text{Exit}} = AV(t)dt = \frac{\pi d^2}{4}[2gh(t)]^{1/2}\,dt \qquad \ldots\text{(a)}$$

With $V(t) = \frac{[2gh(t)]^{1/2}}{2}$ where h is level; and t is time

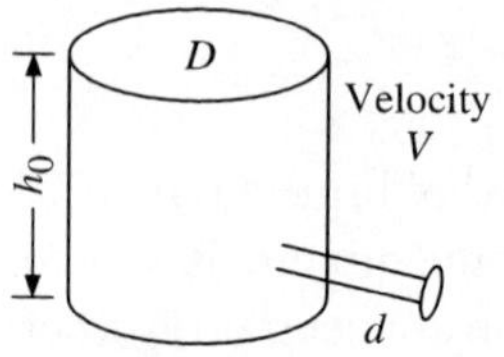

Fig. 1.10

Let the water level in tank at time t be $h(t)$.

The volume of water level loss in tank is

$$dV_{\text{tank}} = -\frac{\pi D^2}{4}dh(t) \qquad \ldots\text{(b)}$$

Equating Eq. (a) and Eq. (b):

$$\frac{\pi d^2}{4}(2gh)^{1/2}dt = \frac{\pi D^2}{4}dh(t) \qquad ...(c)$$

Rearranging $$\frac{dh(t)}{dt} = h(t)^{1/2}\frac{(d^2)}{D^2}(2g)^{1/2} \qquad ...(d)$$

For continuous draining, $dt \to 0$, then Eq. (d) becomes,

$$\frac{dh(t)}{dt} = (2g)^{1/2}\left(\frac{d^2}{D^2}\right)(h)^{1/2}$$. Initial level is h_0.

Solution of the above equation is

Upon integrating on both sides,

$$h^{-1/2}dh = (2g)^{1/2}\left(\frac{d^2}{D^2}\right)dt + c$$

$$2h^{-1/2} = (2g)^{1/2}\left(\frac{d^2}{D^2}\right)t + c \text{ ; at } t = 0,\ h = h_0 \text{ giving}$$

$$c = 2(h_0)^{1/2}$$

Final equation is

$$h(t) = \left[-\left(\frac{g}{2}\right)^{1/2}\left(\frac{d^2}{D^2}\right)t + \sqrt{h_0}\right]^2$$

Time for emptying the tank is $h(t_e) = 0$ gives

$$t_e = \frac{D^2}{d^2}\left(\frac{2h_0}{g}\right)^{1/2}$$

where D = tank diameter; d = pipe diameter; t = time; h = level; g = acceleration due to gravity.

For $D = 12'' = 1'$, $d = 1'' = \frac{1}{12}$ ft, $g = 32.2$; t_e comes as 35.89 seconds.

9. General

Water level in a tapered funnel Fig. 1.11:

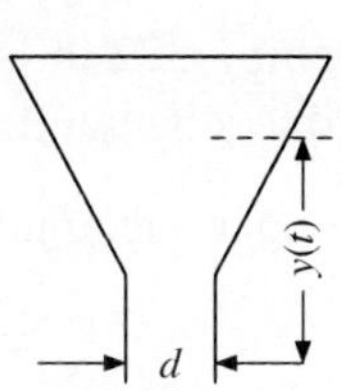

Fig. 1.11

Volume of water leaving the tapered tank during dt time

$$= A_{exit} \times V_e \times dt = dV_{exit}$$

$$V_e = ((2gy(t))^{1/2}; dV_{exit} = \frac{-\pi d^2}{4}(2gy)^{1/2}\,dt \qquad \ldots(a)$$

Total volume of water supplied by funnel during dt time is

$$dV_{funnel} = -\pi r(y)^2\,dy \qquad \ldots(b)$$

$$\frac{y(t)^{\frac{3}{2}}}{\tan^2\theta} - \frac{dy(t)}{dt} = \frac{-d^2}{4}(2g)^{\frac{1}{2}} \qquad \ldots(c)$$

By equating Eqs (a) and (b) with $r(y)$ given in Eq. (c):

$$\frac{d^2}{4}(2gy)^{1/2}\,dt = \frac{gy(t)^2\,dy}{\tan\theta}$$

For continuous operation of $y(t)$, $dt \rightarrow 0$

For $\tan\theta = \tan 45° = 1$; Eq. (c) becomes,

$$\frac{y(t)^2}{(y)^{1/2}}\frac{dy(t)}{dt} = \frac{-d^2}{4}(2g)^{1/2} = c;\ y(t) \text{ at } t = 0 \text{ is } H.$$

Integrating,

$$\frac{2}{5}y^{5/2} = \frac{-d^2}{4}(2g)^{1/2}t + c \qquad \ldots(d)$$

$$c = \frac{2H^{3/2}}{5}$$

Complete solution is

$$y(t)^{5/2} = H^{5/2} - \frac{5d^2}{8}(2g)^{1/2}t$$

For emptying $y_e = 0$;

$$t_e = \frac{8H^{3/2}}{5d^2(2g)^{1/2}} \quad \textbf{Ans.}$$

10. Paratrooper Landing

An armed paratrooper with ammunitions weighing 322 lb jumped with initial velocity from an air plane at an altitude of 10,000 feet with negligible side wind.

Assuming the air resistance, $R(t)$ the paratrooper encountered with is $R(t) = c\,V(t)^2$ where $c = 15$, determine:

(a) Equation for descending velocity of paratrooper

(b) The function of descending velocity, $V(t)$

(c) Time required to land

(d) The impact velocity upon landing.

From equation $\frac{dV(t)}{dt} + P(t)u(t) = g(t)$

(a) $$\frac{dV(t)}{dt} + \frac{c}{m}V(t) = -g; \text{ From } \Sigma F_x = 0 = -w - R(t) - F(t)$$

$$\text{with } w = mg;\ R(t) = cV(t)^2;\ F(t) = ma(t) = m\frac{dV(t)}{dt}$$

With initial condition $V(t)|_{t=0} = V_0$

Paratrooper equation is $\frac{dV(t)}{dt} + \frac{15V(t)^2}{10} = 32.2$

Rearranging: $$10\frac{dV(t)}{dt} = 32.2 - 15V(t)^2 \text{ with } V(0) = 0$$

(b) Solving:

$$V(t) = 4.634\frac{(e^{13.9t} - 1)}{(e^{13.9} + 1)}$$

(c) The descending distance is

$$x(t) = \int_0^t V(t)\,dt = \int \frac{4.634(e^{13.9t} - 1)}{e^{13.9t} + 1}dt$$

(d) Impact velocity: This is at t_g obtained as $x(t_g) = 10{,}000$ ft,

$$V(t_g) = 4.634\frac{(e^{13.9t_g} - 1)}{(e^{13.9t_g} + 1)}$$

11. Force Balance

The flow of solids from the concept of freely moving bodies of constant mass under a constant acceleration is given by

$$V = a\,dt \qquad \text{...(1)}$$

where a is acceleration and dt is time.

Thus, the following forces acting, are the buoyancy effect regulating from displacement of the fluid by the solid and the frictional resistance from the relative motion of the solid and fluid.

Assumptions are : (1) The dust particle is a non-porous and incompressible spherical particle; (2) the accelerating forces are derived from a uniform gravitational field; (3) The particles are freely moving and other particles do not affect the motion of the particle under consideration. Thus, the resulting force tending to move the particle downward is

$$F = ma = mg - wg - F_R \quad \ldots(2)$$

$$F_R^1 = \frac{C_D A \rho V^2}{2} \quad \ldots(3)$$

For a sphere the representative area A may be the projected area, $\frac{\pi D^2}{4}$ and hence

$$F_D = \frac{C_D \pi D^2 r V^2}{g} \quad \ldots(4)$$

Substituting in Eq. (2), we get the steady state maximum falling velocity:

$$\frac{m\,dV}{dt} = mg - wg - F_R^1 \quad \ldots(5)$$

At maximum velocity,

$$\frac{dV}{dt} = 0;\ \frac{dV}{dt} = \frac{(\rho_s - \rho_f)g}{\rho_f} - \frac{3C_D \rho_f V_m^2}{4D\rho_s} \quad \ldots(6)$$

$$\left(\frac{\rho_s - \rho_f}{\rho_f}\right)g = \frac{3C_D \rho_f V_m^2}{4D\rho_s} \quad \ldots(7)$$

$$V_m^2 = \frac{4(\rho_s - \rho_f)}{3\rho_f C_D}$$

$$V_m^2 = \frac{4}{3}\frac{(\rho_s - \rho_f)}{\rho_f C_D} \quad \ldots(8)$$

where C_D is a friction factor,

$$C_D = 4\frac{(\rho_s - \rho_f)gD}{3\rho_f V_m^2}$$

12. Partial Fractions

Model 1: $$\frac{5x-3}{x^2-2x-3} = \frac{A}{(x+1)} + \frac{B}{(x-3)}$$

Factors: $x^2 - 2x - 3 = (x+1)(x-3)$

Here A comes to 2 and B comes to 3.

Model 2: $$\frac{6x+7}{(x+2)^2} = \frac{A}{x+2} + \frac{B}{(x+2)^2}$$

$$A(x+2)^2 + B(x+2) = (6x+7)(x+2)$$

$A(x + 2) + B = 6x + 7$. Giving $A = 6$; $B = -5$

$$\frac{6x+7}{(x+2)^2} = \frac{6}{x+2} - \frac{5}{(x+2)^2} \quad \textbf{Ans.}$$

Model 3: $$\frac{2x^3-4x^2-x-3}{x^2-2x-3} = 2x + \frac{5x-3}{x^2-2x-3} = 2x + \frac{2}{x+1} + \frac{3}{x-3}$$

Model 4: $$\frac{2x+4}{(x^2+1)(x-1)^2} = \frac{Ax+B}{(x^2+1)} + \frac{C}{(x-1)} + \frac{D}{(x-1)^2}$$

Giving $A = 2; C = -2; B = 1; D = 1$

Model 5: $$\frac{x-1}{(x+1)^3} = \frac{A}{x+1} + \frac{B}{(x+1)^2} + \frac{C}{(x+1)^3}$$

Model 1: $$\frac{x^2+1}{(x-1)(x-2)(x-3)} = \frac{A}{x-1} + \frac{B}{x-2} + \frac{C}{x-3}$$

Here equating the coefficients gives, $A = 1, B = -5; C = 5$

$$\frac{x^2+1}{(x-1)(x-2)(x-3)} = \frac{1}{x-1} - \frac{-5}{x-3} + \frac{+5}{x-2} \quad \textbf{Ans.}$$

13. Matrices

Matrix calculus is a powerful mathematical language in connection with linear equations, transformations, systems of differential equations, etc.

$$\text{Matrix } A = [A] = \begin{bmatrix} a_{11} & a_{12} & \cdots & a_{1n} \\ a_{21} & a_{22} & \cdots & a_{2n} \\ \vdots & \vdots & \vdots & \vdots \\ a_{m1} & a_{m2} & \cdots & a_{mn} \end{bmatrix}$$

The numbers $a_{11}, a_{12}, \ldots a_{mn}$ are called elements of the matrix. Horizontal lines are rows and vertical lines are called columns.

Matrix A = Matrix B. The equality refers to matrices which have the same number of rows and same number of columns. The addition and subtraction are defined only for matrices having the same number of rows and the same number of columns.

$$[A] = \begin{bmatrix} -4 & 6 \\ 0 & 1 \end{bmatrix}; [B] = \begin{bmatrix} 5 & -1 \\ 3 & 1 \end{bmatrix}; [A] + [B] = \begin{bmatrix} 1 & 5 \\ 3 & 2 \end{bmatrix}$$

(Adding corresponding elements)

$$[A] - [B] = a_{jk} - b_{jk} = c_{jk} = \begin{bmatrix} -9 & 7 \\ -3 & 0 \end{bmatrix}$$

$$[A] + [B] = [B] + [A]; [A] + ([B] + [C]) = ([A] + [B]) + [C]$$

$$C([A] + [B]) = C[A] + C[B]; (a+b)[A] = a[A] + b[A]$$

$$a(b[A]) = (ab)[A];\ [A] = \begin{bmatrix} a_{11} & a_{12} \\ a_{21} & a_{22} \end{bmatrix};\ A^T = \begin{bmatrix} a_{11} & a_{21} \\ a_{12} & a_{22} \end{bmatrix}$$

$$(A + B)^T = A^T + B^T$$

A matrix having the same number of rows and columns is called a square matrix, and the number of rows is called its order.

1.7 MULTIPLICATION OF MATRICES

The matrices can be multiplied only when the number of columns in the first matrix is equal to the number of rows in the second matrix.

e.g.,
$$\begin{bmatrix} a_1 & b_1 & c_1 \\ a_2 & b_2 & c_2 \end{bmatrix} \times \begin{bmatrix} l_1 & l_2 \\ m_1 & m_2 \\ n_1 & n_2 \end{bmatrix} = \begin{bmatrix} a_1l_1 + b_1m_1 + c_1n_1 & a_1l_l + b_1m_2 + c_1n_2 \\ a_2l_1 + b_2m_1 + c_2n_1 & a_2l_2 + b_2m_2 + c_2n_2 \end{bmatrix}$$

$$AB \neq BA$$

Transpose of a matrix (A) is $A^T : A = \begin{bmatrix} 1 & 2 \\ 4 & 5 \\ 7 & 8 \end{bmatrix};\ A^T = \begin{bmatrix} 1 & 4 & 7 \\ 2 & 5 & 8 \end{bmatrix}$

Inverse of a matrix

$$A^{-1} = \frac{\text{adj } A}{|A|}$$

e.g., Inverse of $A = \begin{bmatrix} 1 & 1 & 3 \\ 1 & 3 & -3 \\ -2 & -4 & -4 \end{bmatrix}$

The determinant of the given matrix is

$$\begin{bmatrix} a_1 & b_1 & c_1 \\ a_2 & b_2 & c_2 \\ a_3 & b_3 & c_3 \end{bmatrix}$$

If A_1, A_2 are the co-factors of a_1, a_2 in the matrix is then $A_1 = b_2\,c_3 - c_2\,b_3$

$$A_1 = +3 \times -4 - (-3 \times -4) = -24;\ A_2 = -(1 \times -4 - 3 \times -4) = -8$$

$$A_3 = (1 \times -3 - 3 \times 3) = -12;\ B_1 = 10;\ B_2 = 2;\ B_3 = 6:$$

$$C_1 = 2;\ C_2\ 2;\ C_3 = 2$$

$$\Delta = a_1A_1 + a_2A_2 + a_3A_3 = 1 \times -24 + 1 \times -8 - 2 \times -12 = -8$$

$$\text{adj } A = \begin{bmatrix} A_1 & A_2 & A_3 \\ B_1 & B_2 & B_3 \\ C_1 & C_2 & C_3 \end{bmatrix} = \begin{bmatrix} -24 & -8 & -12 \\ 10 & 2 & 6 \\ 2 & 2 & 2 \end{bmatrix}$$

Inverse of matrix A is : $\dfrac{adj}{\Delta} = \dfrac{-1}{8}\begin{bmatrix} -24 & -8 & -12 \\ 10 & 2 & 6 \\ 2 & 2 & 2 \end{bmatrix} = \begin{bmatrix} 3 & 1 & \frac{3}{2} \\ \frac{5}{4} & \frac{1}{4} & \frac{3}{4} \\ \frac{1}{4} & \frac{1}{4} & \frac{1}{4} \end{bmatrix}$ **Ans.**

1.8 SOLUTIONS OF LINEAR SYSTEM OF EQUATIONS

1. Cramer's Rule: The system of equations are:

$$a_1x + b_1y + c_1z = d_1;\ a_2x + b_2y + c_2z = d_2;\ a_3x + b_3y + c_3z = d_3$$

$$\Delta = \begin{vmatrix} a_1 & b_1 & c_1 \\ a_2 & b_2 & c_2 \\ a_3 & b_3 & c_3 \end{vmatrix};$$

Then

$$x = \frac{\begin{vmatrix} d_1 & b_1 & c_1 \\ d_2 & b_2 & c_2 \\ d_3 & b_3 & c_3 \end{vmatrix}}{\begin{vmatrix} a_1 & b_1 & c_1 \\ a_2 & b_2 & c_2 \\ a_3 & b_3 & c_3 \end{vmatrix}};$$

$$y = \frac{\begin{vmatrix} a_1 & d_1 & c_1 \\ a_2 & d_2 & c_2 \\ a_3 & d_3 & c_3 \end{vmatrix}}{\begin{vmatrix} a_1 & b_1 & c_1 \\ a_2 & b_2 & c_2 \\ a_3 & b_3 & c_3 \end{vmatrix}};$$

$$z = \frac{\begin{vmatrix} a_1 & b_1 & d_1 \\ a_2 & b_2 & d_2 \\ a_3 & b_3 & d_3 \end{vmatrix}}{\begin{vmatrix} a_1 & b_1 & c_1 \\ a_2 & b_2 & c_2 \\ a_3 & b_3 & c_3 \end{vmatrix}}$$

2. Matrix Method:

$$A = \begin{bmatrix} a_1 & a_2 & a_3 \\ b_1 & b_2 & b_3 \\ c_1 & c_2 & c_3 \end{bmatrix}; X = \begin{bmatrix} z \\ y \\ z \end{bmatrix}; D = \begin{bmatrix} d_1 \\ d_2 \\ d_3 \end{bmatrix}$$

$$X = A^{-1}D = \frac{1}{\Delta}\begin{bmatrix} A_1 & A_2 & A_3 \\ B_1 & B_2 & B_3 \\ C_1 & C_2 & C_3 \end{bmatrix} \times \begin{bmatrix} d_1 \\ d_2 \\ d_3 \end{bmatrix}$$

Example 1.40 Solve the following equations by Cramer's rule.

$$3x + y + 2z = 3;\ 2x - 3y - z = -3;\ x + 2y + z = 4$$

Solution: The equations in matrix form are:

$$a_1x + b_1y + c_1z = d_1$$

$$a_2x + b_2y + c_2z = d_2;\ a_3x + b_3y + c_3z = d_3$$

$$\Delta = \begin{vmatrix} a_1 & b_1 & c_1 \\ a_2 & b_2 & c_2 \\ a_3 & b_3 & c_3 \end{vmatrix} = \begin{vmatrix} 3 & 1 & 2 \\ 2 & -3 & -1 \\ 1 & 2 & 1 \end{vmatrix}$$

$$= 3 \times (-3 \times 1 + 1 \times 2) - 2 \times (1 \times 1 - 2 \times 2) + 1(1 \times - 1 + 3 \times 2) = 8$$

$$(a_1A_1 + a_2A_2 + a_3A_3)$$

$$x = \frac{1}{\Delta}|A| = \frac{1}{8}\begin{vmatrix} d_1 & b_1 & c_1 \\ d_2 & b_2 & c_2 \\ d_3 & b_3 & c_3 \end{vmatrix}$$

$$= -\begin{vmatrix} 3 & 1 & 2 \\ -3 & -3 & -1 \\ 4 & 2 & 1 \end{vmatrix}$$

$$= -\frac{1}{8}[3(-3 + 2) + 3(1 - 4) + 4(-1 + 6)] = 1$$

Similarly,

$$y = \frac{1}{\Delta}\begin{vmatrix} a_1 & d_1 & c_1 \\ a_2 & d_2 & c_2 \\ a_3 & d_3 & c_3 \end{vmatrix} = \frac{1}{8}\begin{vmatrix} 3 & 3 & 2 \\ 2 & -3 & -1 \\ 1 & 4 & 1 \end{vmatrix} = 2$$

$$z = \frac{1}{\Delta}\begin{vmatrix} a_1 & b_1 & d_1 \\ a_2 & b_2 & d_2 \\ a_3 & b_3 & d_3 \end{vmatrix} = \frac{1}{8}\begin{vmatrix} 3 & 1 & 3 \\ 2 & -3 & -3 \\ 1 & 2 & 4 \end{vmatrix} = -1$$

$= -1;\ x = 1;\ y = 2;\ z = -1$ **Ans.**

Matrix Method:

Example 1.41 Solve the following equations by matrix method:

$$3x + y + 2z = 3;\ 2x - 3y - z = -3;\ x + 2y + z = 4$$

Solution:

$$\Delta = \begin{vmatrix} a_1 & b_1 & c_1 \\ a_2 & b_2 & c_2 \\ a_3 & b_3 & c_3 \end{vmatrix} = \begin{vmatrix} 3 & 1 & 2 \\ 2 & -3 & -1 \\ 1 & 2 & 1 \end{vmatrix}$$

Here, $A_1(\text{cofactor of } a_1) = -1\begin{bmatrix} -3 & -1 \\ 2 & 1 \end{bmatrix} = -1;\ A_2 = -3;\ A_3 = 5$

$$B_1 = -3;\ B_2 = 1;\ B_3 = 7;\ C_1 = 7;\ C_2 = -5;\ C_3 = -11$$

Also, $\Delta = a_1A_1 + a_2A_2 + a_3A_3 = 8$

$$\begin{bmatrix} x \\ y \\ z \end{bmatrix} = \frac{1}{\Delta}\begin{bmatrix} A_1 & A_2 & A_3 \\ B_1 & B_2 & B_3 \\ C_1 & C_2 & C_3 \end{bmatrix} \times \begin{bmatrix} d_1 \\ d_2 \\ d_3 \end{bmatrix}$$

$$= \frac{1}{8}\begin{bmatrix} -1 & 3 & 5 \\ -3 & 1 & 7 \\ 7 & -5 & 11 \end{bmatrix} \times \begin{bmatrix} 3 \\ -3 \\ 4 \end{bmatrix}$$

$$= \frac{1}{8}\begin{bmatrix} -3 - 9 + 20 \\ -9 - 3 + 28 \\ 21 + 15 - 44 \end{bmatrix} = \begin{bmatrix} 1 \\ 2 \\ -1 \end{bmatrix}$$

Hence, $x = 1$; $y = 2;\ z = -1$ **Ans.**

Equivalent Matrix:

Two matrices A and B are equivalent if one can be obtained from the other by a sequence of elementary transformations.

Gauss – Jordan Method of Finding Matrix Inverse

$A^{-1} = R_i R_{i-1} \ldots R_2R_1I$; where $R_1, R_2 \ldots$ are unit matrices.

Example 1.42 Find A^{-1} of the matrix A given as: $\begin{bmatrix} 1 & 1 & 3 \\ 1 & 3 & -3 \\ -2 & -4 & -4 \end{bmatrix}$

Solution: $\left[\begin{array}{ccc|ccc} 1 & 1 & 3 & 1 & 0 & 0 \\ 1 & 3 & -3 & 0 & 1 & 0 \\ -2 & -4 & -4 & 0 & 0 & 1 \end{array}\right]$

Writing the same matrix with unit matrix side by side (same order). Operating $R_2 - R_1$ and $R_3 + 2R_1$

$$\begin{bmatrix} 1 & 1 & 3 & \vdots & 1 & 0 & 0 \\ 0 & 2 & -6 & \vdots & -1 & 1 & 0 \\ 0 & -2 & 2 & \vdots & 2 & 0 & 1 \end{bmatrix}$$

Operate $\frac{R_2}{2}$ and $\frac{R_3}{2}$

$$\begin{bmatrix} 1 & 1 & 3 & \vdots & 1 & 0 & 0 \\ 0 & 1 & -3 & \vdots & -1/2 & 1/2 & 0 \\ 0 & -1 & 1 & \vdots & 1 & 0 & 1/2 \end{bmatrix}$$

Operate $R_1 - R_2$ and $R_3 + R_2$

$$\begin{bmatrix} 1 & 0 & 6 & \vdots & 3/2 & -1/2 & 0 \\ 0 & 1 & -3 & \vdots & -1/2 & 1/2 & 0 \\ 0 & 0 & -2 & \vdots & 1/2 & 1/2 & 1/2 \end{bmatrix}$$

Operate $R_1 + 3R_3$; $R_2 - \frac{3}{2}R_3$ and $-\frac{1}{2}R_3$.

$$\begin{bmatrix} 1 & 9 & 0 & \vdots & 3 & 1 & 3/2 \\ 0 & 1 & 0 & \vdots & -3/4 & 1/4 & -3/4 \\ 0 & 0 & 1 & \vdots & -1/4 & -1/4 & -1/4 \end{bmatrix}$$

The inverse of the matrix A is :

$$\begin{bmatrix} 3 & 1 & 3/2 \\ -5/4 & 1/4 & -3/4 \\ -1/4 & -1/4 & -1/4 \end{bmatrix}$$

Characteristic Equation

$$|A - \lambda I| = 0 = \begin{bmatrix} a_{11}-\lambda & a_{12} & a_{13} & \cdots & a_{1n} \\ a_{21} & a_{22}-\lambda & a_{23} & \cdots & a_{2n} \\ \vdots & \vdots & \vdots & \vdots & \vdots \\ a_{n1} & a_{n2} & a_{n3} & \cdots & a_{nn}-\lambda \end{bmatrix} = 0$$

is called the characteristic equation of A. The roots of the characteristic equation are called the characteristic roots or latent roots or eigenvalues of the matrix A.

Example 1.43 Find the eigenvalues and eigenvectors of the matrix.

$$\begin{bmatrix} 1 & 1 & 3 \\ 1 & 5 & 1 \\ 3 & 1 & 1 \end{bmatrix}$$

Solution: The characteristic equation is: $|A - \lambda I| = 0$

$$\begin{vmatrix} 1-\lambda & 1 & 3 \\ 1 & 5-\lambda & 1 \\ 3 & 1 & 1-\lambda \end{vmatrix}, \text{ i.e., } \lambda^3 - 7\lambda^2 + 36 = 0$$

The roots are : $\lambda = -2, 3, 6$.

The eigenvalues (λ) are: –2, 3, 6.

If x, y, z are components of the eigenvector, corresponding to eigenvalues λ, we have

$$|A - \lambda I| x = \begin{bmatrix} 1-\lambda & 1 & 3 \\ 1 & 5-\lambda & 1 \\ 3 & 1 & 1-\lambda \end{bmatrix} \begin{bmatrix} x \\ y \\ z \end{bmatrix} = 0$$

Putting, $\lambda = -2$ gives

$$\begin{bmatrix} 3 & 1 & 3 \\ 1 & 7 & 1 \\ 3 & 1 & 3 \end{bmatrix} \begin{bmatrix} x \\ y \\ z \end{bmatrix} = 0$$

We have $3x + y + 3z = 0$; $7y + z = 0$; $3x + y + 3z = 0$

First and third equation are same. From Eqs. (1) and (2), we get

$$\frac{x}{-20} = \frac{y}{0} = \frac{z}{20} \text{ or } \frac{x}{-1} = \frac{y}{0} = \frac{z}{1}$$

Hence eigenvector is –1, 0, 1. The other eigenvectors corresponding to $\lambda = 3$ and $\lambda = 6$ are : (1, –1, 1) and (1, 2, 1).

Example 1.44 The eigenvalues of a sequence matrix A are the roots of the corresponding characteristic equation.

$$= 5y_1 + 4y_2;\ \dot{y}_2 = y_1 + 2y_2$$

$$\dot{y} = Ay;\ \dot{y} = \begin{bmatrix} \dot{y}_1 \\ \dot{y}_2 \end{bmatrix};\ A = \begin{bmatrix} 5 & 4 \\ 1 & 2 \end{bmatrix}$$

$$\Delta(\lambda) = \begin{bmatrix} (5-\lambda) & 4 \\ 1 & 2-\lambda \end{bmatrix};$$

Solution: Expanding gives,

$$(5 - \lambda)(2 - \lambda) - 1 \times 4 = \lambda^2 - 7\lambda + 6 = (\lambda - 6)(\lambda - 1)$$

Roots are: $\lambda = 6$: $\lambda = 1$ (by factoring)

For $\lambda = 6$; the equations at $-x_1 + 4x_2 = 0$; $x_1 = 4x_2$; $x_1 = \begin{bmatrix} 1 \\ 4 \end{bmatrix}$ eigenvector

For $\lambda = 1; x_1 + x_2 = 0; x_1 = -x_2; x_1 = \begin{bmatrix} 1 \\ -1 \end{bmatrix}$ eigenvector

Symmetric matrix: A real matrix is said to be symmetric if it is equal to its transpose, i.e., $A^T = A$; that is $a_{ij} = a_{ji}$.

Skew symmetric matrix: A real square matrix $A = a_{ji}$ is skew symmetric if $A^T = -A$; $a_{ij} = -a_{ji}$.

Triangle matrix: A square matrix $A = a_{ji}$ whose elements below the principal diagonal are all zero, that is $a_{jk} = 0$ for all $j = k$ is called a triangle matrix.

e.g., $$\begin{bmatrix} 3 & 0 & 0 \\ 0 & 2 & 0 \\ 0 & 0 & 5 \end{bmatrix} \text{ and } \begin{bmatrix} 5 & 1 & 2 \\ 0 & 3 & 6 \\ 0 & 0 & 7 \end{bmatrix}$$

Diagonal matrix: A square matrix $A = a_{jk}$ whose elements above and below the principal diagonal are all zero, that is $a_{jk} = 0$ for all $j \neq k$ is called a diagonal matrix.

e.g., $$\begin{bmatrix} 3 & 0 & 0 \\ 0 & 2 & 0 \\ 0 & 0 & 1 \end{bmatrix} \text{ and } \begin{bmatrix} 4 & 0 & 0 \\ 0 & 5 & 0 \\ 0 & 0 & 8 \end{bmatrix}$$

Unit matrix: A matrix whose diagonal elements are all unity and the other elements are zeroes is called a unit matrix and is denoted by I. For example,

$$I = \begin{bmatrix} 1 & 0 & 0 \\ 0 & 1 & 0 \\ 9 & 0 & 1 \end{bmatrix} \text{ or } I = \begin{bmatrix} 1 & 0 \\ 0 & 1 \end{bmatrix}$$

Zero matrix: A matrix whose elements are all zeroes is called a zero matrix.

Trace of a matrix: The sum of the diagonal elements of a square matrix is called the trace of a matrix.

Null matrix: If all the elements of a matrix are zeroes then it is a null matrix.

1.9 APPROXIMATION AND CONCEPT OF ERROR AND ERROR ANALYSIS

Approximations and Round of Errors

All the numerical methods involve an approximation. The availability of analytical methods allows us to compute the error exactly. For many applied engineering problems, we do not have analytical solutions. Therefore, we cannot compute exactly the errors involved with our numerical methods. We have for approximations or estimates of the errors.

When taking examinations or doing homework problems, you are penalized, not rewarded for your errors. In professional practice errors can be costly and sometimes catastropic. If a structure or device fails, lives can be lost. Although perfection is a desirable goal, it is rarely achieved.

This chapter covers basic topics related to identification, quantification and minimisation of the errors. One of the two major forms of numerical error is round-off error. Round of error is due to the fact that computers can represent only quantities with a finite number of digits. Truncation error is due to the discrepancy introduced by the fact that numerical methods may employ approximations to represent exact mathematical operations and quantities. Errors include blunders, formulation or model errors and uncertain errors.

Significant Figures: The concept of digit or significant figure has been developed to designate the reliability of a numerical value. The significant digits of a number are those that can be used with confidence. They correspond to the number of certain digits plus one estimated digit.

The concept of significant figures has two important implications for our study of numerical methods.

1. Numerical methods give approximate results. To specify how confident we are in our approximate result is in terms of significant figures.
2. Quantities like: π, e, $\sqrt{7}$ represent specific quantities, they cannot be expressed by a limited number of digits.

   For example: $\pi = 3.141592653...$

   Since computers retain only a finite number of significant figures, the omission of the remaining significant figures is called round-off error. The concept of significant figures will have relevance to our definition of accuracy and precision.

Accuracy and Precision: Accuracy refers to how closely a computed or measured value agrees with the true value. Precision refers to how closely individual computed or measured values agree with each other. Inaccuracy is systematic deviation from the truth. Imprecision on the hand is the magnitude of the scatter. The collective term error is used to represent both in the accuracy and the imprecision.

Error Definitions: These include truncation errors. These result when approximations are used to represent exact mathematical procedures and round-off errors. These result when numbers having limited significant figures are used to represent exact numbers.

True value = Approximate value + error

Where E_t is used to designate the exact value of the error.

The subscript t is included to designate that this is the true error.

True fractional relative error = True error/True value

True error = True value – Approximate value

$$\in_t = \text{(True percent relative error)} = \frac{\text{True error}}{\text{True value}} \times 100$$

Calculations of Errors

Problem statement: Suppose we have measured the length of the plot as 9995 cm and the true value is 10,000 cm. Calculate the true error and the percent relative error for the plot.

***Solution*:** The error for measuring the length of the plot: $E_t = 10{,}000 - 9995 = 5$ cm

The percent relative error = $\in_t = (5/10{,}000) \times 100 = 0.05\%$

For numerical methods, the true value will be known only when we deal with functions that can be solved analytically. In real-world applications, we will obviously do not know the true answer a priori. For these situations, an alternative is to normalise the error using the best available estimate of the true value, that is to the approximation itself.

$$\in_a = \frac{\text{Approximate error}}{\text{Current approximate value}} \times 100\%$$

where the subscript *a* refers to the error normalized to an approximate value.

Some numerical methods use an iterative approach to compute answers. For such cases, the error is often estimated as the difference between previous approximate value and current approximate value. The percent relative error is determined according to:

$$\in_a = \frac{(\text{Current approximate value} - \text{previous approximate value}) \times 100}{\text{Current approximate value}}$$

When performing computations, we may not be concerned with the sign of the error. We are interested in whether the percent absolute value is lower than a specified percent value, of tolerance ($\in_s$). For such cases the computation is repeated until,

$$|\in_a| < \in_s$$

It is convenient to relate these errors to the number of significant figures in the approximation.

$$\in_s = (0.5 \times 10^{2-n})\%$$

Error Estimate for Iterative Methods

Problem statement: $e^x = 1 + x + x^2/2! + x^3/3! + \quad .... + x^n/n!$

Add terms until the absolute value of the approximate error estimate $\in_a$ falls below a prescribed error.

Criterion $\in_s$ conforming to three significant figures.

***Solution*:** $\in_s = (0.5 \times 10^{2-3})\% = 0.05\%$

Terms are added to the series until a falls below this level. The first estimate is : 1

The second estimate is: $1 + x$

The third estimate is: $1 + x + x^2/2!$ and so on.

For $x = 0.4$; $e^{0.4} = 1$ (first estimate). True value is: 1.491825

True percent relative error ($\in_t$) = 1.491825 – 1.4/1.491825 × 100 = 6.16%

Approximate estimate of the error ($\in_a$) = 1.4 – 1/1.4 × 100% = 28.6%

Since $\in_a$ is not less than the required value of $\in_s$, so we continue the computation by adding another term, $x^2/2!$ and repeating the error calculations. This process is continued until $\in_a < \in_s$.

Terms	**Result**	$\in_t$ (%)	$\in_a$ (%)
1	1	32.9	----
2	1.4	6.16	28.6
3	1.48	0.793	8
4	1.49066667	0.08433	0.7156
5	1.491733334	0.006167	0.0715
6	1.4918184	0.0004692	0.005704

So after six terms are included, the approximate error ($\in_a$) falls below $\in_s$ = 0.05% and the calculation is terminated.

Flow Diagram for the Error $\in_a$ (EA)

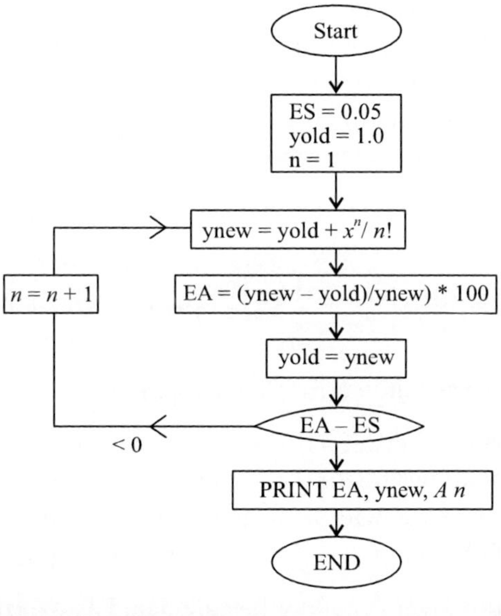

There are many computer programs available nowadays for calculations on computer. They are: 1: Fortran-90, 2: VBA/Excel, 3: MAT LAB, 4: C++

Round-off Errors: Round-off errors originate from the fact that computer retain only a fixed number of significant figures during a calculation. Numbers such as π, e or $\sqrt{7}$ cannot be expressed by a fixed number of significant figures. So they cannot be exactly represented by computers. Computers use a base 2 representation, so they cannot correctly represent base 10 numbers. This discrepancy introduced by omission of significant figures is called as round-off error.

Fractional quantities are represented in computers using floating point form. The number is expressed as : $m.b^e$ where m is mantissa, b is the base of the number of system and e is exponent. For example, 165.92 is 0.16592×10^3 in a floating point base 10 system.

Computer Round-off Errors: Floating point position representation allows both fractions and very large numbers to be expressed, on the computer. It has some disadvantages. Floating point numbers take up more room and take larger or longer to process than integer numbers. Their use introduces a source of error because mantissa holds a finite number of significant figures. That is how a round-off error comes. They are several aspects of floating point representation that have significance regarding computer round of errors.

1. There is a limited range of quantities that may be represented. For integer case, there are large positive and negative numbers that cannot be represented. Attempts to employ numbers outside the acceptable range will result in what is called an overflow error. In addition to large quantities, the foating point representation has added limitation that very small numbers cannot be represented. This is underflow "hole", between zero and first positive number.
2. There are only finite number of quantities that can be represented within the range. The degree of precision is limited. Irrational numbers cannot be represented, exactly. Rational numbers do not exactly match one of the values in the set also cannot be represented precisely. The errors introduced by approximating both these cases are referred as quntising errors. The actual approximation is done in either of two ways. 1. chopping, 2. rounding. For example: π value is = 3.14159265358 is to be stored on a base 10 number system carrying ten significant figures. One method of approximation is to omit or chop off the eleventh and higher terms as $\pi = 3.141592653$ having an associated error $(E_t) = 0.00000000058$.

   This method of retaining only the significant figures was as truncation in computers. It is actually chopping instead of truncation.

   If we round off tenth digit, it gives = 3.141592654. Such rounding reducing the error to $E_t = -0.00000000042...$

   So rounding yields a lower absolute error than chopping.
3. Machine position: The interval between numbers increases as the numbers grow in magnitude. It is this characteristic that allows floating point representation to preserve significant digits. It also means that quantizing errors will be proportional to the magnitude of the number being represented. For cases of chopping: $|\Delta x|/|x| \le \in$ For rounding: $|\Delta x|/|x| \le \in/2$

   where $\in$ is machine epsilon, which cannot be calculated as $\in = b^{1-t}$ where b is the number base and t is number of significant digits in the mantissa. The above quations for chopping and rounding are error bounds.

The magnitude dependence of quantizing errors has a number of practical applications in numerical methods. Most of these relate to the commonly employed operation of testing whether two numbers are equal. This occurs when testing convergence of quantities as well as the stopping mechanism for iterative processes. For these cases, it should be clear that rather than test whether the quantities are equal, it is advisable to test whether their difference is less than an acceptable small tolerance. It should be also evident that normalized rather than absolute difference should be compared, particularly when dealing with numbers of large magnitude.

Commercial computers are larger words and allow numbers to be expressed with more than adequate precision. Computers that use IEEE format allow 24 bits to be sued for the mantissa, which translates into about seven significant base 10 digits of precision with a range of 10^{-38} to 10^{39}. The most common extended precision is double precision, in which the number of words used to store floating point numbers is doubled. It provides about 15 to 16 decimal digits of precision and a range of nearly 10^{-308} to 10^{308}.

1.10 ARITHMETIC MANIPULATIONS

Common Arithmetic Operations: Normalized base 10 numbers are employed to illustrate the effect of round off erosion, simple addition, substraction, multiplication and division.

Addition: Suppose we want to add $0.1653 \times 10^1 + 0.5391 \times 10^{-1}$. The decimal of the mantissa of the second number is shifted to the left a number of places equal to the differences of exponents (1– (–1) = 2).0.5391 × 1 $0.5391 \times 10^{-1} = 0.005391 \times 10^1$. Now adding the two numbers:

$$\begin{array}{l} 0.1653 \times 10^{+1} \\ \underline{0.005391 \times 10^1} \\ 0.170691 \times 10^1 \end{array}$$

and if the result is chopped to 0.1706×10^1, the last two digits of second number are lost from the computation.

Adding a large and a small number: We add a small number 0.0010 to a large number 5000.

$$\begin{array}{ll} 5000 & \rightarrow 0.5000 \times 10^4 \\ 0.0010 & \rightarrow \underline{0.0000001 \times 10^4} \\ & \quad 0.500000 \times 10^4 \end{array}$$

which is chopped to 0.5000×10^4. It shows that the addition is not done.

This type of error occurs in the computer calculation of an infinite series. The initial terms in the series are relatively large in comparison with the later terms.

One way to avoid this error is to sum the series in the reverse order. In this way, each new term is comparable magnitude to the accumulated sum.

Subtraction: Eg: We subtract say 32.85 from 26.52.

$$\begin{array}{ll} 32.85 & \rightarrow 0.3285 \times 10^2 \\ 26.52 & \rightarrow \underline{0.2652 \times 10^2} \\ & \quad 0.0633 \times 10^2 \end{array}$$

To normalize, we shift the decimal one place to the right to give $0.6330 \times 10^1 = 6.330$

A zero added to the end of the mantissa is not significant but is added to fill the empty space created by the shift.

When the numbers are very close as:

$$\begin{array}{r} 0.8984 \times 10^3 \\ -\,0.8983 \times 10^3 \\ \hline 0.0001 \times 10^3 \end{array}$$

which is converted to $0.100 \times 10^0 = 0.1000$. Three non-significant zeroes are added. This introduces a computational error. The lose of significance during substraction of nearly equal numbers is a source of round-off error in numerical methods.

Multiplication and division are straight, not like addition and substraction.

$$0.1642 \times 10^3 \times 0.2553 \times 10^{-2} = 0.04192026 \times 10^1$$

Normalizing the result: 0.4192026 and chopping gives: 0.4192

Substractive cancellation: When substracting two nearly equal floating point numbers this occurs. For example: The roots of quadratic equation, $y = ax^2 + bx + c = 0$. When $b^2 > 4ac$, the difference in numerator is very small. An alternate method is used to minimize the substractive cancellation.

$$x_1 = \frac{-2c}{x_2 = b \pm (b^2 - 4ac)^{1/2}}$$

Smearing: Smearing occurs whenever the individual terms in a summation are larger than the summation itself. This occurs in series of mixed signs like: $e^{-x} = 1 - x + x^2/2! - x^3/3! + ...$

EXERCISES

1. Determine the number of terms necessary to approximate e^{-x} to 8 significant digits. Calculate the value for $x = 0.\ 4\pi$. Write a program to calculate the result.

2. For approximating the square root of any positive number b is formulated as: $x = \dfrac{x + (b/x)}{2}$.

   Write a flow diagram or algorighm and computer program to calculate square root

   (Hint: For square root 6, start $x_1 = 2$, second x is $x_2 = \dfrac{2 + 6/2}{2} = 2.5$ and continue $x_3 = ...$)

3. Determine the eigenvalues for the below given matrix by analytical method. Use iteration method and compare with analytical values.

$$A = \begin{bmatrix} 1 & 2 & 3 \\ 2 & 4 & 6 \\ 3 & 6 & 1 \end{bmatrix}$$

4. Explain the following (a) Real matrix; (b) Symmetric matrix; (c) Triangular matrix; (d) Rank, and (e) Hermitian matrix.

❑❑❑

2

Ordinary Differential Equations: Solutions, Applications and Series Solutions

2.1 REVIEW OF ORDINARY DIFFERENTIAL EQUATIONS

e.g.,
$$\frac{dy}{dx} = (4x + y + 1)^2$$

1. Variable separable method

e.g.,
$$\frac{dy}{dx} = (4x + y + 1)^2; \text{ Let } t = 4x + y + 1;\ \ y = (t - 4x - 1)$$

$$\frac{dy}{dx} = \frac{dt}{dx} - 4;\ \frac{dt}{dx} = t^2 + 4$$

$$\frac{dt}{(t^2+4)} = dx; \text{ Integrating, } \int \frac{dt}{(t^2+4)} = \int dx$$

$$\left(\frac{1}{2}\right)\tan^{-1}\frac{(4x+y+1)}{2} = x + c \textbf{ Ans.}$$

2. Leibnitz method

$$\frac{dy}{dx} + py = Q \text{ where } P \text{ and } Q \text{ are functions of } x.$$

$$\text{Integrating factor} = \int P\, dx$$

$$\frac{dy}{dx} e^{\int P\,dx} + Py e^{\int P\,dx} = Q\, e^{\int P\,dx}$$

$$\frac{d}{dx}\left(e^{\int P\,dx} y\right) = Qe^{\int P\,dx}$$

Integrating,
$$e^{\int P\,dx} y = \int Q\, e^{\int P\,dx} y + \text{constant}$$

e.g.,
$$(x+1)\frac{dy}{dx} - y = e^x(x+1)^2$$

Rearranging: $$-\frac{dy}{dx} - \frac{y}{x+1} = e^x(x + 1)$$

This is of form: $$\frac{dy}{dx} + Py = Q$$

Integrating factor is : $$e^{\int P\,dx} = e^{-\ln(x+1)} = \frac{1}{(x+1)}$$

$$y \times \text{I.F.} = \int e^x (x+1)\,\text{I.F.}\,dx + C$$

$$\frac{y}{x+1} = \int e^x \frac{(x+1)\,dx}{(x+1)} + C = \int e^x + C = e^x + C$$

$$y = e^x(x + 1) + C(1 + x) \textbf{ Ans.}$$

3. Bernoulli equation

$$\frac{dy}{dx} + Py = Qy^n$$

Dividing by y^n throughout and then change of variable converts into Leibnitz form.

e.g., $$\frac{dy}{dx} + \frac{y}{x} = x^3y^3$$

Dividing by y^3 gives

$$y^{-3}\frac{dy}{dx} + \frac{y^{-2}}{x} = x^3$$

Let $y^{-2} = z$; $$\frac{-2y^{-3}dy}{dx} = \frac{dz}{dx}$$

$$\frac{-1}{2}\frac{dz}{dx} + \frac{z}{x} = x^3; \quad \frac{dz}{dx} - \frac{-2z}{x} = -2x^3$$

This is of Leibnitz form.

$$\text{I.F.} = e^{\int \frac{-2}{x}dx} = x^{-2}$$

$$\text{I.F.} \times z = \int(-2x^3x^{-2}dx) + C = -x^2 + C; \; y^{-2}\,x^{-2} = -\frac{2x^2}{2} + C$$

$$(C - x^2)\,x^2\,y^2 = 1 \textbf{ Ans.}$$

2.2 EQUATIONS REDUCIBLE TO EXACT EQUATIONS

$$d\left(\frac{y}{x}\right) = \left(\frac{x\,dy - y\,dx}{x^2}\right)$$

$$d \ln\left(\frac{y}{x}\right) = \frac{x\,dy - y\,dx}{(xy)}$$

$$d \tan^{-1}\left(\frac{y}{x}\right) = \frac{(x\,dy - y\,dx)}{(x^2 + y^2)}$$

4. Linear differential equations

$$\frac{d^n y}{dx^n} + P_1 \frac{d^{n-1} y}{dx^{n-1}} + P_2 \frac{d^{n-2} y}{dx^{n-2}} + \dots + P_n y = X$$

where $P_1, P_2, P_3 \dots P_n$ are functions of x only. The total solution is complimentary function plus the particular integral.

e.g.,
$$\frac{d^2 y}{dx^2} + 4y = 0;\ \frac{d}{dt} = D$$

$$(D^2 + 4)y = 0;\ \text{gives } D = \pm 2$$

The complimentary solution is:

$$y = c_1\, e^{2t} + c_2\, e^{-2t}$$

or
$$y = c_1 \cos 2t + c_2 \sin 2t \ \textbf{Ans.}$$

2.3 RULES FOR FINDING THE PARTICULAR INTEGRAL

General linear differential equation is of form:

$$\frac{d^n y}{dx^n} + k_1 \frac{d^{n-1}}{dx^{n-1}} + k_2 \frac{d^{n-2}}{dx^{n-2}} + \dots + k_n\, y = x$$

in symbolic form $\left(D = \frac{d}{dx}\right)$ it is :

$$(D^n + k_1 D^{n-1} + k_2 D^{n-2} + \dots k_n)y = x$$

$$\text{Particular integral is : P.I.} = \frac{X}{D^n + k_1 D^{n-1} + \dots k_n}$$

Case 1

When
$$X = e^{ax};\ \frac{1}{f(D)} e^{ax} = \frac{1}{f'(a)} e^{ax}$$

If $f'(a) = 0$, then
$$\frac{1}{f(D)} e^{ax} = X \frac{}{f''(a)} e^{ax}$$

e.g.,
$$\text{P.I. of } (D^2 + 5D + 6)y = e^x;\ \text{P.I.} = \frac{e^x}{D^2 + 5D + 6}$$

Putting $D = 1$, giving $= \dfrac{1}{1^2 + 5 \times 1 + 6} e^x = \dfrac{e^x}{12}$

e.g., Find P.I. of $(D + 2)(D - 1)^2 y = e^{-2x}$; P.I. $= \dfrac{1 e^{-2x}}{(D+2)(D-1)^2}$

$$= \frac{1}{(D+2)}\left(\frac{1}{(D-1)^2} e^{-2x}\right) = -\frac{1}{(D+2)}\frac{1}{(-2-1)^2} e^{-2x} = \frac{1}{9(D+2)} e^{-2x}$$

$$= -\frac{1}{9}\frac{x \times 1}{1} e^{-2x} = -\frac{x}{9} e^{-2x}$$

Case 2

When $X = \sin(ax + b)$ or $\cos(ax + b)$; Since $D \sin(ax + b) = a \cos(ax + b)$; $D^2 \sin(ax + b) = -a^2 \sin(ax + b)$; $D^3 \sin(ax + b) = -a^3 \cos(ax + b)$; $D^4 \sin(ax + b) = a^4 \sin(ax + b)$; In general $(D^2)^r \sin(ax + b) = (-a^2)^r \sin(ax + b)$; $\dfrac{1}{f(D^2)} \sin(ax + b) = \dfrac{1}{f(-a^2)} \sin(ax + b)$ provided $f(-a^2) \neq 0$.

If $f(-a^2) = 0$ then $\dfrac{1}{f(D^2)} \sin(ax + b) = \dfrac{x}{f'(-a^2)} \sin(ax + b)$;

If $f'(-a^2) = 0$ then $\dfrac{1}{f(D^2)} \sin(ax + b) = \dfrac{x^2}{f''(-a^2)} \sin(ax + b)$

e.g., Find P.I. of $(D^2 + 1)y = \cos(2x - 1)$

$$\text{P.I.} = \frac{\cos(2x-1)}{(D^2+1)} = \left(\frac{1}{D(-4)+1}\right)\cos(2x-1) = \frac{\cos(2x-1)}{(1-4D)}$$

$$= \left(\frac{(1+4D)\cos(2x-1)}{(1+4D)(1-4D)}\right) = \frac{(1+4D)\cos(2x-1)}{(1-16D^2)}$$

$$= \frac{1+4D\cos(2x-1)}{1-16(-4)}$$

$$= \frac{(1+4D)\cos(2x-1)}{65} = (1 + 4D)\cos(2x - 1)$$

$$= \left(\frac{1}{65}\right)\cos(2x-1) + 4D\cos(2x-1)$$

$$= \left(\frac{1}{65}\right)(\cos(2x-1) - 8\sin(2x-1))$$

Case 3

When $X = x^m$; P.I. $= \left[\dfrac{1}{f(D)}\right] x^m = f(D)^{-1} x^m$

e.g., P.I. of $\frac{d^2y}{dx^2} + \frac{dy}{dx} = x^2 + 2x + 4$; in symbolic form:

$$\text{P.I.} = \left[\frac{1}{(D^2 + D)}\right](x^2 + 2x + 4)$$

$$= \left(\frac{1}{D}\right)(1 + D)^{-1}(x^2 + 2x + 4)$$

$$= \left(\frac{1}{D}\right)(1 - D + D^2)(x^2 + 2x + 4)$$

$$= \left(\frac{1}{D}\right)[(x^2 + 2x + 4) - (2x + 2) + 2]$$

$$= (x^2 + 4)\frac{1}{D} = \left(\frac{x^3}{3}\right) + 4x$$

Case 4

$X = e^{ax} V$ where V is a function of x.

$$\left[\frac{1}{f(D)}\right]e^{ax} V = e^{ax}\left[\frac{1}{f(D + a)}\right]V$$

e.g., particular integral of $(D^2 - 2D + 4)y = e^x \cos x$

$$\text{P.I.} = \left[\frac{1}{(D^2 - 2D + 4)}\right]e^x \text{ ü } x;$$

$$= e^x\left\{\frac{1}{[(D + 1)^2 - 2(D + 1) + 4]}\right\} \cos x \text{ (Replacing } D \text{ by } D + 1)$$

$$= e^x\left[\frac{1}{(-1 + 3)}\right]\cos x = \left(\frac{1}{2}\right)e^x \text{ ü } x \text{ (Putting } D^2 = -1)$$

Case 5

When X is any other function x, then

$$\text{P.I.} = \left[\frac{1}{f(D)}\right]X$$

If $f(D) = (D - m_1)(D - m_2) \ldots (D - m_n)$

Resolving into partial fractions,

$$\frac{1}{f(D)} = \left[\frac{A_1}{(D - m_1)}\right] + \left\{\frac{A_2}{(D - m_2)}\right\} + \ldots + \left[\frac{A_n}{(D - m_n}\right]$$

$$\text{P.I.} = \left\{\left[\frac{A_1}{(D-m_1)}\right]+\left[\frac{A_2}{(D-m_2)}\right]+\left[\frac{A_n}{(D-m_n)}\right]\right\}X$$

$$= A_1 e^{m_1 x}\int X e^{-m_1 x}dx + A_2 e^{m_2 x}\int X e^{-m_2}dx + \ldots + A_n e^{mnx}\int X e^{-mnx}dx$$

e.g., $$\frac{d^2y}{dx^2}-\frac{2dy}{dx}+y = xe^x \sin x$$

In symbolic form: $(D^2 - 2D + 1)y = 2e^x \sin x$; to find complimentary function (C.F.):

$$D^2 - 2D + 1 = 0;\ (D-1)^2 = 0;\ D = 1,\ D = 1;\ \text{C.F.} = (C_1 + C_2x)e^x$$

For particular integral:

$$e^x x \sin\frac{x}{(D-1)^2} = e^x\left[\frac{1}{(D+1-1)^2}\right]x \sin x$$

$$= e^x\left(\frac{1}{D}\right)\left[x(-\cos x) - \int 1-(-\cos x)dx\right]$$

$$= e^x\int[-x\cos x + \sin x]\,dx$$

$$= -e^x(x \sin x + 2\cos x);$$

Complete solution $y = (C_1 + C_2x)e^x - e^x(x \sin x + 2 \cos x)$.

Case 6

Two other methods for finding the particular integrals:

Method of variation of parameters (works for second other differential equations).

Case 7

Method of undetermined coefficients:

For example, $X = 2e^{3x}$, trial solution is: ae^{3x}

For example, for $X = 3 \sin 2x$, trial solution is: $a_1 \sin 2x + a_2 \cos 2x$

For example, $X = 2x^3$, trial solution is: $a_1x^3 + a_2x^2 + a_3x + a_4$.

5. Particular integrals: P.I.

This comes from right hand side terms.

(a) $(D^2 + 5D + 6) = e^x$; P.I. $\dfrac{e^x}{(D^2+5D+6)}$; Putting $D = 1$ (coefficient of x), gives:

$$\frac{e^x}{(1+5+6)} = \frac{e^x}{12}$$

(b) $(D^2 + 1)y = \cos(2x - 1)$; $D^2 = -2^2 = -4$; $\dfrac{1}{[D(-4)+1]}\cos(2x-1)$

$$= \frac{\cos(2x-1)}{(1-4D)} = \left[\frac{(1+4D)}{(1-16D^2)}\right]\cos(2x-1)\ ;\ D^2 = -4;$$

(c) $$\frac{e^x \cos x}{(D^2 - 2D + 4)} = e^x \frac{\cos x}{[(D+1)^2 - 2(D+1) + 4]} = e^x \frac{\cos x}{(D^2+3)}$$

Putting $D^2 = -1^2 = -1$; $e^x \dfrac{\cos x}{(-1+3)} = \left(\dfrac{e^x \cos x}{2}\right)$

6. Linear Legendre equation

$$\frac{(1+x^2)d^2y}{dx^2} + \frac{(1+x)dy}{dx} + y = 2\sin[\log(1+x)]$$

$$D = \frac{d}{dt};\ \frac{(1+x^2)d^2y}{dx^2} = D(D-1)y;$$

$$\frac{(1+x)dy}{dx} = Dy;\ 1 + x = e^t;\ t = \log(1+x)$$

$$D(D-1)y + Dy + y = 2\sin t;\ D^2y - Dy + Dy + y = 2\sin t;\ (D^2+1)y = 2\sin t$$

This is solved by complimentary solution and particular integral.

7. Cauchy equation

$$\frac{x^2d^2y}{dx^2} = D(D-1)y$$

$$\left(\frac{1+4D}{65}\right)\cos(2x-1) = \frac{1}{65}\cos(2x-1)\frac{-8}{65}\sin(2x-1)$$

$$D = \frac{d}{dt};\ \frac{x^2d^2y}{dx^2} = D(D-1)y;\ x = e^t;\ t = \ln x;$$

$$D(D-1)y - Dy + y = t;\ D^2y - 2Dy + y = t;\ (D^2 - 2D + 1)y = t;$$

Solved by complimentary solution and particular integral.

8. To find particular integral of $\dfrac{\tan x}{(x^2+4)}$ cosec x, secant x:

Solution:

$$\frac{\text{ü}\ \ x}{\text{ü}x^2 +} = \frac{\tan x}{[(D+2i)(D-2i)]}$$

$$= \left(\frac{1}{4i}\right)\left[\frac{1}{(D-2i)} - \frac{1}{(D+2i)}\right]\tan x$$

Using $$\frac{1}{D-a}\int X = e^{ax}\int Xe^{-ax}dx = \frac{1}{4i}e^{-2ix}\int \tan x\, e^{-2ix}$$

$$-\frac{1}{4i}e^{-2ix}\int \tan x\, e^{2ix}dx$$

$$= \frac{1}{4i}\left(e^{2ix}\int \tan x\,(2\cos^2 x-1)\right)-\frac{i}{4i}e^{2ix}\int \tan x\sin 2x\,dx$$

$$= \frac{1}{4i}e^{2ix}\int \tan x\,(2\cos^2 x-1)-\frac{i}{4i}e^{+2ix}\int \tan x\sin 2x\,dx$$

$$= \frac{1}{4i}e^{2ix}\int (2\cos^2 x\tan x-\tan x)-\frac{i\sin x}{\cos x}\sin 2x$$

$$= \frac{1}{4i}e^{2ix}(\sin 2x-\tan x)-i(1-\cos 2x)dx$$

$$\cos 2x = 1-2\sin^2 x$$

$$= \frac{1}{4i}e^{2ix}\left(\frac{-\cos 2x}{2}+\log\cos x-i\left(\frac{x-\sin 2x}{2}\right)\right)$$

$$= \frac{e^{2ix}}{4}\left(\frac{i\sin 2x}{2}-\frac{\cos 2x}{2}+\log\cos x-ix\right)$$

Changing + i to $-i$:

$$= \frac{-e^{-2ix}ix}{4}+\ldots$$

$$= \left(\frac{e^{2ix}+e^{-2ix}}{8}\right)\sin 2x\frac{-\cos 2x}{8i}(-e^{-2ix}+e^{2ix})+\ln\frac{\cos x}{4i}\left(\frac{e^{2ix}-e^{-2ix}}{2i\sin 2x}\right)$$

$$-\frac{xi}{4}\left(\frac{e^{2ix}+e^{-2ix}}{2\cos 2x}\right)$$

$$y = C_1\cos 2x+C_2\sin 2x+\frac{\ln\cos x\sin 2x}{2}-\frac{x\cos 2x}{2}$$

9. Particular integral (Variation of parameters method)

$$\frac{d^2y}{dx^2}+a_1(x)\frac{dy}{dx}+a_2(x)y = Q(x) \qquad \ldots(1)$$

Assuming y_1 and y_2 are unknown solutions, 1st complimentary solution:

$$C_1y_1+C_2y_2 \qquad \ldots(2)$$

The particular solution is assumed of the form:

$$I = uy_1+Vy_2 \qquad \ldots(2a)$$

where u and V are functions of x.

$$\frac{dI}{dx} = uy_1' + Vy_2' + (u'y_1' + V'V_2') \quad ...(3)$$

$u'y_1 + V'y_2 = 0$ (taken as) as imposed condition. Further differentiation yields,

$$\frac{d^2I}{dx^2} = (uy_1'' + Vy_2'') + (u'y_1' + V'y_2') \quad ...(4)$$

Putting Eqs. (2a) to (4) in Eq. (1) gives:

$$\frac{d^2y}{dx^2} + a_1\frac{dy}{dx} + a_2y = Q(x) \quad ...(1)$$

$$y = C_1 y_1 + C_2 y_2 + (uy_1 + Vy_2) \text{ (P.I.)}$$

$$y' = C_1 y_1' + C_2 y_2' + \frac{dI}{dx}$$

$$y'' = C_1 y_1'' + C_2 y_2'' + \frac{d^2I}{dx^2}$$

$$\left(C_1 y_1'' + C_2 y_2'' + \frac{d^2I}{dx^2}\right) + a_1\left(C_1 y_1' + C_2 y_2' + \frac{dI}{dx}\right) + a_2(C_1 y_1 + C_2 y_2 + I) = Q(x)$$

C.S is obtained by taking zero on R.H.S.

$$u(y_1'' + a_1y_1' + a_2y_1) + V(y_2'' + a_1y_2' + a_2y_2) + u'y_1' + V'y_2'$$

$$+ C_1(y_1'' + a_1y_1' + a_2y_1) + C_2(y_2'' + a_1y_2' + a_2y_2) = Q(x)$$

This is a solution of homogenous equation in y_1 and y_2.

$$u'y_1' + V'y_2' = Q(x) \quad ...(5)$$

$$u'y_1 + V'y_2 = 0 \quad ...(6)$$

Solving the above equations gives:

$$u' = -\frac{y_2Q(x)}{(y_1y_2' - y_2y_1')}$$

$$V' = \frac{y_1Q(x)}{(y_1y_2' - y_2y_1')}; \; I = uy_1 + Vy_2$$

Example 2.1 P is a function of both x, y as given by:

$$P = Ax + B\left[\frac{(xy - a)}{y}\right] + \frac{C}{(xy)^{1/2}} \quad ...(1)$$

where A, B, C, a are constants. Obtain expression for x, y for max. or min. values of P.

Solution: $$\frac{dP}{dx} \text{ at constant } y = A + B - 0.5\, Cy^{-1/2}\, x^{-3/2} = 0;$$

$$\frac{(A+B)}{(0.5C)} = y^{-1/2}\, x^{-3/2} \qquad \ldots(2)$$

$$\frac{dP}{dy} \text{ at constant } x = 0.5B\, a\, y - 0.5Cx^{-1/2}\, y^{-3/2} = 0;$$

Giving $$\frac{Ba}{C} = x^{-1/2};\ x = \frac{C^2}{(B^2 a^2)} \qquad \ldots(3)$$

Substituting Eq. (3) into Eq. (2) gives:

$$y = \left[\frac{0.25}{(A+B)^2}\right] B^6 A^6 C^{-4}$$

Example 2.2 Discuss the maximum and minimum of $(a) x^3\, y^3(1 - x - y)$ and $(b)\ 2(x^2 - y^2) - x^4 + y^4$.

Solution: $$F = x^3\, y^3 - x^4\, y^3 - x^3\, y^4;\ \frac{dF}{dx} \text{ at constant}$$

$$y = 3x^2\, y^3 - 4x^3\, y^3 - 3x^2\, y^4 = 0$$

$$y^3(3x^2 - 4x^3 - 3x^2 y) = 0;\ y^3 = 0;\ x^2(3 - 4x) = 0;\ x = 0$$

or $$3x^2 - 4x^3 - 3x^2 y = 0;\ 3x^2 - 4x^2 = 3x^2 y;$$

$$3x^2\left(\frac{1-4x}{3}\right) = 3x^2 y;\ \text{Giving } 4x = 3(1 - y);\ \text{if } y = 0, \text{ then } x = \frac{3}{4}$$

$$\frac{d^2F}{dx^2} = 6xy^3 - 12x^2y^3 - 6xy^4;$$

For minimum $\frac{d^2F}{dx^2}$ is positive. For maximum $\frac{d^2F}{dx^2}$ = negative.

$$\frac{dF}{dy} \text{ at constant } x = 3y^2\, x^3 - 3y^2\, x^4 - 4y^3\, x^3 = x^3(3y^2 - 3y^2\, x - 4y^3) = 0$$

At $x = 0$; $$3y^2 - 3y^2x - 4y^3 = 0;\ 3y^2\left(1 - x - \frac{4}{3y}\right) = 0;$$

or $$y = \frac{4}{3(1-x)}$$

Example 2.3 Find the minimum value of $x^m\, x^n\, z^p$ given $x + y + z = a$

Solution: $$\frac{dF}{dx} = mx^{m-1}\, y^n\, z^p + \lambda = 0 \qquad \ldots(1)$$

$$\frac{dF}{dy} = x^m\, ny^{n-1}\, z^p + \lambda = 0 \qquad \ldots(2)$$

$$\frac{dF}{dz} = x^m y^n pz^{p-1} + \lambda = 0 \qquad \ldots(3)$$

Equation (1) – Eq. (2) = $mx^{m-1} y^n z^p - x^m ny^{n-1} z^p = 0$; giving:

$$x^m y^n z^p \left(\frac{m}{x} - \frac{n}{y}\right) = 0; \text{ So } \frac{m}{x} = \frac{n}{y}; y = \frac{nx}{m} \qquad \ldots(4)$$

From Eqs. (2) and (3) $\left(\frac{x^m n y^n z^p}{y}\right) - \left(\frac{pz^p x^m y^n}{z}\right) = 0$; giving:

$$\left(\frac{n}{y} - \frac{p}{z}\right) x^m y^n z^p = 0; \frac{n}{y} = \frac{p}{z}; \frac{py}{n} = z \qquad \ldots(5)$$

From Eq. (3) Eq. (1)

$$x = \frac{nz}{p} \qquad \ldots(6)$$

$$xyz \text{ (Eq. (4) × Eq. (5) × Eq. (6)} = \left(\frac{nx}{m}\right)\left(\frac{py}{n}\right)\left(\frac{mz}{p}\right); x^m = \frac{m^m z^m}{p^m}; y^n = \frac{n^n x^n}{m^n}$$

$$Z^p = \frac{p^p y^p}{n^p}; x^m y^n z^p = z^m x^n y^p m^{m-n} n^{n-p} p^{p-m} \quad \textbf{Ans.}$$

Example 2.4 Homogeneous equations of the form:

$$\frac{dy}{dx} = \frac{f(x, y)}{\phi(x, y)}$$

Put $$y = Vx; \frac{dy}{dx} = V + x\frac{dV}{dx}$$

Separate the variables V and x and integrate.

e.g., solve: $$2xy\frac{dy}{dx} = 3y^2 + x^2; \ y = Vx; \ \frac{dy}{dx} = V + x\frac{dV}{dx}$$

Solution: $$2xVx\left(V + x\frac{dV}{dx}\right) = 3V^2 x^2 + x^2; 2x V^2 x + 2x^3 V\frac{dV}{dx} = 3V^2 x^2 + x^2$$

$$2x^2 V^2 + 2x^3 V\frac{dV}{dx} = 3V^2 x^2 + x^2; 2x^3 V\frac{dV}{dx} = V^2 x^2 + x^2$$

$$2Vx\frac{dV}{dx} = V^2 + 1; \ 2V\frac{dV}{(V^2 + 1)} = \frac{dx}{x}$$

Integrating: $$\ln (V^2 + 1) = \ln x$$

$$e^{\ln(V^2+1)} = e^{\ln x}; V^2 + 1 = x; V = \frac{y}{x}; \left(\frac{y^2}{x^2}\right) + 1 = x + \text{constant. } \textbf{Ans.}$$

10. Linear differential equations: Compimentary solutions

Case 1: If all the roots are real and different:

$$(D - m_1)\ (D - m_2)\ \ldots\ (D - m_n) = 0$$

Then the solution is

$$y = C_1 e^{m_1 x} + C_2 e^{m_2 x} + \ldots + C_n e^{m_n x}$$

Case 2: If two roots are equal ($m_1 = m_2$);

$$y = (C_1 + C_2 x) e^{m_1 x} + C_3 e^{m_3 x} + \ldots$$

Case 3: One pair of roots is imaginary;

$$m_1 = a + b_i;\ m_2 = a - b_i$$

$$y = C_1 e^{(a+bi)x} + C_2 e^{(a-ib)x} + C_3 e^{m_3 x} + \ldots$$

Case 4: If two pairs of imaginary roots are equal;

$$m_1 = m_2 = (a + b_i),\ m_4 = m_3 = a - b_i;$$

$$y = [(C_1 x + C_2) \cos Bx + (C_3 x + C_4) \sin Bx] + \ldots + C_n\, e^{mnx}$$

Differential equations of other types

e.g.,

$$\frac{d^2 y}{dx^2} = f(x)$$

Integrating with respect to x gives:

$$\frac{dy}{dx} = f(x)dx + C_1 = F(x)$$

Again integrating: $y = \int F(x)dx + C_2$ **Ans.**

For $\frac{d^n y}{dx^n} = f(x)$, the solution is obtained by integrating n times.

Equations of form: $$\frac{d^2 y}{dx^2} = f(y)$$

Multiplying both sides by $\frac{2dy}{dx}$, we get

$$\frac{2dy}{dx} \times \frac{d^2 y}{dx^2} = 2f(y)\frac{dy}{dx}$$; Integrating with respective to x:

$$\left(\frac{dy}{dx}\right)^2 = 2\int f(y)\,dy + c = F(y);\ \frac{dy}{dx} = [F(y)]^{1/2}$$

Separating the variables and integrating, we get

$$\int \frac{dy}{[F(y)]^{1/2}} = x + c; \text{ This is the desired solution.}$$

11. Equations not containing *y*

$$F\left(\frac{d^2y}{dx^2}, \frac{dy}{dx}, x\right) = 0; \text{ Put } p = \frac{dy}{dx}; \frac{d^2y}{dx^2} = \frac{dp}{dx} = \frac{dp}{dy} \times \frac{dy}{dx} = \frac{p\,dp}{dy}$$

Giving $$f\left(\frac{p\,dp}{dy}, p, x\right) = 0. \text{ This can be solved.}$$

12. Equations not containing *x*

$$f\left(\frac{d^2y}{dx^2}, \frac{dy}{dx}, y\right) = 0$$

Putting $$\frac{dy}{dx} = p; \frac{d^2y}{dx^2} = \frac{dp}{dx} = \frac{p\,dp}{dy}$$

Giving: $$f\left(P\frac{dp}{dy}, p, y\right) = 0$$

This can be solved.

13. Equations

Whose one solution is known. If $y = u(x)$ is a known solution, then put $y = uV$ in it. This reduces the differential equation to one of order $n - 1$ in $\frac{dV}{dx}$.

A second order differential equation can be solved by reducing it to a first order.

Equations can be solved by changing the independent variable.

$$\frac{d^2y}{dx^2} + P\frac{dy}{dx} + Qy = R \qquad \ldots(1)$$

Let $Z = f(x)$; $$\frac{dy}{dx} = \frac{dy}{dZ} \times \frac{dZ}{dx} \qquad \ldots(2)$$

$$\frac{d^2y}{dx^2} = \frac{d}{dx}\left(\frac{dy}{dZ} \times \frac{dZ}{dx}\right) = \frac{dy}{dZ} \times \frac{d^2Z}{dx^2} + \left(\frac{dZ}{dx}\right)^2 \frac{d^2y}{dx^2} \qquad \ldots(3)$$

Substituting Eqs. (2) and (3) in Eq. (1) : $$\frac{d^2y}{dz^2} + P_1\frac{dy}{dz} + Q_1 y = R_1 \qquad \ldots(4)$$

where $P_1 = \left[\frac{d^2Z}{dx^2} + \frac{P\,dZ}{dx}\right]\left(\frac{dZ}{dx}\right)^2$; $Q_1 = \frac{Q}{\left(\frac{dZ}{dx}\right)^2}$; $R_1 = \frac{R}{\left(\frac{dZ}{dx}\right)^2}$

Equation (4) can be solved by taking $P_1 = 0$, $Q_1 = a$ (constant)

14. Total differential equations

$$P + \frac{Q\,dy}{dx} + \frac{R\,dZ}{dx} = 0 \quad \text{...(1)}$$

Equation (1) can be written as

$$P\,dx + Q\,dy + R\,dz = 0 \quad \text{...(2)}$$

This is integrable if:

$$P\left(\frac{dQ}{dZ} - \frac{dR}{dy}\right) + Q\left(\frac{dR}{dx} - \frac{dP}{dZ}\right) + R\left(\frac{dP}{dy} - \frac{dQ}{dZ}\right) = 0 \quad \text{...(3)}$$

For solving if Eq. (3) is satisfied, take one variable Z as constant, then $dZ = 0$. Then integrate the equation: $P\,dx + Q\,dy = 0$. Replace the constant appearing in its integral by $\phi(z)$. Now differentiate the integral just obtained w.r.t. x, y, z. Finally, compare the result with the given differential equation to determine $\phi(z)$.

Simultaneous total differential equation:

$$P\,dx + Q\,dy + R\,dz = 0;\ P_1\,dx + Q_1\,dy + R_1\,dz = 0;$$

where P, Q, R, P_1, R_1, Q_1 are functions of x, y, z.

$$\frac{dx}{(QR_1 - Q_1R)} = \frac{dy}{(RP_1 - PR_1)} = \frac{dz}{(PQ_1 - QP_1)} \text{ and solve these.}$$

15. Equations of form

$$\frac{dx}{P} = \frac{dy}{Q} = \frac{dz}{R}$$

1. **Method of grouping:** Taking $\frac{dx}{P} = \frac{dz}{R}$, integrate to give $f(x, z) = c$ from $\frac{dy}{Q} = \frac{dZ}{R}$ (see if x is absent or can be removed). Then integrate giving $\psi(y, z) = c'$. The two equations taken together constitute the complete solution required.
2. **Method of multipliers:** By proper choice of multipliers l, m, n, which are not necessarily constants, we write:

$$\frac{dx}{P} = \frac{dy}{Q} = \frac{dz}{R} = \frac{I\,dx + m\,dy + n\,dz}{(IP + mQ + nR)}$$

such that $IP + mQ + nR = 0.$

Then $Pdx + Qdy + Rdz = 0$ can be solved giving integral: $\phi(x, y, z) = c$

Then search for another set of multipliers : λ, μ, γ so that $\lambda P + \mu Q + \gamma R = 0$ giving $\lambda dx + \mu dy + \gamma dz = 0$ which on integrating gives: $\psi(x, y, z) = c$. These two solutions taken together constitute the required solution:

Example 2.5 Reactions take place as shown below:

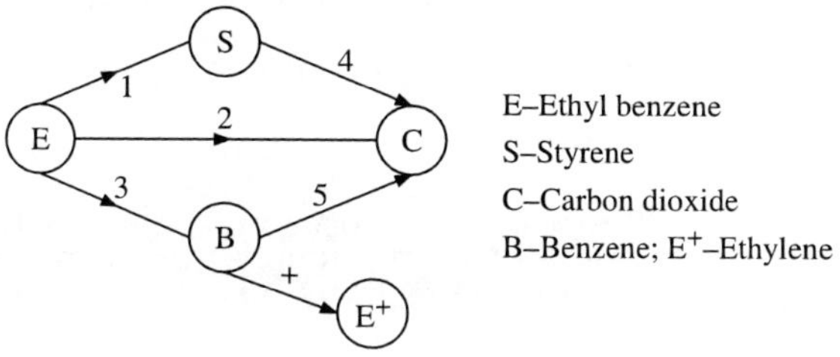

Fig. 2.1

Solution: The rate equations are :

$$\frac{dXE}{dt} = (K_1 + K_2 + K_3)X_E \quad \text{...(1)}$$

$$\frac{dX_S}{dt} = K_1X_E - K_4X_S \quad \text{...(2)}$$

$$\frac{dX_B}{dt} = K_3X_E - K_4X_B \quad \text{...(3)}$$

$$\frac{dX_{E^+}}{dt} = K_5X_{E^\times} \quad \text{...(4)}$$

$$X_C = 1 - X_E - X_S - X_B - X_{E+}$$

$$\frac{dX_C}{dt} = K_2X_E + K_4X_S + \left(\frac{3}{4}\right)K_SX_B$$

Integrating and simplifying Eq. (1):

$$X_E = e^{-K_Et}$$

where $K_E = (K_1 + K_2 + K_3)$. After substituting for X_E in Eqs. (2), (3) and (4).

$$X_S = \left[\frac{K_1}{(K_4 - K_E)}\right](e^{-K_Et} - e^{-K_4t})$$

$$X_B = \left[\frac{K_3}{(K_5 - K_E)}\right](e^{-K_Et} - e^{-K_5t})$$

$$X_{Et} = \left(\frac{K_3}{K_E}\right)(1 - e^{-K_Et})$$

Example 2.6 $\frac{d^2y}{dx^2} + y = \text{cosec } x$

Solution:

$$u' = -\frac{y_2 Q(x)}{(y_1 y_2' - y_2 y_1')}; \; V' = \frac{y_1 Q(x)}{(y_1 y_2' - y_2 y_1')}$$

$$(D^2 + 1) = 0; D = 1i, -1i; y_c = C_1 \cos x + C_2 \sin x; y_1 = \cos x;$$

$$y_2 = \sin x; y'_1 = -\sin x; y'_2 = \cos x$$

$$u' = \frac{(-\sin x \text{ cosec } x)}{(\sin^2 x + \cos^2 x)} = -1; \; v' = \frac{(\cos x \text{ cosec } x)}{(\cos^2 x + \sin^2 x)} = \cot x$$

$$u = -x, \; V = \int V' \, dV = \cot x = \log \sin x;$$

$$\text{P.I.} = -x \cos x + \sin x (\log \sin x)$$

$$y = C_1 \cos x + C_2 \sin x + \sin x \log \sin x - x \cos x.$$

Example 2.7 Clairut's equation:

$$y = px + f(p) \quad \ldots(1)$$

$$\frac{dy}{dx} = P = p + x\frac{dp}{dx} + f'(p)\frac{dp}{dx}; \; [x + f'(p))]\frac{dp}{dx} = 0$$

Giving

$$\frac{dp}{dx} = 0; \text{ i.e., } p = \text{constant} \quad \ldots(2)$$

Eliminating p from Eq. (1) and Eq. (2) :

$$y = cx + f(\text{c}) \quad \ldots(3)$$

Solution of Clairut's equation is obtained on replacing p by c.

e.g., solving: $p = \sin(y - xp)$; $\sin^{-1}p = p(y - xp)$ or $y = px + \sin^{-1}p$ which is Clairut's equation. Its solution $= cx + \sin^{-1}c$.

Example 2.8 Extraction of benzoic acid from benzoic acid – toluene mixture by water: Solvent extraction in two stages (counter current)

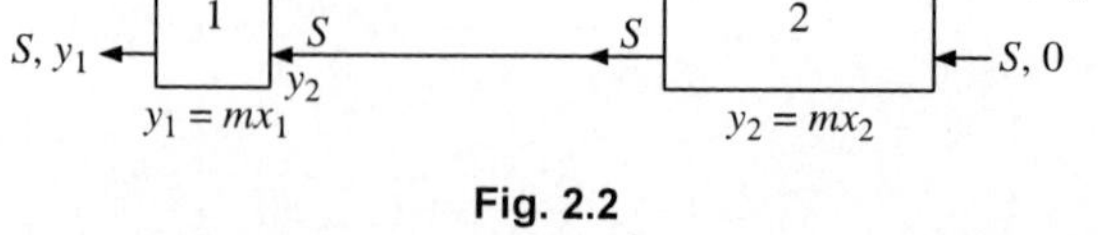

Fig. 2.2

Material balance for stage 1: $RC + Sy_2 = Rx_1 + Sy_1$; For stage 2: $Rx_1 = Rx_2 + Sy_2$

or:

$$RC + mSx_2 = \frac{Ry_1}{m} + Sy_1; \; \frac{Ry_1}{m} = Rx_2 + mSx_2;$$

Eliminating y_1 $$RC + mSx_2 = \left(Rx_2 + mSx_2 + Sy_1\right) = Rx_2 + mSx_2 + \frac{mSy_1}{m}$$

$$= (Rx_2 + mSx_2) + mS\left[\frac{(Rx_2 + mSx_2)}{R}\right]$$

i.e., $$R(RC + mSx_2) = (R + mS)\,(Rx_2 + mSx_2);\; x_2 = \frac{R^2C}{(R^2 + mRS + m^2S^2)}$$

$$y_1 = \frac{mRC(R + mS)}{(R^2 + mRS + m^2S^2)}$$

If $$\alpha = \frac{R}{(mS)};\ \frac{Sy_1}{(RC)} = \frac{mS(R + mS)}{(R^2 + mRS + m^2S^2)}$$

$$E = \frac{Sy_1}{(RC)} = \frac{(\alpha + 1)}{(\alpha^2 + \alpha + 1)} = \frac{(\alpha^2 - 1)}{(\alpha^3 - 1)}$$

Example 2.9 Prove that the limiting velocity of spherical particles in a medium is given by,

$$V_\infty = \frac{g}{(3k)},\ \text{Given } \frac{dr}{dt} = kr$$

Solution: *By force balance:* Gravitational force-drag force/buoyant force = rate of change of momentum

Putting in mathematical terms

$$\left(\frac{4}{3}\right)\pi r^3 \rho g - 6\pi\mu Vr = \frac{d}{dt}\left(\frac{4}{3}\pi r^3 \rho V\right)$$

$$r^3 g - \left(\frac{9}{2}\right)\mu\frac{Vr}{\rho} = 3r^2V\frac{dr}{dt} + r^3\frac{dV}{dt}$$

$$g - \left(\frac{9}{2}\right)\frac{\mu V}{(\rho r^2)} = \left(\frac{3r^2V}{r^3}\right)\frac{dr}{dt} + \left(\frac{r^3}{r^3}\right)\frac{dV}{dt};\ \frac{dr}{dt} = Kr;$$

$$r = \exp(kt + c);$$

$$r = C_0 \exp(Kt);\ \frac{dV}{dt} + 3KV + \left(\frac{9}{2}\right)\frac{\mu V}{(\rho C_1 \exp(2Kt)} = g$$

This is of Leibnitz form,

$$\frac{dy}{dx} + Py = Q;\ \frac{dV}{dt} + V\left\{3K + \frac{9\mu}{[2\rho C_1 \exp(-2Kt)]}\right\} = g$$

Leibnitz equation solution is

$$V\exp\left(\int P dt\right) = \int Q \exp\left(\int P dt\right) + \text{Const.}$$

$$\int P\,dt = \int\left\{\left[\frac{3K+[4.5\mu\exp(-2Kt)}{(\rho C_1)}\right]\right\}dt$$

$$= \left\{3Kt+\left[\frac{4.5\mu\exp(-2Kt)}{pC_1\times-2K)}\right]\right\}$$

$$= 3Kt-\left[\frac{2.25}{(pC_1K)}\right]\exp(-2Kt)\text{ ; I.F. is}$$

$$\exp\left(\int pdt\right) = \exp\left\{3Kt-\left[\frac{2.25e^{-2Kt}}{(pC_1K)}\right]\right\}$$

$$V\times\exp\left\{3Kt-\left[\frac{2.25e^{-2Kt}}{(pC_1K)}\right]\right\} = \int g\exp\left\{3Kt-\left[\frac{2.25}{(\rho C_1K)}e^{-2Kt}\right]\right\}dt+\text{Const.}$$

$$\text{I.F.}*V = \int g\exp(3Kt)+\text{const.} = g\exp\frac{(3Kt)}{3K}+\text{const.}$$

Sine $\exp(-2Kt)$ becomes zero as t becomes ∞;

$$V=\left[\frac{g}{(3K)}\right]\exp(3Kt)\times\exp\left[-3Kt+\frac{(2.25)}{pC_1K)}e^{-2Kt}\right]+\text{const. }\exp(-3Kt+\ldots)\text{; At }t=\infty,\ V=\frac{g}{(3K)}$$

16. Linear differential equations

$$\frac{d^ny}{dx^n}+\frac{P_1d^{n-1}}{dx^{n-1}}+\frac{P_2dx^{n-2}}{dx^{n-2}}+\ldots P_ny = X$$ where P_1, P_2 … P_n are functions of x only or constants.

These are solved using complimentary function and particular integral.

1. $\dfrac{d^2S}{dt^2}+4S=0$; $\dfrac{d}{dt}=D$; $(D^2+4)=0$; $D=2i$; $-2i$

   Complimentary solution is : $S = C_1e^{2it}+C_2e^{-2it}$

   Particular integral comes from the right hand side terms. Here it is zero.

   So $S = C_1e^{2it}+C_2e^{-2it}$; $S = C_3\cos 2t + C_4\sin 2t$.

2. $(D_2+4)y = \tan x$; $y_c = C_1\cos 2x + C_2\sin 2x$; $y'_1 = -2\sin 2x$; $y'_2 = 2\cos 2x$; $y_1 = \cos 2x$; $y_2 = \sin 2x$

$$u' = -\frac{\sin 2x\times\tan x}{(\cos 2x\cos 2x-\sin 2x\times 2\sin 2x)} = -\frac{\sin 2x\tan x}{2}$$

$$V' = \frac{\cos 2x\tan x}{(\cos 2x\times 2\cos x-\sin 2x\times -2\sin 2x)} = \frac{\cos 2x\tan x}{2}$$

$$u = -\int \frac{\sin 2x \tan x}{2dx} = -\frac{x}{2} + \frac{\sin x}{4};\ V = -\int \frac{\cos x \tan x}{2}$$

$$V = -\frac{\cos 2x}{4} + \frac{1}{2} \ln \cos x$$

$$y = C_1 \cos 2x + C_2 \sin 2x + \ln \cos x \sin 2\frac{x}{2} - x \cos \frac{x}{2}.$$

Example 2.10 Mass balance

Q. Two tanks as shown below:

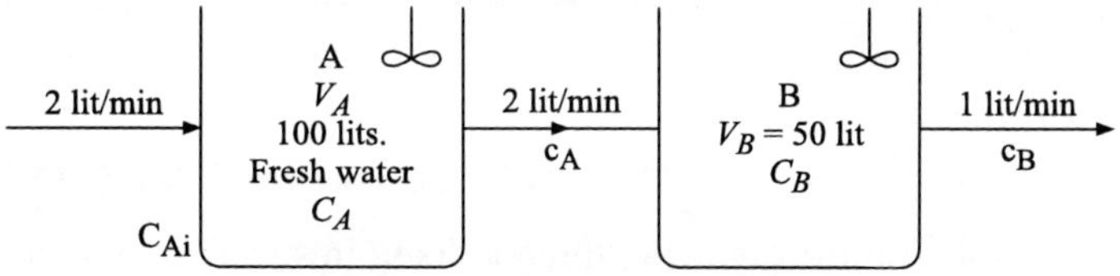

Fig. 2.3

$C_{Ai} = \frac{2}{100} = 0.02$ kg salt/litre of water. Calculate the amount of salt in tank B after one hour?

The initial water in tanks fresh water.

Solution: For tank A: Rate of input of salt-rate of output of salt = Rate of change of accumulation.

Putting in mathematical terms:

$$2 \times 0 - 2 \times C_A = \frac{d}{dt}(V_A \times C_A)$$

Giving: $-2C_A = 100\frac{dC_A}{dt};\ \frac{dC_A}{dt} = -\frac{2}{100dt};\ \ln C_A = -0.02t + \text{const.}$

at $t = 0$; $C_A = 0.02(C_{Ai})$: So constant = ln 0.02

$$C_A(t) = 0.02\, e^{-0.02t}$$

For tank B, by mass balance:

$$2C_A - C_B = \frac{d}{dt}((50+t)C_B)$$

$$2 \times 0.02 \times e^{-0.02t} - C_B = 50\frac{dC_B}{dt} + C_B + t\frac{dC_B}{t} = \frac{dC_B}{dt}(50+t) + C_B$$

$$\frac{dC_B}{dt} + \frac{2C_B}{50+t} = \frac{0.04e^{-0.02t}}{(50+t)};\ \text{This is of } \frac{dy}{dx} + P(x)y = Q(x) \text{ form.}$$

Solution is: $(50+t)^2 \frac{dC_B}{dt} + \frac{2C_B}{(50+t)}(50+t)^2 = 0.04\frac{e^{-0.02t}}{(50+t)}(50+t)^2$

$$C_B(50+t)^2 = \int 0.04\, e^{-0.02t}(50+t)(50+t)\, dt + \text{const.}$$

$$C_B(50+t)^2 = \left(-\left(\frac{50+t}{0.02}\right)e^{-0.02t} - \frac{e^{-0.02t}}{0.0004}\right) \times 0.04 + \text{const.}$$

at $t = 0$: $C_B = 0$; const. = 200

at $t = 60$ minutes : $C_B = \dfrac{0.826}{100}$ kg salt per litre of water.

Amount of salt in tank B after one hour = Volume × Conc. = $(60+50)\dfrac{0.828}{100} = 0.9108$ salt. **Ans.**

17. Flow systems

Two approaches are used in flow systems: (1) Eulerian approach, (2) Langrangian approach.

Eulerian: Changes in small volume element which is fixed in space is studied.

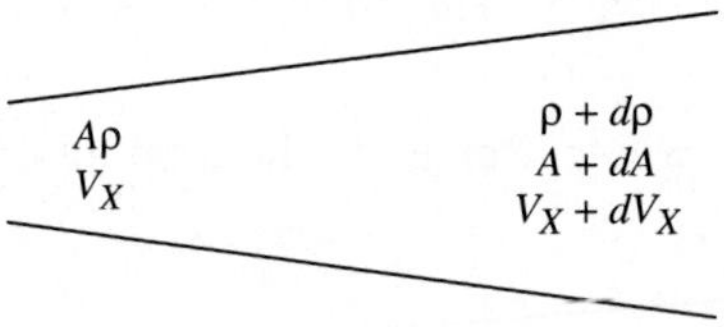

Fig. 2.4

Mass input to dx element – mass output from dx element ± mass change due to reaction = 0 (at steady state)

$$pAV_X - \left(pAV_X + \frac{d}{dx}(pAV_X)\right) = 0$$

Simplifying: $$\frac{d}{dx}(\rho A V_X) = 0;\ \rho A V_X = \text{Const.}$$

which is continuity equation $d(\rho A V_X) = dw = 0$

$$\rho A dV_X + Ad\,\rho V_X + dA\,\rho V_X = 0;$$

Dividing by $\rho A V_x$ throughout:

$$\frac{d\rho}{\rho} + \frac{dA}{A} + \frac{dV_X}{V_X} = 0$$

2.4 LAGRANGE METHOD

Here a small element is studied which moves along with fluid like a person swimming along with element and observing changes in the volume or sitting on the element as it moves in fluid and observing changes in the volume element.

Let T be the elapsed time at which the element is observed. $T = t - t_0$ where t is the absolute time at which the element is observed, t_0 is absolute time at which the element entered the system.

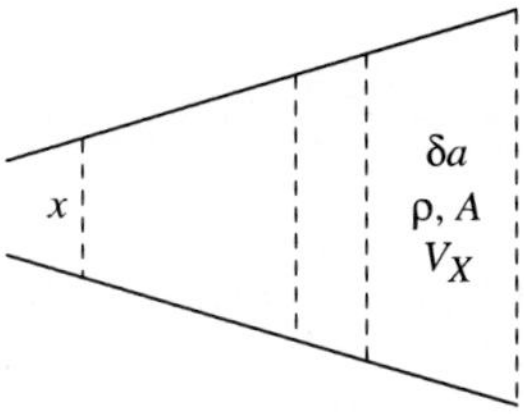

Fig. 2.5

At time T the volume element is: $A\delta a$. A, ρ, V_X refer to the centre of the element. Using the law of conservation of mass, $\frac{d}{dt}(\rho A\,\delta a) = 0$ or $\rho A da = \delta m$ = constant; $V_X = \frac{dx}{dt}$; if $t = t_0$ when $x = 0$; $t - t_0 = \tau = \int_0^x \frac{dx}{V_X}$. At time t, the velocity of the forward force of the element relative to the velocity of the centre of the element is : $\left(V_X + \frac{dV_X}{dx}\frac{\delta a}{2}\right) - V_X$. The relative velocity of the trailing face of the element is: $\left(V_X - \frac{dV_X}{dx}\frac{\delta a}{2}\right) - V_X$. The difference between the two relative velocities is the time rate of change of the length of the element.

$$\frac{dV_X}{dx}\delta a = \frac{d}{dt}(\delta a) \qquad \ldots(1)$$

$$V_X = \frac{dx}{dt} \qquad \ldots(2)$$

$\frac{\text{Eq.}(1)}{\text{Eq.}(2)} = \frac{dV_x}{V_x} = \frac{d(\delta a)}{\delta a}$; from : $\rho A\delta\, a$ = Constant; $\frac{d\rho}{\rho} + \frac{dA}{A} + \frac{d\delta a}{\delta a} = 0$; Using $\frac{d\rho}{\rho} + \frac{dA}{A} + \frac{dV_X}{V_X} = 0$; Giving $\rho A V_X$ = constant. This is continuity equation.

Example 2.11 Rate equation

A compound dissolves in water at a rate proportional to the amount undissolved * difference in concentration of saturated solution and instantaneous concentration. 20 kg of solute is dumped into a tank containing 12 kg of solvent and at the end of 12 minutes the concentration is observed to be $\frac{1}{30}$. Find the amount of solute in solution at any time t, if saturation concentration is $\frac{1}{3}$.

Solution: $\frac{d}{dt}(w_i - w) \propto W\left(\frac{1}{3} - \frac{w}{S}\right)$; $\frac{w}{S}$ is concentration at any time; $w_i - w$ = Amount dissolved in water; w = amount present at time, t (undissolved amount); S – solvent; kg.; w_i = constant.

$$-\frac{dw}{dt} = Kw\left(\frac{1}{3} - \frac{w}{120}\right) = Kw\left(\frac{(40-w)}{120}\right)$$

$$\frac{dw}{dt} = Kw\left(\frac{w-40}{120}\right); \quad \frac{dw}{w(w-40)} = \frac{Kdt}{120}$$

$$\int\frac{Adw}{w} + \int\frac{Bdw}{(w-40)} = \frac{Kdt}{120}; A = -\frac{1}{40}; B = \frac{1}{40}$$

$$\frac{1}{40(w-40)} - \frac{1}{40w} = \frac{Kdt}{120}; \text{ Integrating}$$

$$\frac{1}{40}\ln(w-40) - \left(\frac{1}{40}\right)\ln w = \frac{Kt}{120} + \text{Const.}$$

$$\left(\frac{1}{40}\right)\ln\left(\frac{w-40}{w}\right) = \frac{Kt}{120} + \text{Const. ; At} = t = 0,\ w = 20$$

$$\left(\frac{1}{40}\right)\ln\left(-\frac{20}{20}\right) = \text{Const.; Const.} = 0$$

$$\ln\left(\frac{40-w}{w}\right) = -\frac{40Kt}{120} = -\frac{Kt}{3}$$

at $t = 12$ minutes, $\frac{w}{120} = \frac{1}{30}$; $w = 4$ kg.

For K value: $\ln\left(\frac{40-w}{w}\right) = \ln\left(\frac{40-4}{4}\right) = -\frac{12K}{3}$; $K = -\ln\frac{9}{4} = -0.5493$

$$w = \frac{40}{[1+e^{0.5493\,t/3}]} \textbf{ Ans.}$$

Example 2.12 The compositions of co-existing vapour and liquid phases of a binary mixture are to be determined experimentally by means of Othmer still. The apparatus consists of a boiler connected to a condenser by means of a thermally insulated vapour tube. The condensate collects in a liquid sample trap, the overflow from the trap being returned to the boiler. After boiling has commenced, liquid is condensed in the trap until the trap is filled. At this point the condensate from the trap runs back by the still. For the following conditions obtain an equation relating the difference in composition between vapour leaving the still and the liquid in the trap, to the total amount of liquid condensed (consider both the cases of periods before and after trap is filled). S and T moles of mixture containing x_0 mole fraction of more volatile component are initially charged to the still. The contents of still and trap are completely mixed at all times. When full, the trap holds T moles of liquid. The equilibrium relation is $y^* = mx + b$. Let D represent the total moles of liquid that have been condensed since the state of boiling, Z is mole fraction of more volatile component in the trap, holdups in liquid and vapour phase may be neglected.

Solution: Using Rayleigh's equation (differential equation).

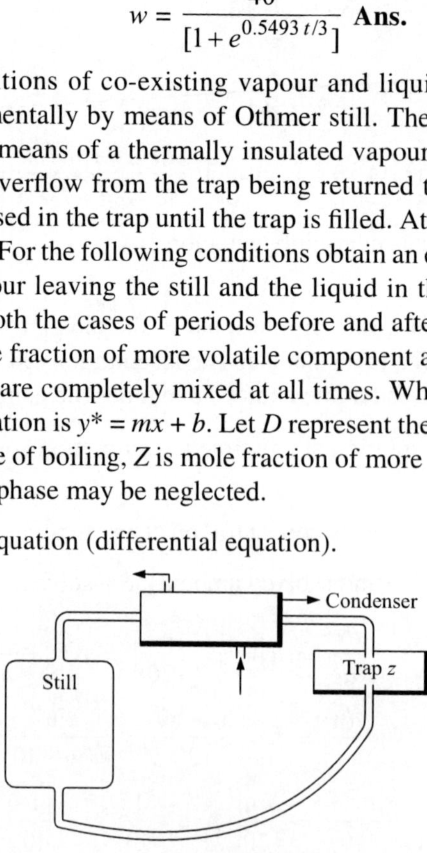

Fig. 2.6

$$\int \frac{dx}{w} = \int \frac{dx_w}{(y^* - x_w)}; \; (y^* - z)$$

to the total amount of liquid condensed. Total amount condensed = D;

$$y^* = f(D); \; w = S + T - D;$$

Before the trap is filled,

$$dw = -dD; \; y^* = m \times xw + b;$$

$$\int \frac{dw}{w} = \int \frac{dx_w}{(y^* - x_w)}$$

$$\int -\frac{dD}{(S + T - D)} = \int \frac{dx_w}{(x_w(m-1) + b)}$$

$$\text{Integrating: } \ln(S + T - D) = \frac{1}{m-1} \ln\left[b + x_w(m-1)\right] + C_1$$

when $D = 0$;
$$x_w = x_0; \; C_1 = \ln(S + T) - \frac{1}{m-1} \ln(b + x_w(m-1));$$

Then the equation becomes:

$$\ln(S + T - D) = \frac{1}{m-1} \ln(b + x_w(m-1)) \ln(S + T)$$

$$- \frac{1}{m-1} \ln(b + x_0(m-1)) \ln \frac{(S + T - D)}{S + T}$$

$$= \frac{1}{m-1} \ln\left[\frac{b + x_w(m-1)}{b + x_0(m-1)}\right];$$

$$y^* = mx_w + b; \; x_w = \left(\frac{(y^* - b)}{m}\right)$$

$$\frac{(S + T - D)^{(m-1)}}{S + T} = \frac{b + x_w(m-1)}{b + x_0(m-1)}; \; b + x_0(m-1)\frac{(S + T - D)}{S + T} = b$$

$$x_w(m-1) = \frac{y^* - b(m-1)}{m}$$

$$y^* - b = m\left(x_0 + \frac{b}{m-1}\right)\left(\frac{S + T - D)}{S + T}\right)^{m-1} - \frac{mb}{m-1}$$

$$y^* = m\left(x_0 + \frac{b}{m-1}\right)\left(\frac{S + T - D}{S + T}\right)^{m-1} - \frac{b}{m-1}$$

$$Dz = (S + T)x_0 - w\,x_w; \; w = S + T - D;$$

$$Dz = (S + T)\,x_0 - (S + T - D)\frac{(y^* - b)}{m};$$

Our aim is to get: $y^* - z$.

$$y^* - z = \left[m\left(x_0 + \frac{b}{m-1} \right)\left(\frac{(S+T-D)}{S+T} \right)^{m-1} - \frac{b}{m-1} \right]$$

$$\left(\frac{(S+T-D)}{mD} + 1 \right) - \frac{b(S+T-D)}{mD} - \frac{(S+T)}{D} * x_0.$$

After trap is filled, $(D_{in} - T)z = -Dy^*_{out} = 0;\ z = \dfrac{Dy^*}{(D-T)};\ y^* - z = y^*\left(1 - \dfrac{D}{D-T}\right)$

Example 2.13 N gm of a solid material was placed in w gm of water at time t_0. The liquid was continuously stirred and maintained at constant temperature. At the end of t_1 seconds, N_1 gm of solid remained undissolved. At the end of a long period of time, N_2 gm of solid remained undissolved. Set up the differential equations required to determine the rate of solution of the solid in terms of N_0, N_1, N_2 and t_0, t_1. Do not integrate the expression.

Data and notes: It is assumed that the rate of solution is proportional to:

1. Surface area of the material.
2. The concentration driving force.
3. The concentration is expressed as gm of solid per gm of water, the original, solid considered of S spheres, each of initial diameter D_0 foot.

Solution: Original solid = S spheres

Volume is $S*\left(\dfrac{4}{3}\right)\pi\left(\dfrac{D_0}{2}\right)^3 \rho_S = N_0$ gm.

Surface area of one sphere = $4\pi\left(\dfrac{D_0}{2}\right)^2;\ N = N_0$ at $t = 0$

$N = N$ at $t = t$; N = gm of solid/gm of water

w = gm of water, constant;

N_2 = equilibrium amount at time t_2

S = number of spheres.

$$\frac{dN}{dt} \propto S\,4\pi\left(\frac{D}{2}\right)^2 \times \frac{(N-N_2)}{w};\ N = S\left(\frac{4}{3}\right)\pi\left(\frac{D}{2}\right)^3 \rho_S$$

$$\left(\frac{D}{2}\right)^2 = \left(\frac{3N}{\rho_S \times S \times 4\pi}\right)$$

$$\frac{dN}{dt} \propto S \times 4\pi\left(\frac{3N}{S \times \rho_S \times 4\pi}\right)^{2/3} \frac{(N-N_2)}{w}$$

$$\frac{dN}{dt} = \frac{K4\pi \times 3^{2/3} N^{2/3}(N - N_2)}{w(S\rho_S\, 4\pi)^{2/3}};\ S = \frac{3N_0}{\rho_S\, 4\pi\left(\frac{D_0}{2}\right)^3}$$

$$\frac{dN}{dt} = \frac{K\, 4\pi 3^{2/3}\left[\rho_S\, 4\pi\left(\frac{D_0}{2}\right)^3\right](N^{2/3} - N^{2/3}N_2)}{w_0(N_0 \times 3)^{2/3}(\rho_S\, 4\pi)^{2/3}}$$

$$\frac{dN}{dt} = \frac{K\pi\, D_0^2(N^{5/3} - N^{2/3}N_2)}{(wN_0)} \quad \textbf{Ans.}$$

Example 2.14 For the figure given below:

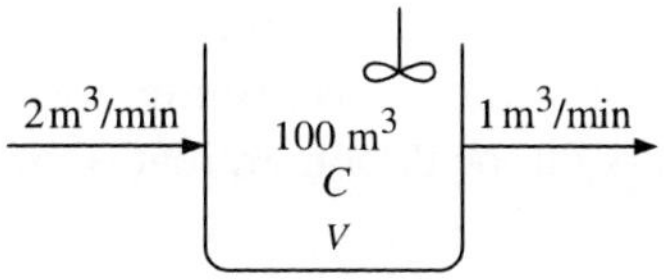

Fig. 2.7

(a) Obtain value of volume (V) in the tank as a function of time.

(b) Obtain concentration C as a function of time.

(c) Calculate C when $V = 120$ m^3.

Solution: (a) By overall balance:

$$2 - 1 = \frac{dV}{dt}$$

$$\frac{dV}{dt} = 1; \text{ Integrating: } V = t + \text{const.}; \text{ at } t = 0,\ V = 100 \text{ m}^3;$$

So Const. = 100; $V = 100 + t$ **Ans.** For (a)

(b) By component balance:

$$2 \times 2 - C \times 1 = \frac{d}{dt}(V \times C) = C\frac{dV}{dt} + V\frac{dC}{dt}$$

$$4 - C = C \times 1 + (100 + t)\frac{dC}{dt}$$

$$4 - 2C = (100 + t)\frac{dC}{dt};\ \frac{dC}{4 - 2C} = \frac{dt}{100 + t}$$

Integrating: $$\frac{-\ln(4 - 2C)}{2} = \ln(100 + t) + \text{const.}; \text{ at } t = 0,\ C = 0$$

$$\text{Const.} = \frac{-\ln 4}{2} - \ln 100;\ \frac{-\ln(4 - 2C)}{2} = \ln(100 + t)\frac{-\ln 4}{2} - \ln 100$$

$$\ln\left(\frac{4}{4-2C}\right) = 2\ln\left(\frac{100+t}{100}\right); \quad \frac{4}{4-2C} = \frac{(100+t)^2}{100}$$

$$C = 2 - \frac{2}{\left(\frac{(100+t)}{100}\right)^2}$$

(c) When $V = 120$; $t = 20$ minutes; $C = 2 - \frac{2}{\left(\frac{120}{100}\right)^2} = 0.61\ \text{kg/m}^3$ **Ans.**

Example 2.15 Product from a plant flows at a rate of 2 tonnes/hour into a storage vessel, holding 20 tonnes of material, and then pumped to another process. Owing to mishap in the plant the material flowing to storage contains 1% of a contaminant. Assuming complete mixing in the vessel (a) if the flow from storage is 2 tonnes/hour and the concentration lasts for one hour, what is the peak contamination in discharge? (b) What is the peak contamination if the discharge is reduced to one tonne per hour?

Solution: Figure 2.8

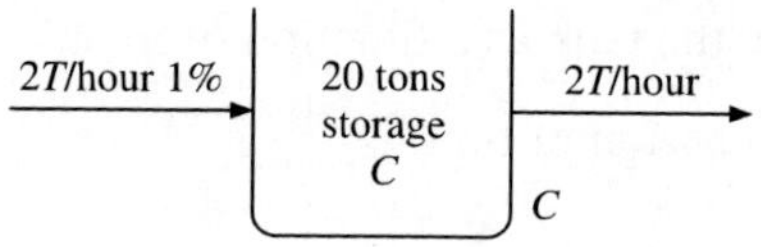

Fig. 2.8

(a) $2 \times 1 - 2 \times C = 20\frac{dC}{dt}$; $20\frac{dC}{dt} = 2 - 2C$; $\frac{dC}{dt} + 0.1C = 0.1$

$$P = 0.1;\ Q = 0.1;\ \int P dt = \int 0.1\, dt = 0.1t$$

$$\text{I.F.} = \exp\left(\int p dt\right) = e^{0.1t};$$

$$C\, e^{0.1t} = \int 0.1\, e^{0.1t} + \text{const.};\ C = e^{0.1t} + \text{const.}$$

at $t = 0$, $C = 0$, so const. = 1

$C = 1 - e^{-0.1t}$; at $t = 1$ hour,

$C = 1 - e^{-0.10} = 0.0952\%$ **Ans.**

(b) $2 \times 1 - 1 \times C = \frac{d}{dt}((20+t)C) = \frac{d}{dt}(20C + (tC))$

$$2 - 2C = 20\frac{dC}{dt} + t\ \frac{dC}{dt} = \frac{dC}{dt}(20+t)$$

$$\frac{dC}{dt} + \frac{2C}{(20+t)} = \frac{2}{(20+t)}$$

Here, $$p = \frac{2}{(20+t)};\ \int p\,dt = 2\ln(20+t)$$

$$\text{I.F.} = \int p\,dt = e^{2\ln(20+t)} = (20+t)^2$$

The solution is $$(20+t)^2 C = \int \frac{2(20+t)^2}{(20+t)}dt + \text{const.}$$

$$C = \frac{(40t+t^2)}{(20+t)^2} + \frac{\text{const.}}{(20+t)^2}$$

at $t = 0$; $C = 0$; So const. = 0;

at $t = 1$ hour, $$C = \frac{40t+t^2}{(20+t)^2};\ C = 0.093\ \%\ \textbf{Ans.}$$

Example 2.16 A small distillation unit is separating propane and butane at 135°C and initially contains 100 kg moles of mixture, whose composition is $x = 0.30$ (x = mole fraction of butane). Additional mixture is fed ($x = 0.30$) at the rate of 5 kg moles/hour. If the total volume of liquid in the still is constant and the concentration of vapour from the still (x_D) is related to x_S, mole fraction of butane remaining in the still is as follows: $x_D = \dfrac{x_S}{(1+x_S)}$. How long will it take for the value of x_S to change from 0.30 to 0.40. What is the steady state value of x_s in the still?

Solution: Figure 2.9

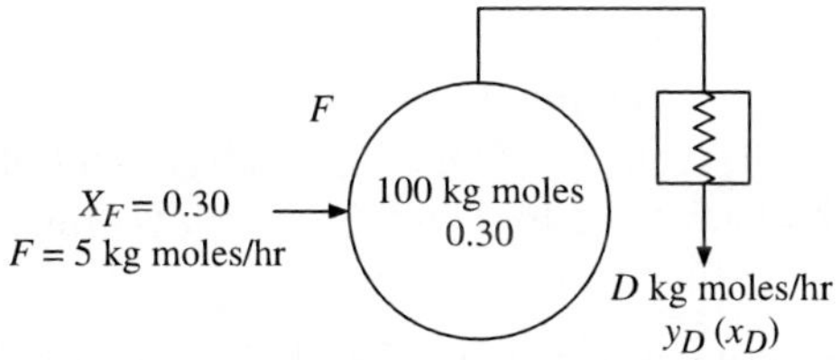

Fig. 2.9

Rate of input – Rate of output = Rate of change of accumulation

$$5 - D = 0\text{: So } D = 5 \text{ kg moles/hour}$$

Component balance (butane):

$$5 \times 0.30 - 5y_D = 100\frac{dx_w}{dt}$$

$$dt = 100\frac{dx_w}{(1.5 - 5y_D)};\ \text{Replacing } x_D \text{ by } x_S$$

$$dt = 100\frac{dx_S}{\left(1.5 - \dfrac{5x_S}{1+x_S}\right)};\ dt = \frac{(100\,dx_S + 100x_S\,dx_S)}{(1.5 - 3.5x_S)}$$

$$dt = \frac{100\,dx_S}{(1.5-3.5x_S)} + \frac{100x_S\,dx_S}{(1.5-3.5x_S)}$$

Integrating, $$\frac{100\,dx_S}{(1.5-3.5x_S)} = \left(\frac{100}{-3.5}\right)\ln(1.5-3.5\,x_S)$$

$$= -28.6\ln(1.5-5x_S)$$

$$\int 100x_s \frac{dx_s}{(1.5-3.5x_s)} = \int -28.6dx_s + \int 42.9\frac{dx_s}{(1.5-3.5x_s)}$$

$$= 28.6x_S - 12.3\ln(1.5-3.5x_s)$$

$$t = -28.6\ln(1.5-3.5x_s) - 12.30\ln(1.5-3.5x_s)$$

$$-28.6x_s + \text{const.};\ \text{At } t = 0,\ x_s = 0.30$$

Giving $$\text{Const.} = 8.58 + 40.9\ln(0.45);$$

$$t = 40.9\ln\left[\frac{0.45}{(1.5-3.5x_s)}\right] + 8.58 - 28.6x_s$$

At $x_s = 0.40$; $$t = 40.9\ln 4.5 - 2.86 = 5.87 \text{ hours}$$

(b) $1.5 - 5y_D = 0;\ y_D = x_D = \dfrac{x_s}{(1+x_s)};\ 1.5 = \dfrac{5x_s}{(1+x_s)};\ x_s = 0.436$ **Ans.**

Example 2.17 Three tanks of 10,000 litres capacity are each arranged so that when water is fed to the first tank an equal quantity of solution overflows from the first tank to the second tank likewise from second to third and from third to some point out of the system. Agitators keep the contents of each tank uniform in concentration. To start let each of the tanks be full of solution of concentration C_0 kg/litre. Run water into the first tank at 50 lpm and let overflows function as described above.

(a) Calculate the time required to reduce the concentration in the first tank to $\dfrac{C_0}{10}$.

(b) Calculate the concentrations in the other two tanks at this time.

Solution: Figure 2.10

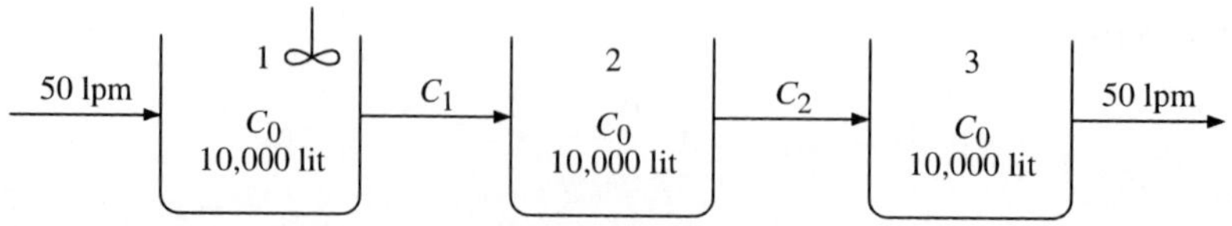

Fig. 2.10

Material balance for tank (1) is

$$50 \times 0 - 50 \times C_1 = 10000\frac{dC_1}{dt};\ 10000\frac{dC_1}{dt} = -50C_1$$

Using separation variable method:

$$200\frac{dC_1}{dt} = -C_1;\ 200\frac{dC_1}{C_1} = -dt;\ \ln C_1 = -\frac{t}{200} + \text{const.};$$

or

$$C_1 = K_1 \exp\left(-\frac{t}{200}\right);$$

where K_1 is a constant; At $t = 0$, $C_1 = C_0$; So $K_1 = C_0$

$$C_1 = C_0 \exp\left(-\frac{t}{200}\right) \quad \ldots(1)$$

Time for $C_1 = \frac{C_0}{10}$ is $\frac{C_0}{10} = C_0 \exp\left(-\frac{t}{200}\right)$; $t = 460.6$ minutes.

For tank (2)

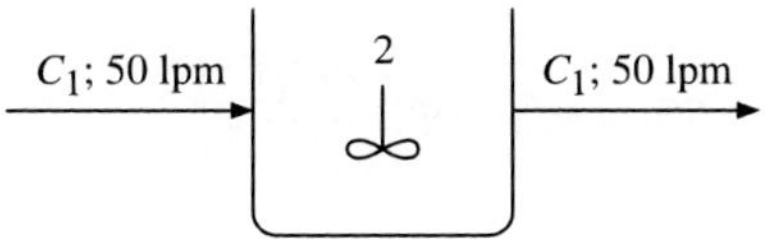

Fig. 2.11

$$50 \times C_1 - 50 \times C_2 = 10{,}000\frac{dC_2}{dt};\ C_1 = C_0\, e^{-\frac{t}{200}}$$

$$50\, C_0\, e^{-\frac{t}{200}} - 50\, C_2 = 10{,}000\frac{dC_2}{dt}$$

$$10{,}000\frac{dC_2}{dt} + 50\, C_2 = 50\, C_0\, e^{-\frac{t}{200}}$$

This is Leibnitz form: $\frac{dy}{dx} + Py = Q(x)$

Here, $P = \frac{1}{200};\ Q = \left(\frac{C_Q}{200}\right) e^{-\frac{t}{200}}$

$$\frac{dc_2}{dt} + \frac{c_2}{200} = \frac{C_0}{200} e^{-\frac{t}{2000}}$$

$$\text{I.F.} = e^{\int P dt} = e^{\frac{t}{200}};\ C_2\, e^{\frac{t}{200}} = \frac{C_0 t}{200} + \text{const.}$$

(Using the following equation)

$$C_2\, e^{\frac{t}{200}} = \int \frac{C_0}{200} e^{\frac{-t}{200}} e^{\frac{t}{200}}\, dt + \text{const.}$$

at $t = 0$, $\qquad C_2 = C_0$, giving const = C_0

So, $$C_2 = C_0\left(e^{-\frac{-t}{200}} + \frac{t}{200}e^{-\frac{-t}{200}}\right) = C_0\, e^{\frac{-t}{200}}\left(\frac{200+t}{200}\right)$$

At $t = 460.6$ minutes; $C_2 = 0.312\ C_0$ **Ans.**

For tank (3)

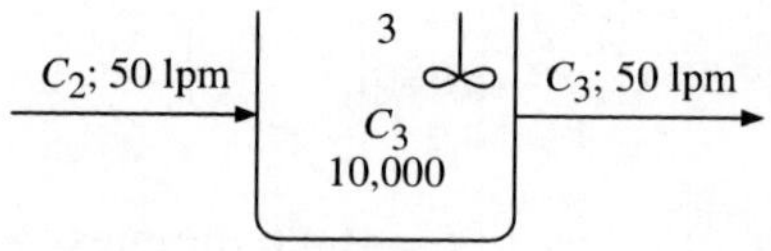

Fig. 2.12

By component balance:

$$50C_2 - 50C_3 = 10{,}000\frac{dC_3}{dt}$$

$$\frac{dC_3}{dt} + \frac{C_3}{200} = \frac{C_0(200+t)}{200 \times 200}e^{-\frac{t}{200}}$$. This is of Leibnitz form

$$\frac{dy}{dx} + Py = Q. \text{ Here } P = \frac{1}{200};\ Q = \frac{C_0(200+t)}{200 \times 200}e^{\frac{-t}{200}}$$

$$\text{I.F.} = e^{\int P dt} = e^{\frac{t}{200}};$$

$$C_3\, e^{\frac{t}{200}} = C_0\int \frac{(200+t)}{200 * 200}e^{\frac{-t}{200}}\, e^{\frac{t}{200}}\, dt + \text{const.}$$

$$C_3\, e^{\frac{t}{200}} = C_0\frac{(200t + 0.5t^2)}{200 * 200} + \text{Const.};\ \text{at } t = 0. C_3 = C_0;$$

Giving constant C_0,

$$C_3 = C_0\, e^{\frac{-t}{200}}\left[1 + \frac{(200t + 0.5t^2)}{200 * 200}\right]$$

At $t = 460.6$ minutes

$$C_3 = C_0 e^{-2.303}\left[1 + \frac{200 \times 460.6 + 0.5 \times 460.6^2}{200 * 200}\right]$$

$C_3 = 0.597C_0$ **Ans.**

Example 2.18 A large meeting room has a volume of 600 m^3. Before the meeting the carbon dioxide content in the room is 0.04%. After the meeting is over the concentration is 0.12%. The room is ventilated by 300 cubic m per hour of fresh air (0.04%). What will be the CO_2 concentration after 6 hours. What must be the ventilation rate of air in order to maintain the concentration in the room at 0.08% carbon dioxide in 6 hours?

Solution: Figure 2.13

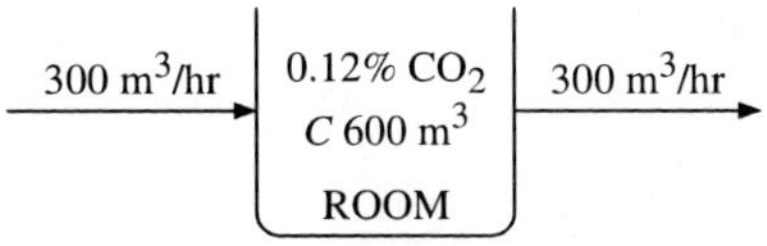

Fig. 2.13

Carbon dioxide balance equation is

$$300 \times 0.04 - 300 \times C = 600\frac{dC}{dt};\ 12 - 300C = 600\frac{dC}{dt}$$

$$\frac{dC}{dt} + 0.5C = 0.02$$

This is Leibnitz equation of the form,

$$\frac{dy}{dx} + Py = Q$$

Here, $P = 0.50$; $Q = 0.02$; I.F. $= \exp\left(\int Pdt\right) = \exp(0.5t)$

The solution is

$$\exp(0.5t) \times C = 0.04 \exp(0.5t) + \text{Const.}$$

At $t = 0$, $C = 0.12$; Const. $= 0.08$

$$C = 0.04 + 0.08 \exp(-0.5t);$$

At $t = 6$ hours, $C = 0.04 + 0.08 \times \exp(-3.0) = 0.044\%$ **Ans.**

(b) $$F \times 0.04 - F^*C = 600\frac{dC}{dt};\ 600\frac{dC}{dt} + FC = 0.04F$$

$$\frac{dC}{dt} + \left(\frac{F}{600}\right)C = \frac{0.04F}{600};\ P = \frac{F}{600};\ \int Pdt = \int \frac{F}{600}dt = \frac{Ft}{600}$$

$$\text{I.F.} = \int \exp(Pdt) = \exp\left(\frac{Ft}{600}\right)$$

The solution is $C\exp\left(\frac{Ft}{600}\right) = \int\left(\frac{0.04\,F}{600}\right)\exp\left(\frac{Ft}{600}\right)dt + \text{Const.}$

$$C\exp\left(\frac{F \times t}{600}\right) = 0.04\exp\left(\frac{Ft}{600}\right) + \text{Const.};\ C = 0.04 + \text{Const.}\exp\left(\frac{-Ft}{600}\right)$$

At $t = 0$, $C = 0.12$, giving const. as 0.08.

At $t = 6$, $C = 0.08$; giving $F = 69.32$ cu.m/hour. **Ans.**

Example 2.19 Diffusion with chemical reaction

A gas is absorbed by a liquid solvent which it reacts chemically. The rate of reaction in liquid is assumed to be proportional to the concentration gradient. The rate of reaction is of first order. The rate of reaction is proportional to the concentration of solute gas in the liquid. Obtain an equation for the concentration of liquid as a function of the distance from the interface.

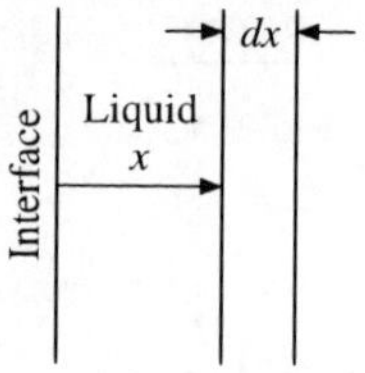

Fig. 2.14

Solution: Taking a small element dx in liquid phase situated at a distance x from the interface:

Rate of input of solute to dx element – Rate of output of solute

From dx element + Change due to reaction in dx element = Rate of change of accumulation of solute in dx element.

$$-D\frac{dC}{dx} - x - \left[D\frac{dC}{dx} + \frac{d}{dx}\left(D\frac{dC}{dx}\right)dx\right] - KC\,dx\,A = 0; \quad \frac{d}{dx}\left(D\frac{dC}{dx}\right)dx - KC\,dx = 0$$

If D is independent of x, $\dfrac{Dd^2C}{dx^2} = KC; \quad \dfrac{d^2C}{dx^2} = \left(\dfrac{K}{D}\right)C = \alpha C$

By linear differential equation methods as C.S. and P.I.:

$$C(D^2 - d) = 0; D = +/\text{- Sq.Rt } (\alpha)$$

$$y_c = C_1 \exp(\sqrt{\alpha})x + C_2 \exp(-\sqrt{\alpha x})$$

$$= A \sinh\sqrt{\alpha x} + B \cosh\sqrt{\alpha x}$$

Example 2.20 Determine x as a function of time and other physical constants.

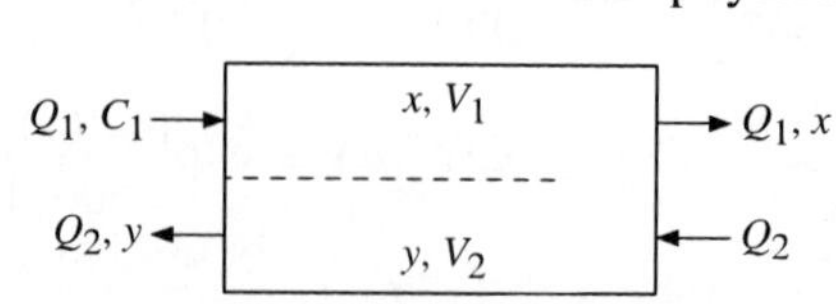

Fig. 2.15

Solute balance: Rate of input of solute to kettle – Rate of output of solute from kettle – Change due to mass transfer = Rate of accumulation of solute in kettle.

$$Q_1C_1 - Q_2y - Q_1x = V_1\frac{dx}{dt} + V_2\frac{dy}{dt}\ ;\ y = mx$$

$$Q_1C_1 - Q_2mx - Q_1x = V_1\frac{dx}{dt} + V_2m\frac{dy}{dt}$$

$$\frac{dx}{dt}(V_1 + V_2m) + x(Q_1 + Q_2m) = Q_1C_1$$

$$\frac{dx}{dt} + \left[\left(\frac{Q_1 + Q_2m}{(V_1 + V_2m)}\right)\right]x = \frac{Q_1C_1}{(V_1 + V_2m)}\ ;\ P = \frac{(Q_1 + Q_2m)}{(V_1 + V_2m)}$$

$$\int Pdt = \left[\frac{(Q_1 + Q_2m)}{(V_1 + V_2m)}\right]t$$

$$\exp\left(\int Pdt\right) = \exp\left[\frac{(Q_1 + Q_2m)}{(V_1 + V_2m)}\right]t$$

Solution is : $x\ \exp\left[\frac{(Q_1 + Q_2m)}{(V_1 + V_2m)}\right]t = \int\left[\frac{Q_1C_1}{(V_1 + V_2m)}\right]\exp\left[\frac{(Q_1 + Q_2m)}{(V_1 + V_2m)}\right]t + \text{Const.}$

$$x\ \exp\left[\frac{(Q_1 + Q_2m)}{(V_1 + V_2m)}\right]t = \left[\frac{(Q_1C_1)}{(Q_1 + Q_2m)}\right]\exp\left[\frac{(Q_1 + Q_2m)}{(V_1 + V_2m)}\right]t + \text{Const.}$$

At $t = 0$; $\quad x = y = 0;\ \text{Const.} = -\frac{Q_1C_1}{(Q_1 + Q_2m)}$

So, $\quad x = \left[\frac{Q_1C_1}{(Q_1 + Q_2m)}\right]\left\{1 - \exp\left[-\frac{(Q_1 + Q_2m)}{(V_1 + V_2m)}\right]t\right\}$

At steady state: $\quad x = \frac{Q_1C_1}{(Q_1 + Q_2m)}$

18. Kinetics fundamentals

$$r_i = \left(\frac{1}{V}\right)\frac{dN_i}{dt} = \frac{\text{Moles of } i \text{ formed}}{(\text{Volume of fluid} \times \text{Time})}$$

$A + B \rightleftarrows R + S$; r_R (Forward) $= K_1\, C_A\, C_B$; r_R (Reverse) $= K_2\, C_R\, C_S$

$aA + bB + cC + \ldots = dD + eE + fF + \ldots$ Gives the relation

$$-\frac{dN_A}{a} = -\frac{dN_B}{b} = -\frac{dN_C}{c} \ldots = \frac{dN_D}{d} = \frac{dN_E}{e} = \frac{dN_F}{f}$$

$$r_A = -KC_A^a\, C_B^b\, C_C^c = \left(\frac{1}{V}\right)\frac{dN}{dt}\ ;\ r_A = -KP_A^a\, P_B^b\, P_C^c$$

For constant volume reactor

$$\frac{dN}{dt} = -KN_A\, C_A^{a-1}\, C_B^b\, C_C^c$$

For constant volume

$$\frac{dN}{dt} = -K'N_A^a\, N_B^b\, N_C^c$$

Example 2.21 Tubular reactor

A first order reversible chemical reaction $A \rightarrow B$ is carried out isothermally in a tubular reactor of length L and cross-sectional area S. The rate of reaction is given by: $= K_1C_A$. A known mixture of A and B and inert are fed to the reactor at a constant rate. Find the steady state concentration of A as function of the length of the reactor.

If the above reactor is connected in series with a constant volume continuous stirred tank reactor (CSTR) to which is also maintained as the same temperature as tubular reactor, find the total conversion and the fraction taking place in CSTR.

Solution: Figure 2.16

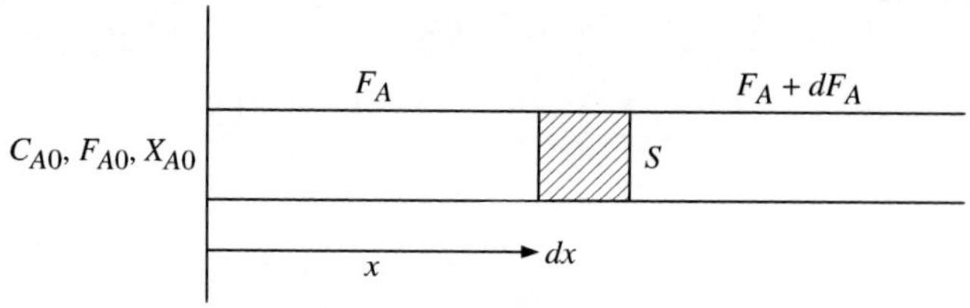

Fig. 2.16

Taking an element of thickness dx located at a distance x from the beginning of the reactor, by mass balance:

Rate of input of component A to dx element, moles/time – Rate of output of component A from dx element, moles/time – Change of component A due to reaction in dx element, moles/time = Rate of accumulation of component A in dx element, moles/time.

$$F_{A0} \times x_{A0}\,(F_A) - (F_A + d_{FA}) - (-r_A \times S \times dx) = 0 \text{ (steady state)}$$

$$F_A - F_A - d_{FA} - r_A\, S\, dx = 0;\; dF_A = dF_{A0}\,(1 - x_A) = -F_{A0}\, dx_A$$

$$F_{A0}\, dx_A = -r_A\, S\, dx;\; \int_{x_{A0}}^{x_{Af}} \frac{dx_A}{-r_A} = S\int_0^x -\frac{dx}{F_{A0}}$$

Here, $$r_A = K_1\, C_A$$

$$\int_{C_{A0}=C_{Ai}}^{C_{Af}} \frac{dC_A}{-K_1\, C_A\, C_{A0}} = S\int_0^x -\frac{dx}{F_{A0}};\; \frac{-1}{K_1 C_{A0}} \ln\left(C_{Af} - C_{A0}\right) = \frac{Sx}{F_{A0}}$$

$$\ln\left(\frac{C_{Af}}{C_{A0}}\right) = -\frac{Sx\,K_1\,C_{A0}}{F_{A0}};\ \frac{C_{A0}}{C_{Af}} = \exp\left(K_1\frac{xSC_{A0}}{F_{A0}}\right)$$

or

$$C_{Af} = C_{A0}\exp\left(-K_1 x\,S\frac{C_{A0}}{F_{A0}}\right) \textbf{ Ans.}$$

For CSTR:

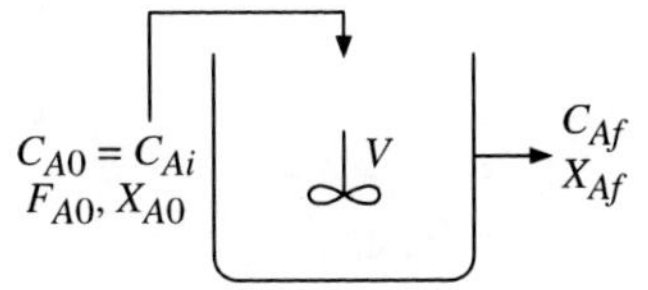

Fig. 2.17

The component balance equation is

Input of component A–output of component A–change of A due to reaction = 0 (steady state)

$$F_{A0}\times C_{A0} - F_{A0}\,C_{Af} + K_1\,V_{CAf} = 0;\ C_{Af} = C_{A0}(1 - x_{Af})$$

$$C_{A0}\times F_{A0} = C_{Af}(K_1\,V + F_{A0});\ \frac{C_{Af}}{C_{A0}} = \frac{F_{A0}}{(F_{A0}+KV)}$$

$$C_{Af} = C_{A0}\times\frac{F_{A0}}{(F_{A0}+K_1V)}$$

Tubular reactor conversion:

$$C_{Af1} = C_{A0}\exp\left(-K_1 x\,S\frac{C_{A0}}{F_{A0}}\right)$$

CSTR conversion:

$$C_{Af2} = C_{Af1}\frac{F_{A0}}{(F_{A0}+K_1V)}$$

$$\text{Fractional conversion} = \frac{C_{Af2}}{(C_{Af1}+C_{Af2})}$$

Example 2.22 It is desired to produce a substance B from raw material A in a continuous stirred tank reactor of effective volume, Vm^3. If $\frac{Q\,\text{m}^3}{\text{minute}}$ of a solution of A of concentration C_0 is fed to the empty reactor and the chemical reaction of A is represented by: $A \underset{K_2}{\overset{K_1}{\rightleftarrows}} B \xrightarrow{K_3} C$ in which all the reactions are of first order. Show that the number of moles of B in the initial discharge from the reactor is given by the solution of the differential equation.

$$\frac{d^2 N_B}{dt^2} + P\frac{dN_B}{dt} + RN_B = S$$

where $P = K_1 + K_2 + K_3$; $R = K_1 K_3$; $S = QC_0 K_1$

Solution: Components are

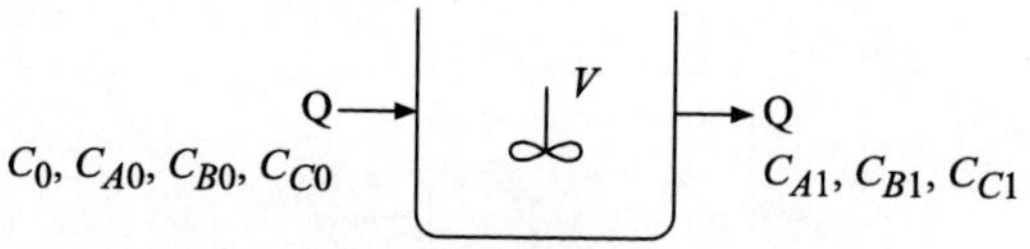

Fig. 2.18

The material balance equations for the components are:

Rate of accumulation in the reactor = input – output – change due to reaction

$$V\frac{dC_{A1}}{dt} = QC_{A0} - QC_{A1} - (VK_1 C_{A1} - VK_2 C_{B2}) \quad \ldots(1)$$

$$V\frac{dC_{B1}}{dt} = QC_{B0} - QC_{B1} - (VK_2C_{B1} + VK_3CB_1 - VK_1 C_{A1}) \quad \ldots(2)$$

$$\frac{VdC_{C1}}{dt} = QC_{C0} - QC_{C1} - (-VK_3C_{B1}) \quad \ldots(3)$$

Differentiating Eq. (2) with respect to time t gives

$$\frac{Vd^2C_{B1}}{dt} = -Q\frac{dC_{B1}}{dt} - VK_2\frac{dC_{B1}}{dt} - VK_3\frac{dC_{B1}}{dt} + VK_1\frac{dCA_1}{dt} \quad \ldots(4)$$

Substituting for $\frac{dC_{A1}}{dt}$ from Eq. (1):

$$V\frac{d^2C_{B1}}{dt} = -Q\frac{dC_{B1}}{dt} - VK_2\frac{dC_{B1}}{dt} - VK_3$$

$$\frac{dC_{B1}}{dt} + K_1(QC_0 - QC_{A1} - VK_1CA_1 + VK_2C_{B1})$$

Substituting for C_{A1} from Eq. (2):

$$VK_1C_{A1} = V\frac{dC_{B1}}{dt} + QC_{B1} + VK_2C_{B1} + VK_3C_{B1}$$

Equation (4) becomes:

$$V\frac{d^2C_{B1}}{dt^2} = -\frac{QdC_{B1}}{dt} - VK_2\frac{dC_{B1}}{dt} - VK_3\frac{dC_{B1}}{dt} + K_1QC_0 + VK_2K_1C_{B1}$$

$$-Q\left(\frac{dC_{B1}}{dt} + \frac{QC_{B1}}{V} + K_3C_{B1} + K_2C_{B1}\right) - K_1\left(Vd\frac{C_{B1}}{dt} + QC_{B1} + VK_2C_{B1} + VK_3C_{B1}\right)$$

Rearranging: $$\frac{Vd^2C_{B1}}{dt^2} = \frac{-VdC_{B1}}{dt}\left(\frac{2Q}{V}+K_1+K_2+K_3\right)$$

$$- VC_{B1}\left(\frac{Q^2}{V^2}+\frac{Q}{V}K_2+\frac{QK_3}{V}+\frac{QK_1}{V}+K_1K_3\right)+K_1Q\,C_0$$

Terms containing $\frac{Q}{V}$ does not come until steady state is reached or reactor is filled completely.

Using $N = V * C$

$$\frac{d^2N_B}{dt^2}+\frac{dN_B}{dt}(K_1+K_2+K_3)+(K_1K_3)N_B$$

$$= QC_0\,K_1 \text{ where } K_1+K_2+K_3 = P;\ K_1K_3 = R;\text{ and } S = QC_0\,K_1$$

Example 2.23 (a) Derive an expression to relate x the moles of C formed with time t for the following reaction between A and B.

$$mA + nB - pC$$

Assume that the rate of formation of C is proportional to the product of concentrations of A and B, initially a moles of A and b moles of B are taken in the reactor, given $\frac{m}{n} = \frac{a}{b}$.

(b) In the reaction $A + 2B = C$, initially 40 moles of A and 80 moles of B are charged to the reactor. If 12 moles of C are formed in 20 minutes, determine the time required for the formation of 15 moles of C (assume that the rate of formation of C is proportional to the product of concentrations of A and B).

Solution: (a) $$mA + nB = pC;\ \left(\frac{m}{p}\right)A+\left(\frac{n}{p}\right)B = C$$

For x moles of C formed:

$$\left[\left(\frac{a-mx}{p}\right)\right]\left(\frac{b-nx}{p}\right) = x_C;\ \frac{dN_C}{dt} = KN_A\,N_B$$

$$\frac{dx_C}{dt} = K\left(\frac{a-mx}{p}\right)\left(\frac{b-nx}{p}\right);\quad \frac{dx}{\left[\left(\frac{a-mx}{p}\right)\left(\frac{b-nx}{p}\right)\right]} = K\,dt$$

$$\frac{dx_C}{\left(\frac{a-mx}{p}\right)\left(\frac{b-mx}{p}\right)} = K\,dt;\quad \frac{dx_C}{ab-x\left(\frac{an}{p}+\frac{bm}{p}\right)+\frac{mn}{p}x^2} = K\,dt$$

This is of form: $\frac{dx_C}{A+Bx+Cx^2} = K\,dt$. This satisfies $B^2 = 4AC$ relation.

Integrating on both sides:

$$Kt + \text{Const.} = \frac{2P^2}{2anp - \dfrac{2an^2x}{p^2b}}$$

at $t = 0$; $x = 0$ giving Const. $= \dfrac{p}{(an)}$

$$Kt + \frac{p}{an} = \frac{p^2}{(anp - mn\,x)}; \; x = \frac{ap}{m} - \frac{p^2a}{m\,(K\tan + p)} \quad \textbf{Ans.}$$

(b) $m = 1; n = 2; p = 1; a = 40; b = 80; \dfrac{m}{n} = \dfrac{a}{b};$

Relation is satisfied.

From part (a), $t = \dfrac{p^2}{K(anp - mnx)} - \dfrac{p}{Kan}$

12 moles of C in 20 minutes, find t for

$$C = 15; K = \frac{1}{\tan}\left(\frac{p^2an}{anp - mnx} - 1\right)$$

$$= \frac{1}{20 \times 40 \times 2}\left[\frac{1 \times 40 \times 2}{40 \times 2 \times 1 - 1 \times 2 \times 2} - 1\right] = \frac{0.43}{1600}$$

$$t = \frac{1}{Kan}\left(\frac{80}{80 - 30} - 1\right) = \frac{12}{0.43} = 28 \text{ min. } \textbf{Ans.}$$

Example 2.24 In the experimental study of the safonification of methyl acetate by sodium hydroxide it is found that 25% of ester is converted to alcohol in 12 minutes when initial concentrations of both ester and caustic are 0.01 m. What conversion of ester would be obtained in one hour if the initial ester concentration is 0.025 m and initial caustic concentration is 0.015 m.

Solution: Methyl Acetate + Sodium Hydroxide – Sodium Acetate + Methyl Alcohol

$$a - x \qquad b - x \qquad x \qquad x$$

Assuming that the rate of reaction is proportional to concentration of reactants and the reaction is first order with respect to each of the reactants, we can write

$$\frac{dx}{dt} = K(a - x)\,(b - x) \text{ where } x \text{ is reaction rate constant,}$$

$$\frac{dx}{(a - x)\,(b - x)} = K\,dt. \text{ In this case, } a = b = 0.01 \text{ m}$$

$$\frac{dx}{(0.01 - x)^2} = K\,dt. \text{ Integrating, } \frac{1}{(0.01 - x)} = Kt + \text{Const.}$$

At $t = 0$; $x = 0$; Giving Const = 100

At $t = \frac{12}{60}$ hr = 0.2 hr.; $x = 0.25 \times 0.01 = 0.0025$ m

$$\frac{1}{(0.01 - 0.0025)} = 0.2\,K + 100; \text{ Giving } K = 166.7$$

In second case: $\frac{dx}{dt} = K(a - x)\,(b - x)$

$$= 166.7\,(0.025 - x)\,(0.015 - x); \quad \frac{dx}{(0.025 - x)(0.015 - x)}$$

$$= 166.7\,dt$$

Arranging in partial fractions:

$$-\frac{dx}{0.01}\left(-\frac{1}{0.015 - x} + \frac{1}{0.025 - x}\right) = -166.7\,dt$$

$$\left(-\frac{1}{0.025 - x} - \frac{1}{0.015 - x}\right)dx = -1.667\,dt. \text{ Integrating on both sides :}$$

ln $(0.015 - x)$ – ln $(0.025 - x) = 1.667t$ + Const.

When $t = 0$; $x = 0$; –ln(0.025) + ln(0.015) = –1.677*0 + Const.

$$\text{Const.} = \ln\left(\frac{0.015}{0.025}\right)$$

$$\ln\left(\frac{0.015 - x}{0.025 - x}\right) = -1.667 + \ln\left(\frac{0.015}{0.025}\right)$$

$$\ln\left(\frac{0.015 - x}{0.025 - x}\right) = \left(\frac{0.015}{0.025}\right)e^{-1.667} = 0.1128$$

$$0.015 - x = (0.025 - x) * 0.1128$$

$$x = \frac{0.01218}{0.8872} = 0.01373$$

$$\% \text{ Conversion based on ester} = \left(\frac{0.01373}{0.025}\right) \times 100 = 55\% \textbf{ Ans.}$$

Example 2.25 A thin rod is extending from a wall at a temperature T_0 into surroundings at a temperature T_S. The long rod is long enough so that the free end is adiabatic. Assuming that there is no temperature gradient in radial direction within the rod (T vs. x). Heat is lost to the surrounding by a heat transfer coefficient h. Assume variation of thermal conductivity K of the material with temperature is negligible. Diameter of the rod is D. Length of rod is L.

Solution: Figure 2.19

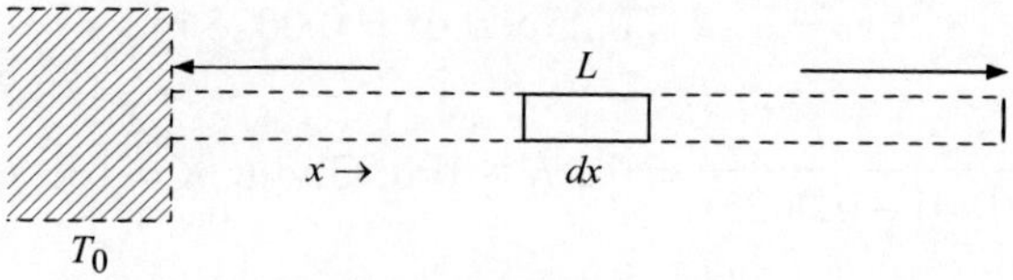

Fig. 2.19

Taking a small element of thickness *dx* situated at a distance *x*.

Input of heat to *dx* element – Output of heat to *dx* element – Output of heat from *dx* element by heat loss. = Rate of heat accumulation in *dx* element

Putting corresponding mathematical terms,

$$-KA\frac{dT}{dx} - * - \left(KA\frac{dT}{dx} + \frac{\partial}{\partial x}\left(KA\frac{dT}{dx}\right)dx - Sh(T - T_S)\right) = \frac{d}{dt}\left(dx\frac{\pi D^2}{4}\rho C_p T\right)$$

At steady state: $\frac{d}{dt}$ term is zero, giving:

$$-KA\frac{dT}{dx} + KA\frac{dT}{dx} + \frac{d}{dx}\left(KA\frac{dT}{dx}\right)dx = Sh(T - T_S)$$

where $S = \pi D\, dx$ and $A = \frac{\pi D^2}{4}$

$$\frac{d}{dx}\left(KA\frac{dT}{dx}\right)dx = \pi\, dx\, h\,(T - T_S);\ \frac{K\pi D^2}{4}\frac{d^2T}{dx^2} = h\pi D\,(T - T_S)$$

$$\frac{d^2T}{dx^2} = \frac{4h}{DK} - (T - T_S)$$

Let $y = T - T_S$; $dy = dT;\ C^2 = \frac{4h}{(DK)}$

$$\frac{d^2y}{dx^2} - C^2 y = 0;\ y_C = C_1 e^{-Cx} + C_2 e^{+Cx};\ \text{At } x = 0,\ T = T_0$$

$$(T_0 - T_S) = C_1 + C_2 \quad \ldots(1)$$

Giving $C_1 = (T_0 - T_S) - C_2$

at $x = L$; $T = T_S;\ 0 = C_1 e^{-\sqrt{\frac{4h}{KD}}L} + C_2 e^{\sqrt{\frac{4h}{KD}}L}$

$$C_2 = \frac{(T_0 - T_S)}{\left(1 - e^{2\sqrt{\frac{4h}{KD}}L}\right)};\ C_1 = -C_2 e^{2\sqrt{\frac{4h}{KD}}L}$$

$$T - T_S = (T_0 - T_S) + \frac{(T_0 - T_S)\, e^{-\sqrt{\frac{4h}{KD}}x}}{\left(1 - e^{2\sqrt{\frac{4h}{KD}}L}\right)} \quad \textbf{Ans.}$$

Example 2.26 Butyl acetate is to be produced in a battery of CSTRs operating in series by reacting 2000 kg/hour of butanol, with 326 kg of glacial acetic acid per hour, containing sufficient sulphuric acid to catalyse the reaction at 100ºC. Under these conditions, the rate of reaction is expressed by the equation,

$$r = KC_A^2;\ K \text{ is } 0.28 \text{ m}^3/(\text{kg.mole.Min})$$

where C_A is concentration kg moles/m^3. If the effective volume of each tank is 10 m^3 and density of reaction mixture is $\left(\frac{48}{62.4}\right) \times 1000 \text{ kg/m}^3$, estimate the number of reaction vessels required, if the concentration of acetic acid in the final discharge should not exceed $\left(\frac{3}{62.4}\right) \times 1000$.

Solution:

$$\frac{V}{F} = \int \frac{dx}{r} = \int \frac{dC_A}{(KC_A^2)}$$

Since

$$r = KC_A^2;\ \frac{V}{F} = \frac{-1}{(KC_A)};\ \frac{V}{(2000 + 326)} \times \left(\frac{48}{62.4}\right) \times 1000$$

$$= -\frac{1}{KC_A}\Bigg|_{326/V}^{(3/62.4)1000}$$

$$= -\frac{1}{0.28}\left(\frac{1}{\left(\frac{3}{62.4}\right) \times 1000} - \frac{1}{\frac{326}{V}}\right)$$

$$-\frac{V}{2326} \times 769.2 = 3.571 \times \left(\frac{1}{48.08} - \frac{1}{\frac{326}{V}}\right)$$

$$= 3.571 \times \left(0.0208 - \frac{V}{326}\right) = (6.656 - V) \times 0.01116$$

$$0.3307\,V = (0.0743 - 0.01116\,V)$$

$$V = 0.22 \text{ m}^3$$

$$N \text{ (number of tanks)} = \frac{0.22}{10} = 1 \textbf{ Ans.}$$

Example 2.27 The reaction of chemical P to chemical Q releases heat

$$P \rightarrow Q + 770 \text{ kJ/mole}$$

Because pure P reacts explosively, the reaction is conducted in dilute water solution. Consider a batch reactor (no flow in or out) initially charged with 1.0 kg of water and 0.12 mole of P at 150ºC.

Thus, concentration of P at time zero is : $P_0 = 0.12$ mole/kg water. The reactor is thermally insulated.

(a) Calculate the temperature in the reactor after P has completely reacted to form Q. Heat capacity of solution can be taken as water.

(b) Obtain mathematical expression for temperature in reactor as a function of (P).

Solution: (a) 770 kJ/mole

For 0.12 mole: $770 \times 0.12 = 92.4$ kJ

$$mC_p\,dT = 1 \times 1 \times (T - 50) = \frac{92.4}{4.18}\text{ K Cal; } T - 50$$

$$= 22.11;\ T = 72.11^\circ\text{C}$$

(b)
$$T = 50 + \frac{770 \times C_P}{4.18}$$

where C_P is concentration of P in mole/kg water.

Example 2.28 P is a function of both x and y given by

$$P = Ax + B\left(\frac{xy - a}{y}\right) + \frac{c}{(yx)^{1/2}}$$

where A, B, C are constants.

Obtain an expression for the values of x, y in terms of constants corresponding to maximum or minimum values of P.

Solution:
$$P = Ax + B\left(\frac{xy - a}{y}\right) + \frac{C}{(xy)^{1/2}} \quad \text{...(1)}$$

$$\frac{dP}{dx} \text{ at constant } y = A + B - 0.5C\,y^{-1/2}\,x^{-3/2} = 0 \quad \text{...(2)}$$

$$\frac{A + B}{0.5C} = y^{-1/2}\,x^{-3/2} \quad \text{...(3)}$$

$$\frac{dP}{dy} \text{ at constant } x = 0.5\,B\,a\,y^{-3/2} - 0.5\,C\,x^{-1/2}\,y^{-3/2} = 0$$

$$\frac{Ba}{C} = x^{-1/2};\ x = \frac{C^2}{(B^2\,a^2)} \quad \text{...(4)}$$

Substituting Eq. (4) in Eq. (3) gives:

$$y = 0.25\,B^6\,A^6\,\frac{C^{-4}}{(A + B)^2} \quad \textbf{Ans.}$$

Example 2.29 The heat capacity of a house is 3300 kJ/ºC, that is 3300 kJ of heat raises the temperature of house by 1ºC. The heater in the house can supply heat at a maximum rate of 5.2×10^4 kJ per hour. If the inside temperature is 5ºC and the outside temperature is –15ºC., the heater at its maximum rate is kept on at 8 PM. Heat escapes from house through conduction through walls and roof. The rate of heat loss, q_{loss} in kJ/hour is proportional to the difference between inside temperature and outside temperature. The rate of heat loss $(q_{loss}) = K(T_i - T_0)$; where K is = 740 kJ/(ºC × hour).

Find out at what time the temperature in the house will be 20ºC by considering heat loss by conduction through walls and roof?

Solution: $5.2 \times 10^4 - 740\,(T - (-15)) = 3300\dfrac{dT}{dt}$

Rearranging: $$\frac{dT}{dt} + \frac{74T}{330} = \frac{409}{33};\ \frac{dy}{dx} + Py = Q$$

Leibnitz solution is :

$$P = \frac{74}{330};\ \int P dt = \frac{37t}{165}$$

$$T * e^{\frac{37t}{165}} = \int \frac{409}{33} \times e^{\frac{37t}{165}dt}\ dt + \text{Const.}$$

Integrating on both sides:

$$TXe^{\frac{37t}{165}} = \int \frac{409}{33} \times e^{\frac{37t}{165}} dt + \text{Const.}$$

At $t = 0$; $T = 5$ºC; Const. = 55.27

$$T = 55.27 - 50.27e\,(-0.224t)$$

For $T = 20$ºC; $e^{-0.224t} = 0.706$; $t = 1.11$ hours

Time = 8 + 1.11 = 9.11 hrs (p.m)

Example 2.30 In a nuclear reaction, the following equations are formulated:

$$\frac{dx}{dt} + \lambda x = 0;\ \frac{dz}{dt} = my;\ x + y + z = \gamma$$

where λ, μ, and γ are constants. $z = \dfrac{dz}{dt} = 0$ when $t = 0$. Show that:

$$z = \gamma + \frac{\gamma}{(\lambda - \mu)}(\mu_e^{-\lambda t} - \lambda e^{-\mu t})$$

Solution: $x(D + \lambda) = 0$; $D = -\lambda$; $x = C_1 e^{-\lambda t}$

where C_1 is a constant.

$$y = \gamma - x - z; \quad \frac{dz}{dt} = \mu(\gamma - x - z) = \gamma\mu - \mu x - \mu z$$

$$= \gamma\mu - \mu C_1 e^{-\lambda t} - \mu z = \frac{dz}{dt} + \mu z = \gamma\mu - \mu C_1 e^{-\lambda t}$$

$$z(D + \mu) = 0; D = -\mu$$

Complimentary solution is $z = C_2 e^{-\mu t}$ where C_2 is constant. Particular integral is

$$\frac{\gamma\mu}{(D+\mu)} = \gamma\left(\frac{1+D}{\mu}\right)^{-1} = \gamma\left(\frac{1-D}{\mu}\right) = \gamma$$

$$z = C_2 e^{-\mu t} - \frac{\mu C_1 e^{-\lambda t}}{(\mu - \lambda)} + \gamma ;$$

when $t = 0$, $z = 0$ giving:

$$C_2 = \frac{C_1\mu}{(\mu - \lambda)} - \gamma \qquad \ldots(1)$$

$$\frac{dz}{dt} = -\mu C_2 + \frac{C_1\mu\lambda}{(\mu - \lambda)} = 0 \qquad \ldots(2)$$

Solving the two Eqs. (1) and (2) simultaneously gives

$$C_1 = \gamma; \text{ and } C_2 = \frac{\gamma\lambda}{(\mu - \lambda)}$$

$$z = \frac{\gamma\lambda}{(\mu - \lambda)} e^{-\mu t} + \gamma - \frac{\gamma\mu e^{-\lambda t}}{(\mu - \lambda)}$$

$$z = \frac{\gamma}{(\lambda - \mu)}(\mu e^{-\lambda t} - \lambda e^{-\mu t}) + \gamma \textbf{ Ans.}$$

Example 2.31 Extraction of benzoic acid from benzoic acid – toluene mixture by water.

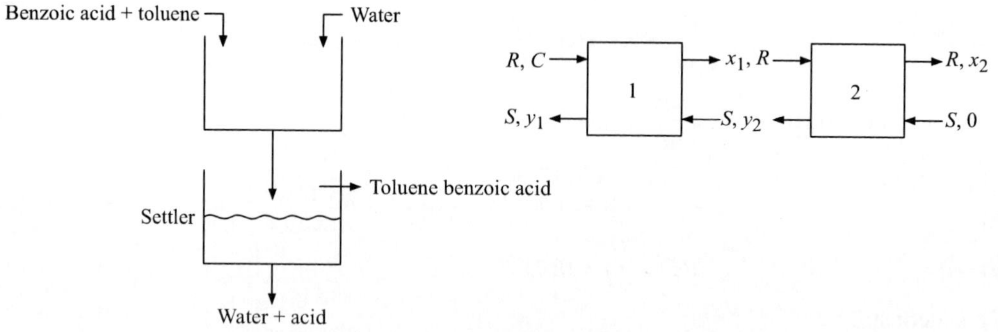

Fig. 2.20

Solution: Material balance for stage 1:

$$RC + Sy_2 = Rx_1 + Sy_1 \qquad \ldots(1)$$

For stage 2:

$$Rx_1 = Rx_2 + Sy_2; RC + mS\,x_2 = R\frac{y_1}{m} + Sy_1$$

$$\frac{Ry_1}{m} = R_2 + mS\,x_2; \text{ Eliminating } y_1;$$

$$RC + mS\,x_2 = (Rx_2 + mS\,x_2) + Sy_1 = Rx_2 + mS\,x_2 + mS\frac{y_1}{m}$$

$$= (Rx_2 + mS\,x_2) + mS\frac{(Rx_2 + mSx_2)}{R};$$

$$R(R + mS\,x_2) = (R + mS)\,(Rx_2 + mS\,x_2)$$

$$x_2 = \frac{R^2C}{R^2 + m\,RS + m^2S^2}; y_1 = \frac{mR\,C\,(RC\,mS)}{R^2 + m\,RS + m^2S^2}$$

If $$\alpha = \frac{R}{(mS)}; \frac{Sy_1}{RC} = \frac{mS\,(R + mS)}{R^2 + m\,RS + m^2R^2}$$

or $$E = \frac{\alpha + 1}{(\alpha^2 + \alpha + 1)} = \frac{\alpha^2 - 1}{\alpha^3 + 1}$$

Example 2.32 Mass transfer with first order chemical reaction

A liquid is in contact with a well mixed gas containing a substance A to be absorbed. Near the surface of the liquid there is a film of thickness across which A diffuses steadily while being consumed by a first order homogenous chemical reaction with rate constant K_1. At the gas-liquid interface, the liquid solution is in equilibrium with the gas and its concentration is C_{Ai}. At the other side of the film, its concentration is virtually zero. Assuming dilute solutions, derive an expression for the ratio of absorption flux with chemical reaction to the corresponding flux without chemical reaction.

Solution: From equation of continuity, assuming that the flux due to bulk motion is negligible, for dilute solutions, we obtain

$$\frac{d^2C_A}{dz^2} - K_1\frac{C_A}{D_{AB}} = 0; C_A = C_{Ai}\left(\frac{1-z}{\delta}\right); N_A = -D_{AB}\frac{dC_A}{dz} = C_{Ai}\frac{D_{AB}}{\delta} \qquad \ldots(1)$$

For one dimensional steady state diffusion, including a first order homogenous equation reaction, neglecting the bulk motion contribution, the equation of continuity, becomes:

$$\frac{dN'_A}{dt} = -K_1C_A; N'_A = -D_{AB}\frac{dC_A}{dz}; \frac{d^2C_A}{dz^2} - \frac{K_1C_A}{D_{AB}} = 0$$

This equation is solved with boundary conditions.

$$C_A(z) = C_{Ai}\,\sinh\frac{\left(\sqrt{\dfrac{K_1}{D_{AB}}}\,(\delta - z)\right)}{\sinh\left(\sqrt{\dfrac{K_1}{D_{AB}}}\,\delta\right)}$$

Evaluating N'_A from the expression for the concentration profile,

$$N'_A = -D_{AB}\frac{dC_A}{dz} = C_{Ai}(K_1\,D_{AB})^{1/2}\,\frac{\cosh\left(\sqrt{\dfrac{K_1}{D_{AB}}}\,(\delta - z)\right)}{\sinh\left(\sqrt{\dfrac{K_1}{D_{AB}}}\,\delta\right)}$$

The absorption flux at the gas-liquid interface is ($z = 0$) is then

$$\left.\frac{N'_A}{N_A}\right|_{z=0} = C_{Ai}(K_1\,D_{AB})^{1/2}\coth\left(\sqrt{\frac{K_1}{D_{AB}}}\,\delta\right)$$

$$\left.\frac{N'_A}{N_A}\right|_{z=0} = \frac{C_A(K_1 D_{AB})^{1/2}\coth\left(\dfrac{K_1}{D_{AB}}\right)^{1/2}\delta}{C_{Ai}\dfrac{D_{AB}}{\delta}}$$

$$= \sqrt{\frac{K_1\,\delta^2}{D_{AB}}}\left(\coth\sqrt{\frac{K_1\,\delta^2}{D_{AB}}}\right)$$

Defining $\quad D_a = K_1\dfrac{\delta^2}{D_{AB}}$ (Damkohler number);

$$\left.\frac{N'_A}{N_A}\right|_{z=0} = (D_a)^{1/2}\coth\sqrt{D_a}$$

when reaction is slow, D_a is small and expanding coth by series:

$$\frac{N'_A}{N_A}\text{ at } z = 0 = \sqrt{D_a}\left(\frac{1}{\sqrt{D_a}} + \frac{\sqrt{D_a}}{3} + \frac{3/\sqrt{D_a}}{45} + \ldots\right)$$

$$= \left(1 + \frac{D_a}{3}\right).$$

When reaction is fast: D_a is large and $\cot h \to 1$.

$$\frac{N'_i}{N_i}\text{ at } z = 0 = (D_a)^{1/2}\ \textbf{Ans.}$$

Example 2.33 Oxygen transport in muscle

August Krough proposed that oxygen was transported from capillaries to surrounding tissues by passive diffusion. He formulated a simple geometric mould model to describe the phenomena. The Krough model is a simplification of a capillary bed, a group of parallel identical units, each containing a central capillary bed and a surrounding cylinder of tissue. The goal is to predict the amount of oxygen (concentration) as a function of position within the tissue cylinder. Each tissue cylinder is assumed to be supplied with oxygen exclusively by the capillary within it. Oxygen diffusion in axial direction is neglected as evidenced by the fact that oxygen concentration gradient is steeper in radial direction than in axial direction.

Solution:

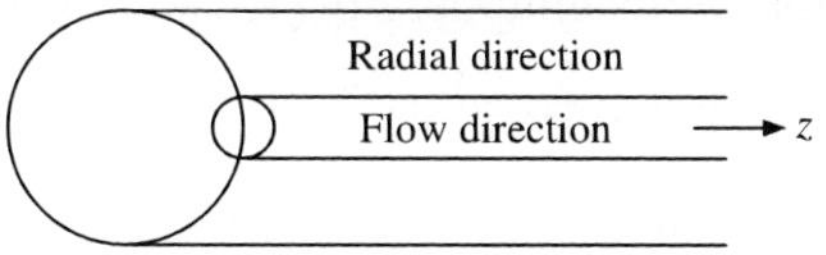

Fig. 2.21

Applying Fick's law of diffusion in radial direction, $R_C \le r_S \le R_t$ and conservation of mass leads to the equation for oxygen diffusion, in muscle tissue (Mcguire and Secomb, 2001):

$$D_{AB}\left(-\frac{1}{r}\frac{d}{dr}\left(r\frac{dC_A}{dr}\right)\right) = R_A(C_A) \qquad ...(1)$$

where C_A is concentration of oxygen in tissue, expressing concentration in terms of partial pressure, $P_a = HC_A$ where H is Henry's constant.

Equation. (1) becomes:

$$K\left(\frac{1}{r}\frac{d}{dr}\left(r\frac{dP_A}{dr}\right)\right) = M(P_A) \qquad ...(2)$$

$$R_C \le r \le R_t$$

K is Krough constant (coefficient) = D_{AB} eff, (H and $M(P_A)$ is the oxygen consumption rate per unit volume of tissue cylinder.

$$M(P_A) = \frac{M_0\, P_A}{P_0 + P_A} \qquad ...(3)$$

where M_0 is oxygen demand and P_0 is partial pressure of oxygen (Michaelis–Menten Kinetics model). The kinetics equation becomes first order if P_a is smaller than P_A and zero order for P_a larger than P_A. Assuming first order here, $M(P_A) = M_0$.

The two boundary conditions to solve Eq. (2) are:

1. $P_A(R_C) = P_{A,\,b}$
2. $\frac{dP_A}{dr}$ at $R_C = 0$

First B.C. becomes: by neglecting oxygen diffusion. Resistance of the capillary wall. $P_{A,\,b}$ is average partial pressure of oxygen in blood. Second B.C. is assumed that no oxygen is exchanged across the outer boundary of the tissue cylinder. The complete solution of Eq. (8) is (with zero order kinetics):

$$P_A(r) = P_{A,\,B} + M_0 \frac{(r^2 - R_C^2)}{4K} - \frac{M_0\, R_t^2}{2K} \ln\left(\frac{r}{R_C}\right) \qquad \ldots(4)$$

Equation (4) can predict whether under a given set of conditions there are regions where partial pressure of oxygen is less than 1 torr in the tissue.

When P_0 and P_a are of same order of magnitude, Eqs. (2) and (3) in dimensionless form are:

The solution is obtained by shooting numerical method.

$$\frac{d^2u_1}{d\eta} + \left(\frac{1}{\eta}\right)\frac{du_1}{d\eta} = u_1 \frac{D_a}{(\theta + u_1)}$$

where $u_1 = \dfrac{pA}{PA}$; $\eta = \dfrac{r}{R_t}$; $\theta = \dfrac{P_0}{P_{A,b}}$; $D_a = \dfrac{M_0 R_t^2}{(KP_{AB})}$;

with boundary conditions:

1. $\eta = \dfrac{R_c}{(R_t\, u_t)} = 1.0$

2. At $\eta = 1.0$, $\dfrac{du_t}{d\eta} = 0$

Example 2.34 For the reaction $A \rightarrow C \rightarrow B$. The initial number of species of A are 10^5. The fraction left after one minute is : $A = 0.50$; B is 0.30 and C is 0.20. Calculate the rate constants of A, B and C.

Solution: Let N_A be the number of moles of A, N_B is the number of moles of B and N_C is the number of moles of C.

Assuming first order rate equation:

$$-\frac{dN_A}{dt} = K_1 N_A; \quad -\frac{dN_B}{dt} = K_2 N_A; \quad -\frac{dN_C}{dt} = K_3 N_A$$

Giving solutions of: $N_A = N_{A0} \exp(-K_1 t)$; $N_B = N_{A0} \exp(-K_2 t)$;

$N_C = N_{A0} \exp(-K_3 t)$; At $t = 1$ minute

$0.50 = \exp(-K_1)$;

Giving $K_1 = 0.70$; $0.30 = \exp(-K_2)$

Giving $K_2 = 0.36$; $0.20 = \exp(-K_3)$

Giving $K_3 = 0.22$ **Ans.**

Example 2.35 For the reaction $A \xrightarrow{K_1} B$ obtain the expression of V_A versus time and constants. If A is increased by 25% in feed, how conversion varies with time?

Solution: Let other components be inert components along with A component. $N_B = 0$; at $t = 0$.

Let N be the total number of moles at any time. $N_{A0} + N_I = N$. For constant volume reactor,

$$C_A = \frac{N_A}{V}$$

$$C_B = \frac{N_B}{V}; r = K_1C_A - K_2C_B$$

$$-\frac{dC_A}{dt} = K_1C_A - K_2C_B; N_A + N_B + N_I = N$$

$$-\left(\frac{1}{V}\right)\frac{dN_A}{dt} = K_1\frac{N_A}{V} - K_2\frac{N_B}{V}; N_B = N - N_I - N_A$$

$$-\frac{dN_A}{dt} = K_1N_A - K_2(N - N_I - N_A)$$

$$-\frac{dN_A}{dt} = -(K_1 + K_2)N_A - K_2(N - N_I); N - N_I = N_{A0}$$

$$\frac{dN_A}{dt} + (K_1 + K_2)N_A = K_2N_{A0}$$

This is of Leibnitz form.

Here, $$P = K_1 + K_2; Q = K_2N_{A0}$$

$$\text{I.F.} = \exp\left(\int Pdx\right) = \exp[(K_1 + K_2)t]$$

The solution is

$$N_A \exp[(K_1 + K_2)t] = \int K_2(N - N_I)\exp[K_1 + K_2)t]dt = \text{Const.}$$

At $t = 0$; $$N_A = N_{A0}$$

$$N_A \exp[(K_1 + K_2)t] = \frac{N_2(N - N_I)}{(K_1 + K_2)}\exp[(K_1 + K_2)t] + N_{A0} - K_2\frac{(N - N_I)}{(K_1 + K_2)}$$

Combining the terms

$$N_A = \frac{N_{A0}\,K_2}{(K_1 + K_2)}\{1 - \exp[-(K_1 + K_2)t]\} + N_{A0}\exp[-(K_1 + K_2)t]$$

When A component is increased by 25%, $N_{A0} + N_I = N = \text{Constant}$;

$$N_{A0} = 1.25\ N_{A0}$$

$$N_A = N - 1.25\, N_{A0}$$

$$N_A = N - 1.25\, N_{A0};\; N_A = \frac{K_2 \times 1.25\, N_0}{(K_1 + K_2)}\{1 - \exp[-(K_1 + K_2)t]\} + 1.25 N_{A0} \exp[-(K_1 + K_2)t]$$

Example 2.36 Let T be the temperature at any point from bolt to the surface. Let reference temperature be surrounding temperature. By heat balance: Find temperature at any point from bolt to surface.

Solution: Heat gained by bolt = Heat lost by bolt to the surroundings – Heat lost to outer covering by conduction = Heat accumulation (zero) at S.S.

1. Heat lost by bolt to the surroundings
2. Heat lost to outer covering by conduction
3. Heat accumulation in bolt (= 0 since steady state)

Putting corresponding mathematical terms:

1. Is: $mCp(T_h - T_s) = \dfrac{\pi d^2}{4} \times t \times C_P (T_h - T_s)\rho_B$ ...(1)
2. Is: $h\pi d^2 \dfrac{(T_h - T_s)}{4}$
3. Is: $K^*d^*t^*\dfrac{dT}{dx} = 0$; d^*t = area. Let x be the distance from the centre of the bolt to outer covering.

   T_h is constant (inside temperature), T_s is constant (surrounding temperature). Substituting in the above Eq. (1)

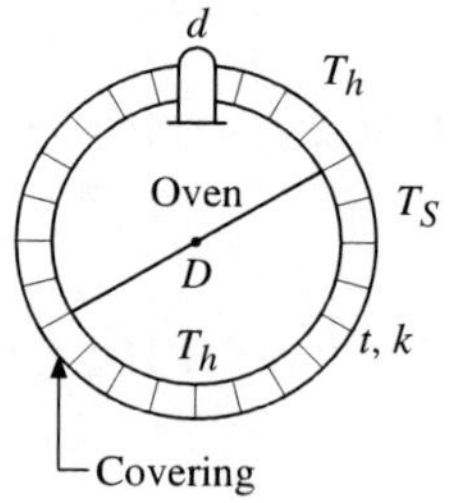

Fig. 2.22

$$\pi d^2 t\, \rho_b \frac{(T_h - T_s)}{4} - h\pi d^2 \frac{(T_h - T_s)}{4} + K^*d^*t \frac{dT}{dt} = 0$$

Cancelling common terms:

$$Kdt \frac{dT}{dx} = -\pi d^2 t\, \rho_b \frac{(T_h - T_s)}{4} + h\pi d^2 \frac{(T_h - T_s)}{4}$$

$$\frac{dT}{dx} = (T_h - T_s)\pi_d \frac{\left(\frac{h}{t} - P_B\right)}{(4K)};$$

Integrating, $$T = \pi(T_h - T_s)\frac{\left(\frac{h}{t} - p_b\right)x}{(4K)} + \text{Const.}$$

At $x = 0$, $T = T_h$;

Giving Const. as: $$T_h\text{: } T = (T_h - T_s)\frac{\left(\frac{h}{t} - p_b\right)x}{(4K)} + T_h \textbf{ Ans.}$$

Example 2.37 A sphere has initial temperature T_0 and later is kept at T_1 temperature in the outer surface. Find the temperature in the sphere as a function of time and radius, *r*.

Solution: T_0 = initial temperature of surface,

T_1 = later surface temperature;

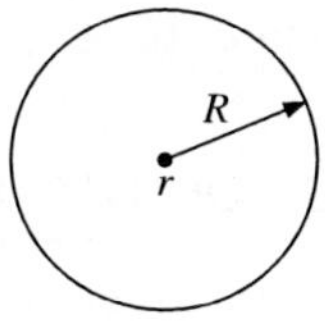

Fig. 2.23

$$\frac{dT}{dt} = C\nabla^2 T \text{ in spherical coordinates.}$$

$$C = \frac{K}{C_p\rho}$$

$$\frac{dT}{dt} = C\left[\left(\frac{1}{r^2}\right)\frac{d}{dr}\left(\frac{r^2 dT}{dr}\right) + \left(\frac{1}{r^2\sin^2\theta}\right)\frac{dT}{d\phi}\right]$$

$$+\left[\frac{1}{r^2\sin\theta}\frac{\partial}{\partial\theta}\left(\sin\theta\frac{dT}{d\theta}\right)\right] = \frac{dT}{dt} \quad \ldots(1)$$

Neglecting temperature variation in θ and θ directions:

$$\nabla^2 T = \frac{d^2T}{dr^2} + \left(\frac{2}{r}\right)\frac{dT}{dr} = \left(\frac{1}{C}\right)\frac{dT}{dt} \quad \ldots(2)$$

With boundary conditions: at $t = 0$, $T = T\,0$ at $r = R$; For all t greater than zero, $T = T_1$ at $r = R$

Using separation variable method;

$$T = f(r)T(t) \text{ gives: } TF''(r) + \left(\frac{2}{r}\right)Tf'(r) = \left(\frac{1}{C}\right)f(r)\,T'$$

$$\frac{f''(r)}{f(r)} + \frac{2f'(r)}{rf(r)} = \left(\frac{1}{C}\right)\frac{T'}{T} = -a^2 \text{ where } a \text{ is a const.} \quad \ldots(3)$$

$$r^2 f''(r) + r^2 f'(r) + a^2 r^2 f(r) = 0 \quad \ldots(4)$$

$$T' + Ca^2 T = 0 \quad \ldots(5)$$

Equation (4) has Bessel solutions.

$$f(r) = C_1 r^{-1/2} J^{1/2}(ar) + C_2 r^{-1/2} J^{-1/2}(ar) \text{ if } a \neq 0; f(r) = \frac{C_3 + C_4}{r} \text{ if } a = 0;$$

$$T = C_5 \exp(-Ca^2 t) \text{ if } a \neq 0; T = C_6 \text{ if } a = 0$$

$$f(r) = \left(\frac{1}{r}\right)\left(\frac{2}{\pi}\right)^{1/2} \exp(-aCt)(A \sin ar + B \cos ar) + \left(\frac{C}{r}\right) + D$$

Constants are evaluated using boundary conditions.

SERIES SOLUTIONS

Example 2.38 Solve using series method (Picard's method):

$$\frac{dx}{dt} = x \cos t; x(0) = 1$$

Solution: $x \cos t; x(0) = 1$

$$dx = x \cos t \, dt; \text{ with } x(0) = 1$$

Integrating,
$$x_1 = 1 + \int_0^t \cos t \, dt * 1 = 1 + \sin|_0^t = 1 + \sin t$$

$$x_2 = 1 + \int_0^t (1 + \sin t) \cos t \, dt$$

$$= 1 + \int_0^t \cos t \, dt + \int \sin t \cos t \, dt$$

$$x_2 = 1 + \left(\sin t - \frac{\cos 2t}{4}\right)\Bigg|_0^t = 1 + \sin t - \frac{\cos 2t}{4} + \frac{1}{4}$$

$$x_3 = 1 + \int_0^t \left(1 + \frac{1}{4}\sin t - \frac{\cos 2t}{4}\right) \cos t \, dt$$

$$x_3 = 1 + \int_0^t \left(\cos t + \frac{1}{4}\cos t + \sin t \cos t - \frac{\cos 2t}{4}\cos t\right) dt$$

$$x_3 = 1 + \sin t + \frac{\sin t}{4} - \frac{\cos 2t}{4} - \frac{\sin 3t}{24} + \frac{1}{8}\sin t + \frac{1}{4}$$

$$x_3 = 1 + \sin t + \left(\frac{1}{4} - \frac{\cos 2t}{4}\right) + \left(\frac{1}{8}\sin t - \frac{\sin 3t}{24} + \frac{1}{4}\sin t\right)$$

Summarising for generalization:

$$x_0 = 1$$

$$x_1 = 1 + \sin t$$

$$x_2 = 1 + \sin t + \left(\frac{1}{4} - \frac{\cos 2t}{4}\right)$$

$$x_3 = 1 + \sin t + \left(\frac{1}{4} - \cos\frac{2t}{4}\right) + \left(\frac{3}{8}\sin t - \sin\frac{3t}{4}\right)$$

Example 2.39 Obtain series solution (Picard's method)

For: $$\frac{dy}{dt} = ty;\ y(0) = 1$$

Solution: $\int dy = \int ty\, dt$. Using $y(0) = 1$

$$\int dy = \int_0^t t\, dt + y(0)$$

$$y_1 = 1 + \int_0^t t\, dt = 1 + \frac{t^2}{2};\ y_1 = 1 + \frac{t^2}{2}$$

$$y_2 = 1 + \int_0^t t\, y_1\, dt = 1 + \frac{t^2}{2} + \frac{t^4}{8}$$

$$y_2 = 1 + \frac{t^2}{2} + \frac{t^4}{8}$$

$$y_3 = 1 + \int_0^t t\, y_2\, dt = 1 + \int_0^t t\left(1 + \frac{t^2}{2} + \frac{t^4}{8}\right) dt$$

$$= 1 + \int_0^t \left(t + \frac{t^3}{2} + \frac{t^5}{8}\right) dt$$

$$y_3 = 1 + \frac{t^2}{2} + \frac{t^4}{2.4} + \frac{t^6}{2.4.6}$$

$$y = 1 + \frac{t^2}{2} + \frac{t^4}{2.4} + \frac{t^6}{2.4.6.} + \ldots. + \frac{t^{2n}}{2.4.6\ldots 2n}$$

Example 2.40 For the second and higher order equations:

Frobenius method:

$$y = a_0x^m + a_1x^{m+1} + a_2x^{m+2} + ... + a_nx^{m+n}$$

For $m = 0$;
$$y = a_0 + a_1x + a_2x^2 + ... + a_nx^n$$

Solve
$$4x\frac{d^2y}{dx^2} + \frac{2dy}{dx} - y = 0 \quad ...(1)$$

Solution:
$$y = a_0 + a_1x + a_2x^2 + a_3x^3 + ...$$

$$y' = a_1 + 2a_2x + 3a_3\, x^2 + ...$$

$$y'' = 2a_2 + 6a_3x + ...$$

Substituting for y', y'' and y in Eq. (1) gives

$$4x(2a_2 + 6a_3x + ...) + 2(a_1 + 2a_2x + a_3\, 3x^2 + ...) - (a_0 + a_1x + a_2\, x^2 + ...) = 0$$

Collecting the coefficients of x^0, x^1, x^2 ....

$$(2a_2 - a_0) + (4a_2 + 8a_0 - a_1)x + (24a_3 + 6a_3 - a_2)x^2 + ... + (2n(2n-1)\,(2a_2 - a_1)$$
$$a_n - a_{n-1})x^{n-1} - a_nx^n = 0$$

Equating to zero:
$$2a_1 - a_0 = 0;\; a_1 = \frac{a_0}{2}$$

$$12a_2 - a_1 = 0;\; a_2 = \frac{a_1}{12} = \frac{a_0}{24}$$

$$30a_3 - a_2 = 0;\; a_3 = \frac{a_2}{30} = \frac{a_0}{720}$$

$$a_n = \frac{a_{n-1}}{2n(n-1)} = \frac{a_0}{2n!}$$

The final equation is

$$y = a_0\left(1 + \frac{x}{2!} + \frac{x^2}{4!} + \frac{x^3}{6!} + \frac{x^n}{2n!} + ...\right)$$

a_0 is obtained from the initial value of y at $x = 0$.

Example 2.41 For second order and higher order differential equations:

$$y = x^m(a_0 + a_1x + a_2x^2 + a_3\, x^3 + ... + a_nx^n)$$

Solve $9x(1-x)\frac{d^2y}{dx^2} - 12\frac{dy}{dx} + 4y = 0$ ...(1)

Solution:

$$y' = m\,a_0 x^{m-1} + (m+1)x^m\,a_1 + (m+2)\,(m+1)a_2\,x^m + ...$$

$$y'' = m(m-1)a_0\,x^{m-2} + (m+1)m\,a_1 x^{m-1} + (m+1)\,(m+2)a_2 x^m + ...$$

Substituting for: y, y', y'' in Eq. (1) gives

$$9x(1-x)\,(m(m-1)a_0\,x^{m-2} + (m+1)m\,a_1\,x^{m-1} + (m+1)\,(m+2)\,a_2\,x^m - 12m\,a_0\,x^{m-1} - 12(m+1)\,x^m\,a_1 - \ldots) = 0$$

Lower power of x in Eq. (1) is: x^{m-1}. Its coefficients equating to zero gives:

$$a_0(9m(m-1) - 12m) = 0;\ (3m-7)3m = 0;\ \text{Giving: } m = 0;\ \frac{7}{3}.$$

$$m = 0 \text{ or } m = \frac{7}{3}$$

Equating the coefficients of x^m, x^{m+1}, ... to zeroes:

$$a_1 = \frac{(3m+1)}{(3(m+1))a_0} = a_2 = \frac{(3m+4)a_1}{3(m+2)}$$

$$= \frac{(3m+4)\,(3m+1)a_0}{3(m+2)\,3(m+1)}$$

When $m = 0$; $$a_1 = \left(\frac{1}{3}\right)a_0\,;\ a_2 = \frac{1.4.a_0}{3.6}\,;\ a_3 = \frac{1.4.7.a_0}{3.6.9}$$

When $m = 0$; $$y_1 = a_0\,x^2\left(1 + \frac{x}{3} + \frac{1.4}{3.6}\right) + \frac{1.4.7}{3.6.9}x^3 + \frac{7}{3}\ ...$$

When $m = \frac{7}{3}$; $$y_2 = a_0 x^{7/3}\left(1 + \frac{8x}{10} + \frac{8.11x^2}{10.13} + ...\right)$$

$$y = C_1\,y_1 + C_2\,y_2$$

$$y = C_1\left(1 + \frac{1}{3}x + ...\right) + C_2 x^{7/3}\left(1 + \frac{8}{10} + ...\right)$$

C_1 and C_2 values are calculated using Boundary conditions.

2: For equal m's:

$$C_1 y_1 + C_2\left(\frac{dy_1}{dm}\right) \text{ at } m_1.$$

For m's differing by an integer, $C_1 y_1 + C_2\left(\frac{dy_1}{dm}\right)$ m_2; m_1 is less than m_2.

Forms of series solution:

The equation of second order differential equation gives two roots which may be:

1. Distinct and not differing by an integer. The complete solution is:

$$Y = C_1 y(m_1) + C_2 y(m_2)$$

2. When the roots are equal (m_1, m_2) (m_1 is less than m_2) are distinct and differ by an integer, then the solution is:

$$Y = C_1 y(m_1) + C_2 \frac{dy}{dm}(m_1)$$

3. Bessel equation: The form is :

$$x^2 \frac{d^2y}{dx^2} + x\frac{dy}{dx} + (x^2 - n^2)y = 0 \qquad \text{...(1)}$$

Using series solution of

$$y = a_0 x^m + a_1 x^{m+1} + a_2 x^{m+2} + ...$$

$$y' = a_0 m\, x^{m-1} + a_1(m+1)x^m + (m+2)a_2\, x^{m+1} + ...$$

$$y'' = a_0(m-1)m\, x^{m-2} + a_1\, m(m+1)x^{m-1} + ...$$

Substituting for, y, y', y'' in Eq. (1) gives:

$$a_0(m^2 - n^2)x^m + a_1[(m+1)^2 - n^2]x^{m+1} + \{a_2[(m+2)^2 - n^2] + a_0\}x^{m+2} + ... = 0$$

Equating the coefficient of x^m, we obtain the equation as $m^2 - n^2 = 0$ (as $a_0 \neq 0$) when $m = n$ or $-n$.

$$a_1 = a_3 = a_5 = a_7 = a_9 = ... = 0$$

$$a_2 = -\frac{a_0}{[(m+2)^2 - n^2]};\ a_4 = -\frac{a_2}{[(9m+4)^2 - n^2]} \text{ etc.}$$

Substituting $a_1, a_3, ..., a_0, a_2$ values.

$$y = a_0 x^m \left\{1 - \frac{x^2}{(m+2)^2 - n^2} + \frac{x^4}{[(m+2)^2 - n^2][(m+4)^2 - n^2]} ....\right\}$$

For $m = n$, we obtain:

$$y_1 = a_0 x^n \left\{1 - \frac{x^2}{[4(n+1)]} + \frac{x^4}{[4^2.2(n+1)(n+2)} - ....\right\} \qquad \text{...(2)}$$

For $m = -n$, we obtain:

$$y_2 = a_0 x^{-n} \left\{1 - \frac{x^2}{[4(-n+1)]} + \frac{x^4}{[4^2.2(-n+1)(-n+2)} - ....\right\} \qquad \text{...(3)}$$

Case 1: When n is not an integral or zero: Complete solution is

$$y = C_1 y_1 + C_2 y_2$$

If $a_0 = \left(\frac{1}{2n}\right)\overline{|(n+1)}$ then Eq. (2) becomes as $J_n(x)$, Bessel function of first kind and order n.

$$J_n(x) = \left(\frac{x}{2}\right)^n \left\{ \frac{1}{\overline{|(n+1)}} - \frac{1}{1!}\overline{|(n+2)}\left(\frac{x}{2}\right)^2 + \left(\frac{x}{2}\right)^4 \frac{1}{\left[2!\overline{|(n+2)}\right]} - \left(\frac{x}{2}\right)^6 \frac{1}{\left[3!\overline{|(n+4)}\right]} + ... \right\}$$

For $n > 0$.

Case 2: For, $n = 0$; $y_1 = y_2$. This is Bessel equation of order zero. Complete solution is : $y = A J_n(x) + B\,Y_n(x)$; $Y_n(x) = \int \frac{dx}{[x J_n(x)^2]} . J_n(x)$. $Y_n(x)$ Newmann function or Bessel equation/function of second kind of order n.

Example 2.42 $y' - 2xy = 0$; $y(x) = a_0 + a_1x + a_2x^2 + ...$

$$y' = a_1 + 2a_2x + 3a_3x^2 + ...$$

Solution: The equation: $y' - 2xy = 0$ becomes: $(a_1 + a_2\,2x + 3a_3x^2 + ...) - 2(a_0x + a_1x^2 + ...) = 0$

Collecting the powers of x

$$a_1 + (2a_2 - 2a_0)x + (3a_3 - 2a_1)x^2 + ... = 0$$

Equating each term to zero gives

$$a_1 = 0;\ 2a_2 - 2a_0 = 0;\ 3a_3 - 2a_1 = 0$$

Giving:

$$a_1 = 0;\ a_2 = a_0;\ a_3 = 0$$

$y(x) = \left(\sum a_n x^n\right)$. Substituting in $y' - 2xy = 0$ gives

$$0 = \left(\sum_{n=0}^{\infty} a_n x^n\right) - 2x\left(\sum a_n x^n\right)$$

$$= \left(\sum n a^n x^{n-1}\right) - \sum 2a_n x^{n+1}$$

$$= \sum (n+1)a_{n+1}x^n - \sum 2a_{n-1}x^n$$

$$= a_1 + \sum (n+1)a_{n+1} - 2a_{n-1})x^n$$

We have $a_1 = 0$; $\qquad a_{n+1} = \frac{2a_{n-1}}{(n+1)}$

We see from this, $a_{2K+1} = 0$ for all K. On the other hand

$$a_{2K} = \frac{2a_{2K-2}}{(2K)};\ \frac{1}{K}a_{2K-2/K} = \frac{a_0}{K!}$$

Therefore, we have $\qquad y(x) = \sum \frac{1}{K!}a_0x^{2K} = a_0\left[\sum \frac{1}{K!}x^{2K}\right]$

We recognize that: $$\sum_{K=0}^{\infty}\frac{x^{2K}}{K!}=\sum_{K=0}^{\infty}\frac{(x^2)^K}{K!}=e^{x^2}$$

So the solution is $$y(x)=e^{x^2} \quad \textbf{Ans.}$$

Example 2.43 $y''-\sin x^*y=y(\pi)=1;\ y'(\pi)=0$

$$y(t)=\sum a_0\, t^n\ ;\ \sin t=\sum(-1)^n\frac{t^{2n+1}}{(2n+1)!}$$

Solution: Original equation becomes

$$0=\sum a_n n(n-1)t^{n-2}+\left(t-\frac{t^3}{3!}+....\right)*(a_0+a_1t+a_2t^2+....)$$

$$=2a_2+6a_3t+12a_4\,t^2+....+a_0t+a_1t^2+\left(a_2-\frac{a_0}{6}\right)t^3+....$$

$$=2a_2+(6a_3-a_0)t+(12a_4+a_1)t^2+....$$

Equating the coefficients for zero gives

$$2a_2=0;\ 6a_3+a_0=0;\ 12a_4+a_1=0;\ 20a_5+a_2-\frac{a_0}{6}=0$$

Giving: $$a_0=1;\ a_1=0;\ a_2=0;\ a_3=-\frac{1}{6};\ a_4=0;\ a_5=\frac{1}{120}$$

Substituting in the original equation, $y''-\sin t*y=0$ gives,

$$=\sum a_n n(n-1)t^{n-2}+\left(t-\frac{t^3}{3!}+\frac{t^5}{5!}\right)(a_0+a_1t+a_2t^2+....)$$

$$=2a_2+6a_3t+12a_4t^2+20\,a_5t^3+30a_6t^4+....$$

$$+a_0t+a_1t^2+\left(a_2-\frac{a_0}{6}\right)t^3+....$$

Giving (equating coefficients of powers to zero):

$$30a_6+a_3-\frac{a_4}{6}=0;\ a_3=-\frac{1}{6};\ a_6=\frac{1}{180};\ a_1=0$$

$$y(t)=1-\frac{t^3}{6}+\frac{t^5}{120}+\frac{t^6}{180}+....$$

$$y(x)=1-\frac{1}{6}(x-\pi)^3+\frac{(x-\pi)^5}{120}+\frac{(x-\pi)^6}{180}+.... \quad \textbf{Ans.}$$

Example 2.44 $y''-xy'+2y=\cos x$

Solution: $$y(x)=\sum a_n\,x^n$$

$$\cos x = \sum \frac{(-1)^n x^{2n}}{2n!}$$

$$\sum \frac{(-1)^n x^{2n}}{2n!} = \sum a_n n(n-1)x^{n-2} + \sum 2a_n x^n - \sum na_n x^n$$

Collecting the power of x gives

$(2a_2 + 2a_0) + \Sigma(a_{n+2}(n+2)(n+1) - n\, a_n + 2a_n)x^n$

$$x^0 : 1 = 2a_2 + 2a_0$$

$$x^1 : 0 = 6a_3 + a_4$$

$$x^2 : \frac{1}{2} = 12\, a_1$$

$$a_3 = \frac{-a_1}{6}$$

$$a_2 = -a_0 \times \frac{1}{2}$$

$$y(x) = a_0 + a_1 x + \left(\frac{1}{2} - a_0\right)x^2 - a_1 \frac{x^3}{6} + ...$$

$$y(x) = \left(\left(\frac{1}{2}\right)x^2 + ...\right) - a_0(1 - x^2 + ...) + a_1\left(x - \frac{x^3}{6} + ...\right)$$

Example 2.45 An equation of from:

$\frac{x(b-x)d^2y}{dx^2} + (b-2x)\frac{dy}{dx} - K(b-x) = 0$ with boundary conditions:

at $x = 0$; y remains finite. ...(1)

At $x = b - a$; $T = T_L$; $y = T_B - T_A$ where T_B is the temperature of the pipe contents, comes in heat transfer.

Solution: Using the Frobenius method of substitutions

$$y = \sum_{0}^{\infty} a_n x^{n+c} (a_0 \neq 0) \qquad ...(2)$$

Equation (1) is (expanding): $xb\frac{d^2y}{da^2} - x^2\frac{d^2y}{dx^2} + b\frac{dy}{dx} - 2x\frac{dy}{dx} - Kb + Kx = 0$ ...(3)

Using Eq. (2) in Eq. (1) gives

$$b\sum a_n (n+c)(n+c-1)x^{n+c-1} - \sum a_n (n+c)(n+c-1)x^{n+c}$$

$$+ b\sum a_n (n+c)x^{n+c-1} - 2\sum a_n (n+c)x^{n+c} - Kb\sum a_n x^{n+c} + K\sum a_n x^{n+c+1} = 0$$

The equation obtained by equating the coefficients of: x^{c-1} is

$$ba_0c(c-1) + ba_0c = 0$$

$$c^2 = 0 \qquad \text{...(4)}$$

Equation (4) gives a solution with equal roots. This is of case, with zero roots (equal roots).

$$y = Au(x, c_1) + b\left(\frac{du}{dc}\right)_{c=c_1}.$$

The second solution has no physical significance.

Equating the next power of x which is x^c,

$$ba_1 - Kba_0 = 0$$

$$a_1 = Ka_0 \qquad \text{...(5)}$$

The recurrence relation is

$$ba_{r+1}(r+1)r - a_r r(r-1) + ba_{r+1}(r+1) - 2a_r r - K\,b_{ar} + Ka_{r-1} = 0$$

Giving: $$ba_{r+1}(r+1)^2 = a_r(r(r+1) + Kb) - Ka_{r-1}$$

This relates three coefficients:

$$a_{r+1}, a_r, a_{r-1}.$$

It is difficult to relate a_r in terms of a_0.

With $T_A = 60$; $T_B = 212$ and alpha (α) = 5^0 gives

$$K = \left(\frac{2}{220}\right) \times 0.0872 = 0.1043 \qquad \text{...(6)}$$

Equation (5) and (6) become

$$a_1 = 0.1043\, a_0$$

$$a_{r+1}(r+1)^2 = a_r(2r(r+1) + 0.1043) - 0.2086a_{r-1} \qquad \text{...(7)}$$

For $r = 1$: $$a_2 \times 2^2 = a_0 \times 0.1043 \times 4.1043 - 0.2086a_0$$

$$a_2\, 2^2 = 0.2195a_0$$

$$a_2 = 0.0549a_0$$

Using Eq. (7):

For $r = 2$: $$a_3 3^2 = 0.0549a_0(12.1043) - 0.2086 \times 0.1043a_0$$

$$a_3 3^2 = 0.6424a_0$$

$$a_3 = 0.0714a_0$$

In Eq. (7), the important terms are: $r = 1$ and $r = 2$. The later of the series approximate to: $a_{r+1}(r+1)^2 = a_r 2r(r+1)$

or $$a_{r+1} = \frac{2r}{r+1} a_r$$

Evaluating further coefficients from Eq. (7) gives: $a_4, a_5, \ldots$)

$$a_4 = 0.1068$$

$$a_5 = 0.171$$

$$a_6 = 0.285$$

$$a_7 = 0.488$$

$$y = a_0(1 + 0.1043x + 0.0549x^2 + 0.0714x^3 + 0.1068x^4 + 0.171x^5 + 0.285x^6 + 0.488x^7 + \ldots) \quad \ldots(8)$$

To evaluate a_0, using the boundary condition:

At $x = \frac{1}{3}$;

$$y = 212 - 60 = 152$$

$$152 = a_0\left(\frac{0.1043}{3} + \frac{0.0549}{9} + \frac{0.0714}{27} + \frac{0.1068}{81} + \frac{0.171}{243} + \frac{0.285}{729}\right)$$

Giving $$a_0 = 145.3 \quad \ldots(9)$$

Final equation in series is

$$y = 145.3(1 + 0.1043x + 0.549x^2 + 0.0714x^3 + 0.1068x^4 + 0.171x^5 + 0.285x^6 + 0.488x^7 + \ldots)$$ **Ans.**

Example 2.46 An equation of the form

$$\frac{d^2T}{dx^2} - \frac{u\rho C_p}{KA}\frac{dT}{dx} + \frac{\pi Dh(T_w - T)}{KA} = 0 \text{ comes in heat transfer.} \quad \ldots(1)$$

Solution: Putting $t = T_w - T$; Eq. (1) becomes,

$$\frac{d^2t}{dx^2} - \frac{u\rho C_p}{KA}\frac{dt}{dx} - \frac{4\,ht}{Kd} = 0 \quad \ldots(2)$$

Inserting numerical values gives

$$\frac{d^2t}{dx^2} - 6880\frac{dt}{dx} - 3000x^{-1/2}t = 0 \text{ (since } h = 5x^{-1/2}) \quad \ldots(3)$$

Substituting $x = z^2$ to remove $\frac{1}{2}$ power,

Equation (3) becomes

$$\frac{zd^2t}{dz^2} - (1+13760z^2)\frac{dz}{dt} - 12000z^2t = 0 \qquad \ldots(4)$$

Using series method:

$$t = \sum_{n=0}^{\infty} a_n z^{n+c} \qquad \ldots(5)$$

By equating the coefficients, $c = 0$ or $c = 2$; using $c = 2$, $a_1 = 0$, $a_2 = 3440a_0$

$$a_3 = 800a_0;\ a_4 = 7.9 \times 10^6 a_0$$

$$t = a_0(3440z^4 + 800z^5 + 7.9 \times 10^6 z^6 + \ldots.)$$

Using $z = x^{1/2}$ (since replaced)

$$t = a_0(3440x^2 + 800x^{5/2} + 7.9 \times 10^6 x^3 + \ldots.)$$

a_0 is evaluated by the boundary condition at $x = 0$; $T = 70$, $T_w - T = t$, $T_w = 600\,°F$; $t = 70 - 600 = -530$. The series is not convergent; so it is not working for this equation. This can be solved by Bessel series solution.

EXERCISES

1. A tank contains 50 gallons of fresh water. Brine containing 2 kg of salt/gallon of water flows into the tank at a rate of 2 gallons/minute and the mixture is kept uniform by stirring and runs out at the same flow rate. How long will it take for the quantity of salt in the tank to increase from 40 to 80 kg. (Ans: 27 min. 28 sec.)
2. A tank initially contains 50 m^3 of pure water. Starting at time $t = 0$, brine containing two kg of dissolved salt per m^3 flows into the tank at the rate of 1 m^3 per minute. The mixture is kept uniform by stirring and the well stirred solution flows out of the tank at the same rate. Calculate the amount of salt present in the tank at the end of 25 minutes?
3. A mixture of benzene and toluene is purified in a batch distillation column unit from an essentially non-volatile impurity. Intially, the still is charged with feed stock of given composition and the feed is supplied at a constant rate. If the still is heated so that the total moles of the liquid is constant. Derive an expression for the time required for benzene content of the overhead product to decrease by 50%.
4. Two thin wall metal pipes of outer diameter d and joined by flanges of diameter d_f and thickness t are carrying steam at temperature T. If the conductivity of the flange metal is K and the exposed surfaces of the flanges are at a temperature T_f, derive an expression for the rate of heat loss from pipe and the proportion that leaves the rim of the flanges?
5. The following elementary reaction takes place in a vessel, at a constant volume and at constant temperature.

$$A \underset{K_3}{\overset{K_1}{\rightleftarrows}} B \overset{K_2}{\rightleftarrows} C$$

Derive an expression for the concentration of A as a function of time. Assume $C_{BO} = C_{CO} = 0$.

6. A tank contains 100 litres of brine in which 40 kg of salt is dissolved. Fresh water runs into the tank at the rate of 8 litres per minute and the mixture is uniform by continuous stirring and runs out at the same rate. How long will it be before 20 kg of salt is left in the tank.

7. The method used in the laboratory to produce CO_2 may be represented by the reversible reactions.

$$CaCO_3 + 2HCl \rightleftarrows CaCl_2 + H_2CO_3 \rightleftarrows CO_2 + H_2O$$

Most of the CO_2 is liberated as gas. If the forward and backward speed coefficients of the first and second reactions are K_1, K_2, K_3 and K_4 respectively, and the reaction is begun with, n moles of $CaCO_3$ (chalk) and n_2 moles of HCl. Set up the differential equation for n moles of chalk left at any time after the beginning of the reaction. Outline the solution.

8. In a chemical reaction the amount of x of a substance at time t satisfies the equation.

$$\frac{dx}{dt} = K(3 - x)(6 - x)$$

where K is a constant. If $x = 0$, when $t = 0$ and $x = 1$ when $t = 10$. Find the value of K? Also find x when $t = 30$.

9. A continuous stirred tank reactor system consists of M tanks in series. L cubic cm of a solution per second is fed to each tank. The feed to the first tank contains a concentration of C_0 moles/litre of component A. Each tank is well stirred. Reaction takes place as $A \rightarrow B$. The rate at which A reacts is given by expression.

$$-\frac{dw_n}{dt} = (KV\,C)_n \text{ moles/liter}$$

If the reaction rate constant K is same for each tank, show that the highest concentration of B which may be produced by the tank system occurs when $V_1 = V_2 = \ldots.\ V_n$. Provided that the total volume $= \sum_{i=1}^{n} V_n$ remains constant.

10. The diffusivity for a binary gas mixture is to be obtained experimentally by measuring the rate of inter diffusion of two gases originally combined in the two ends of a hollow cylinder. A thin diaphragm separating the gases divides the cylinder into two sections of equal volume. The diaphragm is suddenly removed and the gases are allowed to diffuse for a measured time. The diaphragm is then replaced and the gas is analysed. Assuming convection currents are negligible, derive an expression which permits the calculation of diffusivity from experimental data.

11. The following consecutive reactions occur in a batch vessel at a constant volume.

$$A \underset{K_2}{\overset{K_1}{\rightleftarrows}} B \underset{K_4}{\overset{K_3}{\rightleftarrows}} C$$

Assume that one mole of A is present at the start, show that N_A, the moles of A at anytime t is given by the solution of the equation,

$$\frac{d^2N_A}{dt^2} + P\frac{dN_A}{dt} + QN_A - R = 0;$$

where $P = K_1 + K_2 + K_3 + K_4$

$Q = K_1K_3 + K_2K_4 + K_1K_4$

$R = K_2K_4$

12. Solve the differential equation: $(1 + x^2)\frac{dy}{dx} + xy = 1$

13. Solve: $4x\frac{d^2y}{dx^2} + 2\frac{dy}{dx} - y = 0$. Using series solution of

$$y = a_0 + a_1 x + a_2 x^2 + \ldots + a_n x^n.$$

14. A supply of hot air is to be obtained by drawing cool air through a heated cylindrical pipe. The pipe is of 4 inch diameter and 4 ft long and is maintained at a temperature of $T_w = 600$°F. throughout, its length. The properties of air are: Specific heat = 0.24 BTU/lb°F. Thermal conductivity of air = 0.20 BTU/hr.ft.°F, density of air = 0.050 lb/ft^3, flow rate of air = 1000 ft^3/hour; inlet temperature of air = 70°F; overall heat transfer coefficient = $5x^{-1/2}$BTU. (hr. ft^2. °F)

Where x is the distance from pipe inlet in feet. Find the temperature of exit gas as a function of x by approximate method.

15. A copper fin of L feet long in triangular in cross section. w–feet thick at base, tapers to a line at the end. Base temperature is T_b. Surrounding temperature is T_S. Heat transfer coefficient (heat loss) by convection is h BTU/hr sq. ft. °F). Determine T (temperature of fin T) as a function of L, x and given parameters. Angle of fin is θ°.

16. A wetted wall column is to be used for the absorption of a gas (A). The gas is consumed in liquid phase by a pseudo first order reaction in terms of gas concentration. Develop an expression giving the point absorption rate and effective penetration depth as a function of distance from liquid phase.

17. A system consists of N stirred tanks each of volume V ft^3 arranged in a cascade. If each tank initially contains pure water and a salt stream of concentration x_0 lb/ft^3 is fed to the first tank at a rate of R ft^3/hour. Calculate the output concentration from the last tank as a function of time and other constant terms. The stirring is 100%.

18. A tank with 2000 kg capacity initially contains 800 kg of water. A 50% by weight of solution of ethyl alcohol and water is added at the rate of 200 kg per hour, and the mixed solution is

withdrawn from the tank at 100 kg/hour. What is the composition of liquid in the tank just as it becomes full? Assume through mixing in the tank.

19. N_0 gm of a solid material was placed in w gm of water, at time t_0. The liquid was continuously stirred and maintained at a constant temperature. At the end of t_1 seconds, N_1 gm of solid remains undissolved. At the end of a very long period of time, N_2 gm of solid remained undissolved. Set up the differential equation required to determine the rate of solution of the solid in terms of N_0, N_1, N_2 and t_0, t_1.

    Data and notes: It is assumed that the rate of solution is proportional to (a) the surface area of the material and (b) the concentration driving force when the concentration is expressed as gm of solid per gram of water. The original solid consisted of S spheres, each of internal diameter D_0 feet.

20. For the reaction: $A \xrightarrow{K_1} B \xrightarrow{K_2} C$. Formulate the differential equation for the number of moles of (N_A) with time. Assume constant volume reaction and initially only A is present. Outline the solution.

21. A closed kettle of total surface area A sq.ft. is heated through the surface by surrounding steam T_S ºF. The kettle is charged with M lb of liquid of specific gravity of C BTU/lb.ºF at a temperature of T_0 ºF. If the process is controlled by heat transfer coefficient, h BTU/(hr.sq.ft. ºF), how does the temperature of the liquid vary with time?

    (Ans: $T = T_s - (T_S - T_0)\exp(-hA_0/MC)$)

22. A reaction mixture is flowing through a pipe of cross-sectional area, A at a constant velocity. By reason of concentration gradient, there is also present the diffusion of the reactant in the axial direction, in accordance with Fick's law. The rate is $-DA\dfrac{dC}{dt}$ where D is diffusivity. Obtain the differential equation and hence find out the general solution assuming a first order chemical reaction having reaction velocity constant as K.

23. A tank of cross-sectional area A, m^2 at water level has an orifice of B m^2 in the bottom. If h metres is the depth of water in the tank at any time, sec (t), the rate of flow from the tank is given, by $A\dfrac{dh}{dt} = -0.6\,B\,(2gh)^{1/2}$ where $g = 9.81$ m/sec^2. Find the time required to empty the cylindrical tank 3 m long, 2 metres diameter through a hole of 75 mm. in diameter.

    (a) If the tank is originally full and its axis is vertical and (b) if the tank is half-full and its axis is horizontal and the hole at the base.

24. The equation of steady radial heat conduction with cylindrical symmetry is given by: $r\dfrac{d^2T}{dr^2} + \dfrac{dT}{dr} = 0$ where T is the temperature, at a distance r from the axis of the cylinder. The heat flux Q per unit time over a unit length l of the cylindrical surface is given by: $Q = -2\pi rlK$

$\frac{dT}{dr}$ where K is a constant. For a radial flow between two co-axial cylinders, $T = T_1$ at $r = r_1$ and $T = T_2$ at $r = r_2$ where r_1 is less than or equal to r_2.

$$\text{Prove: } T = \left(T_1 \ln\left(\frac{r_2}{r_1}\right) + T_2 l\left(\frac{r}{r_1}\right) \times \ln\left(\frac{r_2}{r_1}\right)\right)$$

25. Two thin metal pipes of 2 cm diameter (external) are joined by flanges of 1″ (inch) thick and 8 cm diameter are carrying steam at 100°C. If conductivity of the flange metal is K gm Cal/(sec cm^2 °C) and the exposed surface loses heat to the surroundings at 30ºC according to heat transfer coefficient h gm.Cals/(sec.cm^2.ºC). Show the temperature distribution in the flange is: $r\frac{d^2T}{dr^2} + \frac{dT}{dr} + \frac{hr}{K}(T - 30) = 0$ where T is temperature in ºC and r is the radial distance coordinate in cm. Indicate the boundary conditions required for solutions of the equation.

26. A certain chemical dissolved in water at a rate proportional to the product of amount undissolved and the difference between the concentration of saturated solution and the concentration of actual solution. When 30 gm of chemicals are agitated with 100 gm of water, 10 gm dissolve in 2 hours. How much will dissolve in 5 hours. Saturated solution has concentration as 50 gm per 100 gm of solvent.

27. An elevated horizontal cylindrical tank of 1 metre and 2 metres long is insulated with asbestos lagging of thickness 1.4 cm and is employed as a maturing vessel for batch chemical process. Liquid at 95ºC is charged into the tank and allowed to mature for 5 days. Calculate the final temperature by forming a differential equation assuming that heat loss through supports is negligible and the thermal capacity of the lagging can be neglected. Use the following data. Liquid film coefficient of heat transfer (h_1) = 150 W/m^2.ºC.

    Thermal conductivity of asbestos (K) = 0.20 W/m.ºC

    Surface coefficient of heat transfer by convection and radiation (h_c) = 10 W/m^2.ºC; Density of liquid (ρ) = 1000 kg/cu.m; Heat capacity of liquid (S) = 2500 J/Kg.ºC; Atmospheric temperature (t) is assumed to vary as: $t = 10 + 10 \cos\left(\frac{\cos \pi\theta}{12}\right)$ where θ is in hours.

    Atmospheric temperature at the time of charging = 20°C

28. Solve: $\frac{d^2y}{dx^2} - 3\frac{dy}{dx} + 2y = xe^{3x} + \sin 2x.$

29. $\frac{dy}{dx} = \frac{(x^2 + y^2)}{(xy^2)}.$

30. Apply Cholesky factorizing method to solve the equations:

$3x + 2y + 7z = 4$

$2x + 3y + z = 5$

$3x + 4y + z = 7$

31. Solve the equation by series method? Use Frobenius method. The differential equation is

$$9x(1-x)\frac{d^2y}{dx^2} - 12\frac{dy}{dx} + 4y = 0$$

32. Solve: $\dfrac{d^3z}{dx^3} - 2\dfrac{d^2z}{dx\,dy} = 2e^{2x} + 3x^2y.$

33. A system consists of N stirred tanks each of volume θ ft^3 arranged in a cascade. If each tank initially contains pure water and salt stream of concentration x_0 lbs/ft^3 is fed to the first tank at a rate of R ft^3/hour, calculate the output concentration from the last tank as a function of time and other constant terms. The stirring is perfect.

34. Solve the following differential equation:

$$\frac{d^2y}{dx^2} + x\frac{dy}{dx} + 5y = e^{-t}\sin t$$

Given $y(0) = 0$ and $y'(0) = 1$

❑❑❑

3

Partial Differentiation, Partial Differential Equations and Applications

3.1 PARTIAL DIFFERENTIATION

(a) $w = f(x_1, x_2, x_3, \ldots, x_n)$

$$\frac{dw}{dt} = \frac{df}{dt} = \frac{df}{dx_1} \times \frac{dx_1}{dt} + \frac{df}{dx_2} \times \frac{dx_2}{dt} + \ldots.$$

(b) $w = f(x)$

$$x = f(u)$$

$$\frac{dw}{du} = \frac{df}{dx} \times \frac{dx}{du}$$

$$\frac{d^2w}{dx^2} = \frac{d}{du}\left(\frac{df}{dx} \times \frac{dx}{du}\right)$$

$$= \frac{df}{dx} \times \frac{d^2x}{du^2} + \frac{dx}{du} \times \frac{d}{du}\left(\frac{df}{dx}\right)$$

$$\frac{d}{du}\left(\frac{df}{dx}\right) = \frac{d^2f}{dx^2} \times \frac{dx}{du}$$

$$\frac{d^2w}{dx^2} \text{ becomes } = \frac{df}{dx} \times \frac{d^2x}{du^2} + \left(\frac{dx}{du}\right)^2 \times \frac{d^2f}{dx^2}$$

(c) $w = x^n$; $x = u^2 - 2$

$$\frac{dw}{dx} = \frac{df}{dx} = nx^{n-1}$$

$$\frac{dx}{du} = 2u;$$

$$\frac{dw}{du} = \frac{df}{dx} \times \frac{dx}{du} = nx^{n-1}.2u = n(u^2 - 2)^{n-1}2u$$

$$\frac{d^2 f}{dx^2} = \frac{d}{dx}\left(\frac{df}{dx}\right) = n(n-1)x^{n-2}$$

$$\frac{d^2 x}{du^2} = 2$$

$$\frac{d^2 f}{du^2} = \frac{d^2 w}{du^2} = \frac{df}{dx} \times \frac{d^2 x}{du^2} + \left(\frac{dx}{du}\right)^2 \frac{d^2 f}{dx^2}$$

$$\frac{d^2 f}{dx^2} = nx^{n-1} \times 2 + (2u)^2 n(n-1)x^{n-2} \quad \textbf{Ans.}$$

(d) $\dfrac{d^2 w}{du\, dv} = \dfrac{d^2 f}{du\, dv} = \dfrac{df}{dx} \times \dfrac{d^2 x}{du\, dv} + \dfrac{dx}{du} \times \dfrac{dx}{dv} \times \dfrac{d^2 f}{dx^2}$

$$\frac{d^2 w}{dv^2} = \frac{df}{dx} \times \frac{d^2 x}{dv^2} + \left(\frac{dx}{dv}\right)^2 \left(\frac{d^2 f}{dx^2}\right)$$

(e) $w = f(x, y);\ x = F_1(u);\ y = F_2(u)$

$$\frac{dw}{du} = \frac{df}{du} = \left(\frac{df}{dx}\right)\left(\frac{dx}{du}\right) + \left(\frac{df}{dy}\right)\left(\frac{dy}{du}\right)$$

$$\frac{d^2 w}{du^2} = \frac{d}{du}\left(\frac{df}{dx} \times \frac{dx}{du} + \frac{df}{dy} \times \frac{dy}{du}\right)$$

$$= \frac{df}{dx} \times \frac{d^2 x}{du^2} + \frac{dx}{du} \times \frac{d}{du}\left(\frac{df}{dx}\right) + \frac{df}{dy} \times \frac{d^2 y}{du^2} + \frac{dy}{du}\frac{d}{du}\left(\frac{df}{dy}\right)$$

$$\frac{d}{du}\left(\frac{df}{dx}\right) = d\frac{(df/dx)}{dx} \times \frac{dx}{du} + \frac{d}{dy}\left(\frac{df}{dx}\right) \times \frac{dy}{dx}$$

$$= \frac{d^2 f}{dx^2} \times \frac{dx}{du} + \frac{d^2 f}{dy\, dx} \times \frac{dy}{du}$$

$$\frac{d}{du}\left(\frac{df}{dy}\right) = \frac{d(df/dy)}{dx} \times \frac{dx}{du} + \frac{d}{dy}\left(\frac{df}{dy}\right) \times \frac{dy}{du}$$

$$= \frac{d^2 f}{dx\, dy} \times \frac{dx}{du} + \frac{d^2 f}{dy^2} \times \frac{dy}{du}$$

$$\frac{d^2 w}{du^2} = \frac{df}{dx} \times \frac{d^2 x}{du^2} + \frac{d^2 f}{dx^2} \times \left(\frac{dx}{du}\right)^2 + \frac{2d^2 f}{dx\, dy} \times \frac{dx}{du} \times \frac{dy}{du}$$

$$+ \frac{df}{dy} \times \frac{d^2 y}{du^2} + \frac{d^2 f}{dy^2} \times \left(\frac{dy}{du}\right)^2$$

(f) $w = x^2 + y^2;\ x = u^3;\ y = u^2$

$$\frac{df}{dx} = \frac{dw}{dx} = 2x$$

$$\frac{df}{dy} = 2y$$

$$\frac{dx}{du} = 3u^2$$

$$\frac{dy}{du} = 2u$$

$$\frac{dw}{du} = 2x \times 2u + 2$$

$$\frac{d^2 f}{dx^2} = 2$$

$$\frac{d^2 f}{du^2} = 2y$$

$$\frac{d^2 f}{dx\,dy} = 0$$

$$\frac{d^2 x}{du^2} = 6u$$

$$\frac{d^2 y}{du^2} = 2$$

$$\frac{d^2 w}{dx^2} = 2x \times 6u + 2 \times (3u^2)^2 + 2 \times 0 \times 3u^2 \times 2u + 2y \times 2 + 2 \times 2u^2.$$

$$\frac{d^2 w}{dx^2} = 6u^4 + 4u^2 + 6ux + 4y$$

(g) $w = f(x, y);\ x = f_1(u);\ y = f_2(v)$

$$\frac{df}{du} = \frac{dw}{du} = \frac{df}{dx} \times \frac{dx}{du} + \frac{df}{dy} \times \frac{dy}{du}$$

$$\frac{dw}{du} = \frac{df}{dx} \times \frac{dx}{du}$$

$$\frac{dw}{dV} = \frac{df}{dy} \times \frac{dy}{dV}$$

$$\frac{d^2 w}{du^2} = \frac{d}{dw}\left(\frac{dw}{du}\right) = \frac{df}{dx} \times \frac{d^2 x}{du^2} + \frac{dx}{du} \times \frac{d}{du}\left(\frac{df}{dx}\right) = \frac{d}{du}\left(\frac{df}{dx}\right) = \frac{d^2 f}{dx^2} \times \frac{dx}{du}$$

$$\frac{d^2w}{du^2} = \frac{df}{dx} \times \frac{d^2x}{du^2} + \left(\frac{dx}{du}\right)^2 \frac{d^2f}{dx^2}$$

$$\frac{d^2w}{dV^2} = \frac{df}{dy} \times \frac{d^2y}{dV^2} + \frac{d^2f}{dy^2}\left(\frac{dy}{dV}\right)^2$$

$$\frac{d^2w}{du\,dV} = \frac{dx}{du} \times \frac{dy}{dV} \times \frac{d^2f}{dx\,dy}$$

(h) $w = f(x, y); x = F_1(u, v); y = F_2(u, v)$

This is cylindrical coordinate conversion.

$$\frac{dw}{du} = \frac{df}{dx} \times \frac{dx}{du} + \frac{df}{dy} \times \frac{dy}{du}$$

$$\frac{dw}{dv} = \frac{df}{dx} \times \frac{dx}{dv} + \frac{df}{dy} \times \frac{dy}{dv}$$

$$\frac{d}{du}\left(\frac{dw}{du}\right) = \frac{d^2w}{du^2}$$

$$= \frac{df}{dx} \times \frac{d^2x}{du^2} + \frac{dx}{du} \times \frac{df}{dx} + \frac{df}{dy} \times \frac{d^2y}{dx^2} + \frac{dy}{du} \times \frac{d}{du}\left(\frac{df}{dy}\right)$$

$$\frac{d}{du}\left(\frac{df}{dx}\right) = \frac{d^2f}{dx^2} \times \frac{dx}{du} + \frac{d^2f}{dx\,dy} \times \frac{dy}{du}$$

$$\frac{d}{dv} = \left(\frac{d}{dx} \times \frac{dx}{dv} + \frac{d}{dy} \times \frac{dy}{dv}\right)$$

$$\frac{d}{du}\left(\frac{df}{dy}\right) = \frac{d^2f}{dy\,dx} \times \frac{dx}{du} + \frac{d^2f}{dy^2} \times \frac{dy}{du}$$

$$\frac{d}{du} = \left(\frac{dx}{du} \times \frac{d}{dx} + \frac{d}{dy} \times \frac{dy}{du}\right)$$

Substituting the above equation of $\frac{d}{du}\left(\frac{df}{dx}\right)$ gives:

$$\frac{d^2w}{du^2} = \frac{df}{dx} \times \frac{d^2x}{du^2} + \frac{d^2f}{dx^2} \times \left(\frac{dx}{du}\right)^2 + \frac{2d^2f}{dx\,dy} \times \frac{dx}{du} \times \frac{dy}{du}$$

$$+ \frac{df}{dy} \times \frac{d^2y}{dx^2} + \frac{d^2f}{dy^2} \times \left(\frac{dy}{du}\right)^2$$

$\frac{d^2w}{dv^2}$ = (u is replaced by V in the above equation).

$$\frac{d^2w}{du\,dv} = \frac{df}{dx} \times \frac{d^2x}{du\,dv} + \frac{dx}{dv} \times \frac{d}{du}\left(\frac{df}{dx}\right) + \frac{df}{dy} \times \frac{d^2y}{du\,dv} + \frac{dy}{du} \times \frac{d}{du}\left(\frac{df}{dy}\right)$$

$$\frac{d^2w}{du\,dv} = \frac{df}{dx} \times \frac{d^2x}{du\,dv} + \frac{d^2f}{dx^2} \times \frac{dx}{du} \times \frac{dx}{dv} + \frac{d^2f}{dx\,dy} \times \frac{dy}{du} \times \frac{dx}{dv} + \frac{df}{dy} \times \frac{d^2y}{du\,dv} + \frac{dy}{dv} \times \frac{dx}{du} + \frac{d^2f}{dy^2} \times \frac{dy}{dv} \times \frac{dy}{du}$$

Conversion of coordinates comes under this transformation.

Equation of Continuity in Three Dimensions

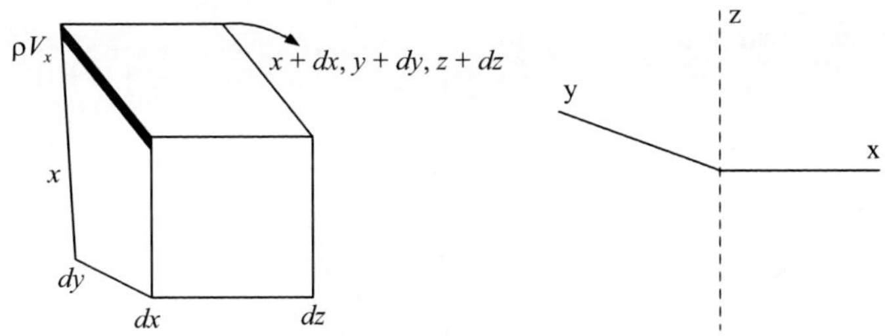

Fig. 3.1

Rate of mass input to dx element – rate of mass output from dx element = Rate of mass accumulation in dx element

(Through dx element)$dy\ dz\left(\rho V_x|_x - \rho V_x|_{x+dx}\right) = \dfrac{dx\,dy\,dz\,d\rho}{dt}$

(Through dy element)$dz\ dx\left(\rho V_y|_y - \rho V_y|_{y+dy}\right)$

(Through dz element)$dx\ dy\left(\rho V_z|_z - \rho V_z|_{z+dz}\right)$

Dividing by $dx\ dy\ dz$ throughout:

$$\frac{d\rho}{dt} = -\left[\frac{d}{dx}(\rho V_x) + \frac{d}{dy}(\rho V_y) + \frac{d}{dz}(\rho V_z)\right]$$

$$\frac{D\rho}{dt} = -(\nabla \cdot \rho V)$$

$$\frac{D\rho}{dt} = -\rho(\nabla \cdot V)$$

where $\dfrac{D}{Dt} = \left(\dfrac{d}{dt} + Vx\dfrac{d}{dx} + Vy\dfrac{d}{dy} + Vz\dfrac{d}{dz}\right)$

3.2 UNSTEADY STATE HEAT CONDUCTION EQUATION (ONE-DIMENSIONAL)

x-direction:

Rate of heat input at x-Rate of heat output at $x + dx$ ± Heat gain/Heat loss = Rate of heat accumulation in dx element.

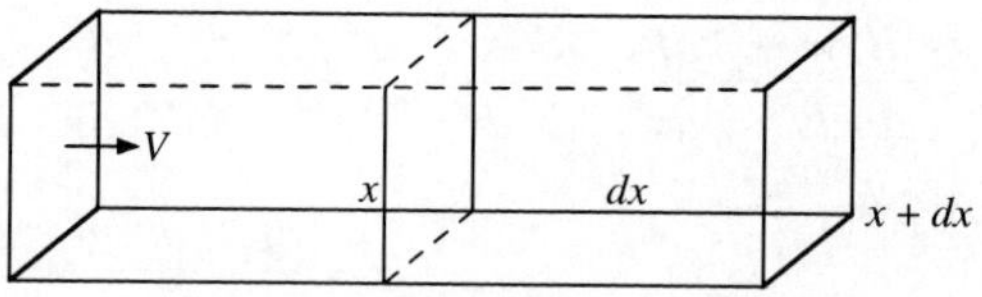

Fig. 3.2

T = temperature

ρ = density of solid

A = area of solid surface normal to x direction

t = time

x = distance

K = thermal conductivity.

Putting mathematical terms:

$$-KA\frac{dT}{dx} - \times - \left[\frac{KAdT}{dx} + \frac{d}{dx}\left(\frac{KAdT}{dx}\right)dx\right] + 0 = \frac{d}{dt}(A\,dx\,\rho CpT)$$

Simplifying:
$$\frac{KAd}{dx}\left(\frac{dT}{dx}\right)dx = A\,dx\rho Cp\frac{dT}{dt}$$

$$KA\frac{d^2T}{dx^2} = A\rho Cp\frac{dT}{dt}$$

$$\frac{dT}{dt} = \left(\frac{K}{Cp\rho}\right)\frac{d^2T}{dx^2}$$

$$\frac{dT}{dt} = \frac{C^2d^2T}{dx^2}$$

3.3 UNSTEADY STATE HEAT CONDUCTION EQUATION (TWO-DIMENSIONAL)

A thin plate becomes a two-dimensional case, t is time, T is temperature.

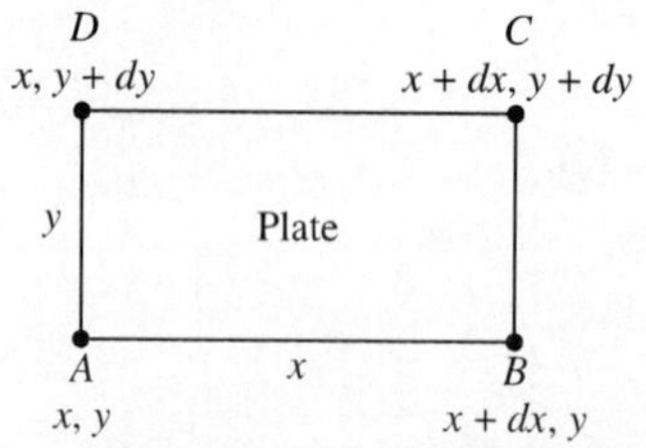

α = thickness of plate

Fig. 3.3

Putting in mathematical terms:

Rate of heat input to plate by conduction – Rate of heat output from plate by conduction ± Rate of heat gain/Rate of heat loss = Rate of heat accumulation in plate.

Putting in mathematical terms:

From side AB: $-K\alpha\, dx \left.\frac{dT}{dy}\right|_{y} + K\alpha \left.\frac{dT}{dy}\right|_{y+dy} + 0 = Cp\rho\, \alpha\, dx\, dy \frac{dT}{dt}$

From side AD: $-K\alpha\, dy \left.\frac{dT}{dx}\right|_{x} + K\alpha\, dy \left.\frac{dT}{dx}\right|_{x+dx} + 0$

Dividing by $dx\, dy\alpha$ throughout: α = Thickness of plate

$$\alpha K\, dx\, dy \left\{ \left[\left.\frac{dT}{dx}\right|_{x+dx} - \left.\frac{dT}{dx}\right|_{x} \right] \Big/ dx + \left[\left.\frac{dT}{dy}\right|_{y+dy} - \left.\frac{dT}{dy}\right|_{y} \right] \Big/ dy \right\}$$

$$\frac{dT}{dt} = \left(\frac{K}{Cp\rho}\right)\left[\frac{d^2T}{dx^2} + \frac{d^2T}{dy^2}\right]$$

$$\frac{dT}{dt} = C^2\left[\frac{d^2T}{dx^2} + \frac{d^2T}{dy^2}\right]$$

3.4 UNSTEADY STATE HEAT CONDUCTION EQUATION (THREE-DIMENSIONAL)

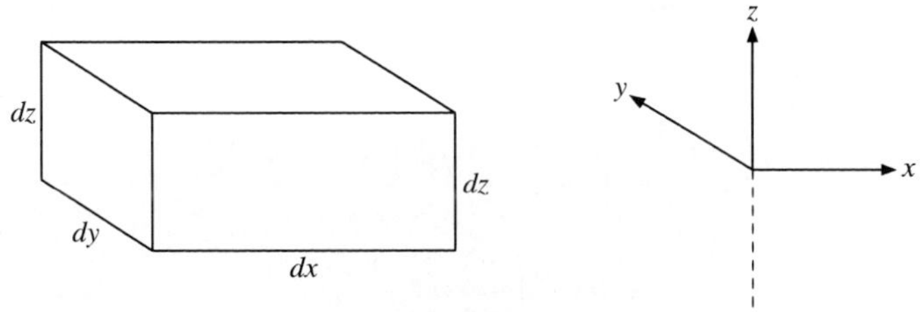

Fig. 3.4

Using unsteady state heat balance:

Heat input – Heat output ± Heat loss/Heat gain = Rate of heat accumulation

In x-direction: $-K\, dy\, dz \left.\frac{dT}{dx}\right|_{x} + K\, dy\, dz \left.\frac{dT}{dx}\right|_{x+dx} + 0$

In y-direction: $-K\, dx\, dz \left.\frac{dT}{dy}\right|_{y} + K\, dx\, dz \left.\frac{dT}{dy}\right|_{y+dy} + 0$

In z-direction: $-K\, dx\, dy \left.\frac{dT}{dz}\right|_{z} + K\, dx\, dy \left.\frac{dT}{dz}\right|_{z+dz} + 0 = \rho Cp\, dx\, dy\, dz \frac{dT}{dt}$

Dividing by $dx\, dy\, dz$ throughout:

$$K\left\{\left[\left(\left.\frac{dT}{dx}\right|_{x+dx} - \left.\frac{dT}{dx}\right|_{x}\right)\Big/dx\right] + \left[\left(\left.\frac{dT}{dy}\right|_{y+dy} - \left.\frac{dT}{dy}\right|_{y}\right)\Big/dy\right]\right.$$

$$\left. + \left[\left(\left.\frac{dT}{dz}\right|_{z+dz} - \left.\frac{dT}{dz}\right|_{z}\right)\Big/dz\right]\right\} = \rho\, Cp\frac{dT}{dt}$$

$$\frac{d^2T}{dx^2} + \frac{d^2T}{dy^2} + \frac{d^2T}{dz^2} = \left(\frac{\rho Cp}{K}\right)\frac{dT}{dt}$$

or

$$\frac{dT}{dt} = \left\{\frac{K}{Cp\rho}\left(\frac{d^2T}{dx^2} + \frac{d^2T}{dy^2} + \frac{d^2T}{dz^2}\right)\right\}$$

$$\frac{dT}{dt} = C^2\nabla^2 T$$

3.5 CONVERSION OF CARTESIAN COORDINATES TO CYLINDRICAL COORDINATES

Converting $\frac{d^2T}{dx^2} + \frac{d^2T}{dy^2} + \frac{d^2T}{dz^2}$ to cylindrical coordinates. In cylindrical coordinates the relations are:

$$x = r\cos\theta$$

$$y = r\sin\theta$$

$$z = z$$

$$r = \left(x^2 + y^2\right)^{1/2}$$

$$\theta = \tan^{-1}\left(\frac{y}{x}\right)$$

$$x = f(r, \theta)$$

$$\frac{dT}{dx} = \frac{dT}{dr}\times\frac{dr}{dx} + \frac{dT}{d\theta}\times\frac{d\theta}{dx}$$

$$\frac{dr}{dx} = \frac{x}{(x^2 + y^2)^{1/2}} = \cos\theta$$

$$\frac{d\theta}{dx} = -\frac{y}{r^2} = -\frac{\sin\theta}{r}$$

$$\frac{dT}{dx} = \frac{dT}{d\theta}\left(-\frac{\sin\theta}{r}\right) + \frac{dT}{dr}\times\cos\theta$$

$$\frac{d}{dx} = \left(\cos\theta\frac{d}{dr} - \frac{\sin\theta}{r}\frac{d}{d\theta}\right)$$

$$\frac{d^2T}{dx^2} = \frac{d}{dx}\left(\frac{dT}{dx}\right) = \left(\cos\theta\frac{\partial}{\partial r} - \frac{\partial}{\partial\theta}\frac{\sin\theta}{r}\right)\left(-\frac{\sin\theta}{r}\frac{dT}{d\theta} + \cos\theta\frac{dT}{dr}\right)$$

$$= \cos\theta\left(\frac{\sin\theta}{r^2}\frac{dT}{d\theta} - \frac{\sin\theta}{r}\frac{d^2T}{d\theta\,dr}\right) + \cos\theta\left(\cos\theta\frac{d^2T}{dr^2}\right)$$

$$-\frac{\sin\theta}{r}\left(-\frac{dT}{d\theta}\frac{\cos\theta}{r} - \frac{\sin\theta}{r}\frac{d^2T}{d\theta^2} - \sin\theta\frac{dT}{dr} + \cos\theta\frac{d^2T}{d\theta\,dr}\right)$$

$$= \cos^2\theta\frac{d^2T}{dr^2} + \frac{\sin\theta\cos\theta}{r^2}\left(\frac{dT}{d\theta}\right) - \frac{\sin\theta\cos\theta}{r}\left(\frac{d^2T}{d\theta\,dr}\right)$$

$$+\frac{\sin\theta\cos\theta}{r^2}\left(\frac{dT}{d\theta}\right) + \frac{\sin^2\theta}{r^2}\left(\frac{d^2T}{d\theta}\right) + \frac{\sin^2\theta}{r}\left(\left(\frac{dT}{dr}\right) - \frac{\sin\theta\cos\theta}{r}\left(\frac{d^2T}{d\theta\,dr}\right)\right)$$

$$= \cos^2\theta\frac{d^2T}{dr^2} + \frac{2\sin\theta\cos\theta}{r^2}\left(\frac{dT}{d\theta}\right) - \frac{2\sin\theta\cos\theta}{r}\left(\frac{d^2T}{d\theta\,dr}\right)$$

$$+\frac{\sin^2\theta}{r^2}\left(\frac{d^2T}{d\theta^2}\right) + \frac{\sin^2\theta}{r}\left(\frac{dT}{dr}\right)$$

Similarly, $y = f(r, \theta)$

$$\frac{dT}{dy} = \frac{dT}{dr}\times\frac{dr}{dy} + \frac{dT}{d\theta}\times\frac{d\theta}{dy}$$

$$\frac{d\theta}{dr} = \sin\theta$$

$$\frac{d\theta}{dy} = \frac{\cos\theta}{r}$$

$$\frac{d}{dy} = \left(\sin\theta\frac{d}{dr} + \frac{\cos\theta}{r}\frac{d}{d\theta}\right)$$

$$\frac{d^2T}{dy^2} = \frac{d}{dy}\left(\frac{du}{dy}\right)$$

$$= \left(\sin\theta\frac{d}{dr} + \frac{\cos\theta}{r}\frac{d}{d\theta}\right)\left(\sin\theta\frac{dT}{dr} + \frac{\cos\theta}{r}\times\frac{dT}{d\theta}\right)$$

$$= \sin^2\theta\frac{d^2T}{dr^2} + \frac{2\sin\theta\cos\theta}{r}\frac{d^2T}{dr\,d\theta} + \frac{\cos^2\theta}{r^2}\frac{d^2T}{d\theta^2}$$

$$+\frac{\cos^2\theta}{r}\frac{dT}{dr} - \frac{2\sin\theta\cos\theta}{r^2}\frac{dT}{d\theta}$$

$$\frac{d^2T}{dz^2} = \frac{d^2T}{dz^2}$$

$$\frac{d^2T}{dx^2} + \frac{d^2T}{dy^2} + \frac{d^2T}{dz^2} = \frac{d^2T}{dr^2} + \frac{1}{r^2}\frac{d^2T}{d\theta^2} + \frac{1}{r}\frac{dT}{dr} + \frac{d^2T}{dz^2}$$

3.6 UNSTEADY STATE HEAT CONDUCTION EQUATION IN A CYLINDER

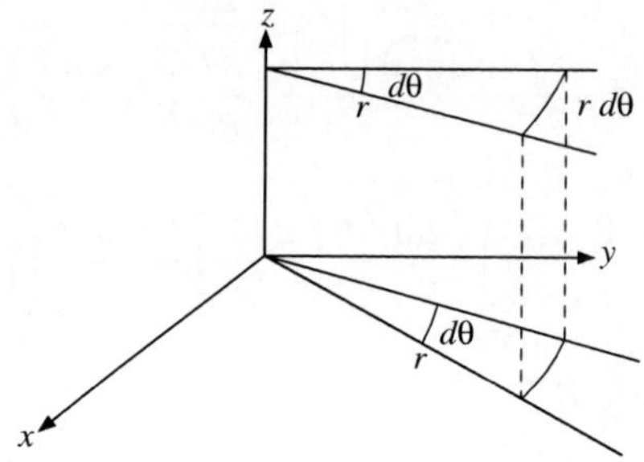

Fig. 3.5

Varying coordinates	*Constant*	*Change in arc length*	
dr	θ, z	dr	
$d\theta$	r, z	$r\,d\theta$	
dz	θ, r	dz	
Coordinate	Area	Area	Area normal to coordinates
r	dAr	$d\theta\,dz$	$r\,d\theta\,dz$
θ	$dA\theta$	$dr\,dz$	$dr\,dz$
Z	dAz	$dr\,d\theta$	$r\,dr\,d\theta$

Unsteady state heat conduction equation is:

Heat input rate – Heat output rate ± Heat gain/Heat loss = Rate of heat accumulation

$$(r\text{-direction})\left[q_r - \left(q_r + \frac{dq_r}{dr}\right) + 0\right] + (\theta - \text{direction})\left[q_\theta - \left(q_\theta + \frac{dq_\theta}{d\theta}d\theta\right) + 0\right]$$

$$+ (z\text{-direction})\left[q_z - \left(q_z + \frac{dq_z}{dz}dz\right) + 0\right]$$

$$= rdr\,d\theta\,dz\rho\,Cp\,\frac{dT}{dt}$$

Simplifying gives:

$$-\frac{dq_r}{dr} \times dr - \frac{dq_\theta}{d\theta} \times d\theta - \frac{dq_z}{dz}dz = rdr\,d\theta\,dz\rho\,Cp\,\frac{dT}{dt} \quad \ldots(1)$$

We have

$$q_r = -\frac{KAr\,dT}{dr} = -\frac{Krd\theta dz\,dT}{dr}$$

$$q_\theta = -\frac{KA\theta\,dT}{d\theta} = -\frac{K\,dr\,dz\,dT}{rd\theta}$$

$$q_z = -\frac{K\,Az\,dT}{dz} = -\frac{K\,rd\theta\,dr\,dT}{dz}$$

Subsituting in Eq. (1) gives

$$\frac{dr\,d\theta\,dz\,Kd}{dr}\left(\frac{rdT}{dr}\right) + \frac{dr\,dz\,d\theta\;Kd}{r\quad d\theta}\left(\frac{dT}{d\theta}\right) + \frac{rdr\,d\theta\,dz\,K\,d}{dz}\left(\frac{dT}{dz}\right) = rdrd\theta dz\rho Cp\frac{dT}{dt}$$

Dividing by r throughout,

$$\left(\frac{1}{r}\right)\frac{dT}{dr} + \frac{d^2T}{dr^2} + \left(\frac{1}{r^2}\right)\frac{d^2T}{d\theta^2} + \frac{d^2T}{dz^2} = \left(\frac{\rho Cp}{K}\right)\frac{dT}{dt}$$

Rearanging:

$$\frac{d^2T}{dr^2} + \left(\frac{1}{r}\right)\frac{dT}{dr} + \left(\frac{1}{r^2}\right)\frac{d^2T}{d\theta^2} + \frac{d^2T}{dz^2} = \left(\frac{\rho Cp}{K}\right)\frac{dT}{dt}$$

$$\alpha\nabla^2 T = \frac{dT}{dt}$$

3.7 UNSTEADY STATE HEAT CONDUCTION EQUATION IN SPHERICAL COORDINATES

In sphere,

$$x = r\cos\theta\sin\phi$$

$$y = r\sin\theta\sin\phi$$

$$z = \cos\phi$$

The volume element in spherical coordinates is

$dx\,dy\,dz$ (rectangular coordinates) = $dr\,d\theta\,d\phi$ in lengths.

Coordinate	*Coordinate changing*	*Coordinate held constant*	*Change in arc length*
R	dr	θ, ϕ	dr
θ	$d\theta$	r, ϕ	$(r\sin\phi)d\theta$
ϕ	$d\phi$	r, θ	$rd\phi$
Area symbol	*Coordinate*	*Area*	*Area element in length unts*
dAr	r	$d\theta\,d\phi$	$r^2\sin\phi\,d\theta\,d\phi$
$dA\theta$	θ	$d\phi\,dr$	$rd\theta\,dr$
$dA\phi$	ϕ	$d\theta\,dr$	$r\sin\phi\,d\theta\,dr$

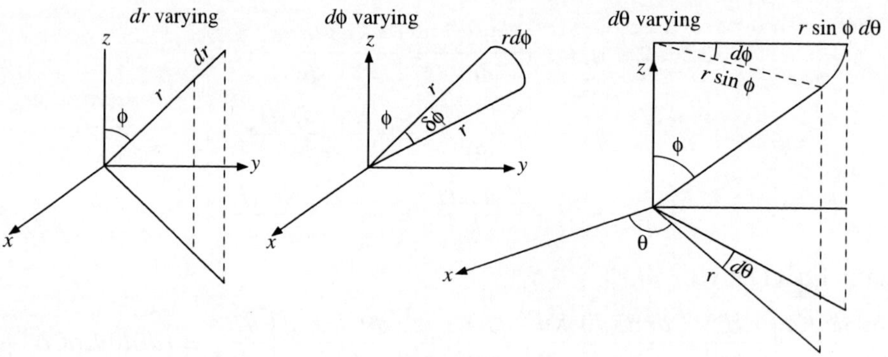

Fig. 3.6

The unsteady state heat conduction equation is:

Rate of heat input – Rate of heat output ± Heat gain/Heat loss = Rate of heat accumulation.

(ln r-direction) $\left[q_r - \left(q_r + \frac{dq_r}{dr} \times dr\right) + 0\right]$ + (ln θ direction) $\left[q_\theta - \left(q_\theta + \frac{dq_\theta}{d\theta} \times d\theta\right) + 0\right]$ + (in ϕ direction) $\left[q_\phi - \left(q_\phi + \frac{dq_\phi}{d\phi} \times d\phi\right) + 0\right] = r^2 \sin\phi\, d\theta\, d\phi\, dr \frac{pCp\, dT}{dt}$

Simplifying gives:

$$-\frac{dq_r}{dr} \times dr - \frac{dq_\theta}{d\theta} \times d\theta - \frac{dq_\phi}{d\phi} \times d\phi$$

$$= r^2 \sin\phi\, d\phi\, d\theta\, dr \frac{\rho CP\, dT}{dt} \qquad \text{...(1)}$$

We know

$$q_r = -K\left(\frac{dT}{dr}\right) dAr = -K(r^2 \sin\phi\, d\phi\, d\theta)\frac{dT}{dr} \qquad \text{...(2)}$$

$$q_\theta = -K\left(\frac{1}{r\sin\phi}\right)\frac{dT}{d\theta} \times dA\theta$$

$$= -\frac{K\, d\phi\, dr}{\sin\phi} \times \left(\frac{dT}{d\theta}\right) \qquad \text{...(3)}$$

$$q_\phi = -K\left(\frac{1}{r}\right) dA\phi \times \frac{dT}{d\phi} = -K \sin\phi\, d\theta\, dr \frac{dT}{d\phi} \qquad \text{...(4)}$$

Substituting Eqs. (2), (3), (4) into, Eq. (1) gives

$$K \sin\phi\, d\phi\, d\theta \frac{d}{dr}\left(\frac{r^2\, dT}{dr} \times dr\right) + \left(\frac{dr\, d\phi}{\sin\phi}\right)$$

$$\times \frac{d}{d\theta}\left(\frac{KdT}{d\theta}\right) d\theta + d\theta\, dr \frac{Kd}{d\phi}\left(\frac{\sin\phi\, dT}{d\phi}\right) d\phi$$

$$= r^2 \sin\phi \, d\phi \, d\theta \, dr \frac{C p\rho \, dT}{dt}$$

Giving: $$\frac{1}{r^2}\frac{d}{dr}\left(\frac{r^2\, dT}{dr}\right) + \frac{1}{r^2 \sin^2\phi}\frac{d}{d\phi}\left(\frac{\sin\phi\, dT}{d\phi}\right) + \frac{1}{r^2\sin^2\phi} \times \frac{d^2T}{d\theta^2}$$

$$= \frac{1}{C^2}\frac{dT}{dt}$$

or $$\frac{dT}{dt} = C^2\nabla^2 T$$

Example 3.1 A rectangular block of metal is subjected to temperature variation in all directions but the surface at $y = 0$ is thermally insulated. Express this condition by an equation.

Solution: It is convenient to use Cartesian coordinates in this problem. Heat can pass through the faces by conduction only (since metal).

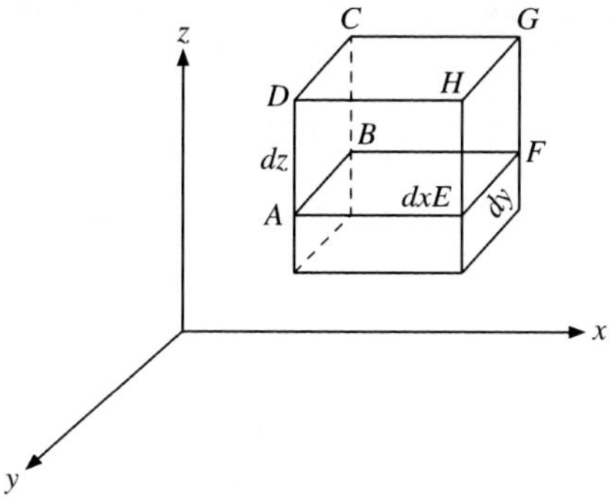

Fig. 3.7

The faces are:	In	Out
In x-direction:	$ABCD$	$EFGH$
In y-direction:	$ADHE$	$BCGF$
In z-direction:	$ABFE$	$DGHC$

Input in y-direction ($ADHE$) = zero (since insulated)

Output in y-direction ($BCGF$) = $-\left[K\left(\frac{dT}{dy}\right) dx\, dz + \frac{d}{dy}\left(\left(\frac{KdT}{dy}\right) dx\, dz\right) dy\right]$

Input rate in x-direction ($ABCD$) = $-K\left(\frac{dT}{dx}\right) dy\, dz$

Output rate in x-direction ($EFGH$) = $-\left[K\left(\frac{dT}{dx}\right) dy\, dz + \frac{d}{dx}\left(\left(\frac{KdT}{dx}\right) \times dy\, dz\right) dx\right]$

Input rate in z-direction ($ABFE$) = $-\frac{KdT}{dz} \times dy\, dx$

Output r ate in z-direction ($DGHC$) = $-\left[\frac{KdT}{dz} \times dy\, dx + \frac{d}{dz}\left(\frac{KdT}{dz} \times dy\, dx\right) dz\right]$

Using the unsteady state heat balance by conduction in the element:

Heat input rate – Heat output rate ± Heat change = Rate of heat accumulation. Cancelling the common terms,

$$K\left(\frac{d^2T}{dx^2}\right) dy\, dz\, dx + K\left(\frac{d^2T}{dy^2}\right) dx\, dy\, dz + K\left(\frac{d^2T}{dz^2}\right) dx\, dy\, dz + \left(\frac{dT}{dy}\right) dx\, dz\, K$$

$$= \rho Cp\, dx\, dy\, dz\, \frac{dT}{dt}$$

Cancelling $dx\, dy\, dz$ throughout

$$K\left(\frac{d^2T}{dx^2} + \frac{d^2T}{dy^2} + \frac{d^2T}{dz^2}\right) dy + \frac{KdT}{dy} = \rho Cp\, dy \frac{dT}{dt}$$

Taking the limit as $y \rightarrow 0$, then the equation becomes with $\frac{dT}{dy} = 0$ at $y = 0$. This is the required boundary condition.

Physical significance of the below given partial differential equations.

(1) $\frac{du}{dt} = \frac{ad^2u}{dx^2} - bu$

This is unsteady state heat conduction equation in one dimension, with heat loss by radiation or convection.

(2) $\frac{du}{dt} = \frac{a\, d^2u}{dx^2} - \frac{b\, du}{dx}$

This is also unsteady state heat conduction equation with heat loss by conduction.

(3) $\frac{d^2u}{dx^2} + \frac{d^2u}{dy^2} + f(x, y) = 0$

This is steady state heat conduction equation in two dimensions (x, y) with a heat source $f(x, y)$ or sink.

3.7.1 Conversion of Coordinates by Vectors Method

Cylindrical Coordinates

$$x = r\cos\theta$$

$$y = \sin\theta$$

$$z = z$$

where θ = angle between r and x

r = radius of the cylinder.

$$(ds)^2 = (dx)^2 + (dy)^2 + (dz)^2 \text{ in Cartesian coordinates}$$

$$= (dr)^2 + (d\theta)^2 + (dz)^2 \text{ in lengths}$$

$$= (dr)^2 + (rd\theta)^2 + (dz)^2 = (dr)^2 + r^2(d\theta)^2 + (dz)^2$$

$$(ds)^2 = h_1^2\, du_1^2 + h_2^2\, dv^2 + h_3^2\, dw^2$$

where u, v, w are curvilinear variables (r, θ, z) of h_1, h_2, h_3. h_1, h_2 and h_3 are scalar variables (magnitudes).

Comparing:

$$h_1 = 1$$

$$h_2 = r$$

$$h_3 = 1$$

$$\text{Volume element } (dV) = h_1h_2h_3\, du\, dv\, dw = rd\, rd\theta\, dz$$

In spherical coordinates:

$$x = r\sin\theta\cos\phi$$

$$y = r\sin\theta\sin\phi$$

$$z = r\cos\phi$$

$$(ds)^2 = h_1^2du^2 + h_2^2dv^2 + h_3^2dw^2$$

$$= (dx)^2 + (dy)^2 + (dz)^2$$

$$= (dr)^2 + (d\theta)^2 + (d\phi)^2 \text{ in lengths}$$

$$= dr^2 + (rd\theta)^2 + (r\sin\theta\, d\phi)^2$$

So,

$$h_1 = 1$$

$$h_2 = r$$

$$h_3 = r\sin\theta$$

$$dV = dx\, dy\, dz$$

$$= h_1\, h_2\, h_3\, dr\, d\theta\, d\phi$$

$$= r^2\sin\theta\, dr\, d\theta\, d\phi$$

$$\nabla f = \left[\left(\frac{A}{h_1}\right)\frac{df}{du} + \left(\frac{B}{h_2}\right)\frac{df}{dv} + \left(\frac{C}{h_3}\right)\frac{df}{dw}\right]$$

$$\nabla \times F = \begin{bmatrix} \dfrac{A}{(h_2\, h_3)} & \dfrac{B}{(h_3\, h_1)} & \dfrac{C}{(h_1\, h_2)} \\ \dfrac{d}{du} & \dfrac{d}{dv} & \dfrac{d}{dw} \\ h_1 f_1 & h_2 f_2 & h_3 f_3 \end{bmatrix}$$

$$\nabla^2 F = \frac{1}{(h_1\, h_2\, h_3)}\left[\frac{d}{du}\left(\frac{h_2\, h_3}{h_1} \times \frac{df}{du}\right) + \frac{d}{dv}\left(\frac{h_3\, h_1}{h2} \times \frac{df}{dv}\right) + \frac{d}{dw}\left(\frac{h_1\, h_2}{h_3} \times \frac{df}{dw}\right)\right]$$

1. In Spherical Coordinates

$$h_1 = 1$$

$$h_2 = r$$

$$h_3 = r \sin \theta$$

$$\nabla^2 f = \left(\frac{1}{r^2 \sin \theta}\right)\left\{\frac{d}{dr}\left(r^2 \sin \theta \times \frac{df}{dr}\right) + \frac{d}{d\theta}\left(\frac{r \sin \theta}{r} \times \frac{df}{d\theta}\right) + \frac{d}{d\phi}\left(\frac{r}{r \sin \theta} \times \frac{df}{d\phi}\right)\right\}$$

$$= \left(\frac{1}{r^2}\right)\frac{d}{dr}\left(\frac{r^2 df}{dr}\right) + \frac{1}{r^2 \sin \theta} \times \frac{d}{d\theta}\left(\sin \theta \frac{df}{d\theta}\right) + \frac{1}{r^2 \sin^2 \theta}\frac{d^2 f}{d\theta^2}$$

$$\nabla^2 f = \frac{d^2 f}{dr^2} + \left(\frac{1}{r^2}\right)\frac{d^2 f}{d\theta^2} + \left(\frac{1}{r^2 \sin^2 \theta}\right)\frac{d^2 f}{d\phi^2} + \left(\frac{2}{r}\right)\frac{df}{dr} + \frac{\cos \theta}{r^2} \times \frac{df}{d\theta}$$

2. In Cylindrical Coordinates

$$\nabla^2 f = \left(\frac{1}{h_1\, h_2\, h_3}\right)\left[\frac{d}{du}\left(\frac{h_2 h_3}{h_1} \times \frac{df}{du}\right) + \frac{d}{dv}\left(\frac{h_3\, h_1}{h_2} \times \frac{df}{dv}\right) + \frac{d}{dw}\left(\frac{h_1\, h_2}{h_3} \times \frac{df}{dw}\right)\right]$$

$$u = r$$

$$v = \theta$$

$$w = z$$

$$h_1 = 1$$

$$h_2 = r$$

$$h_3 = 1$$

$$x = r \cos \theta$$

$$y = r \sin \theta$$

$$z = z$$

$$\nabla^2 f = \left(\frac{1}{r}\right)\left[\frac{d}{dr}\left(\frac{r\,df}{dr}\right) + \frac{d}{d\theta}\left(\frac{1}{r}\times\frac{df}{d\theta}\right) + \frac{d}{dz}\left(\frac{rdf}{dz}\right)\right]$$

$$\nabla^2 f = \frac{d^2 f}{dr^2} + \left(\frac{1}{r}\right)\frac{df}{dr} + \left(\frac{1}{r}\right)\frac{d^2 f}{d\theta^2} + r\frac{d^2 f}{dz^2}$$ **Ans.**

PARTIAL DIFFERENTIAL EQUATIONS

1. Fourier Series

The Fourier series expansion is

$$f(x) = \frac{a_0}{2} + a_1 \cos x + a_2 \cos 2x + a_3 \cos 3x + ...$$

$a_n \cos nx + b_1 \sin x + b_2 \sin 2x + b_3 \sin 3x + \ldots + b_n \sin nx.$

$$f(x) = \frac{a_0}{2} + \Sigma a_n \cos nx + \Sigma b_n \sin nx$$

For x varying from α to $\alpha + 2\pi$, the coefficients are given by

$$a_0 = \frac{1}{\pi}\int_{\alpha}^{\alpha+2\pi} f(x)\,dx$$

$$a_n = \frac{1}{\pi}\int_{\alpha}^{\alpha+2\pi} f(x)\cos nx\,dx$$

$$b_n = \frac{1}{\pi}\int_{\alpha}^{\alpha+2\pi} f(x)\sin nx\,dx$$

If $\alpha = 0$; x varies from 0 to 2π

$$a_0 = \frac{1}{\pi}\int_0^{2\pi} f(x)\,dx$$

$$a_n = \frac{1}{\pi}\int_0^{2\pi} f(x)\cos nx\,dx$$

$$b_n = \frac{1}{\pi}\int_0^{2\pi} f(x)\sin x\,dx$$

If $\alpha = -\pi$; then x varies from $-\pi$ to $+\pi$.

$$a_0 = \frac{1}{\pi}\int_{-\pi}^{\pi} f(x)\,dx$$

$$a_n = \frac{1}{\pi}\int_{-\pi}^{\pi} f(x)\cos nx\, dx$$

$$b_n = \frac{1}{\pi}\int_{-\pi}^{\pi} f(x)\sin nx\, dx$$

2. Discontinuous Functions

$$f(x) = \phi(x),\ \alpha < x < C$$

$$= \psi(x),\ C < x < 2\pi + \alpha \text{ where } C \text{ is the point discontinuity}$$

$$a_0 = \frac{1}{\pi}\left[\int_{\alpha}^{C} \phi(x)\,dx + \int_{c}^{\alpha+2\pi} \psi(x)dx\right]$$

$$a_n = \frac{1}{\pi}\left[\int_{\alpha}^{C} \phi(x)\cos nx\,dx + \int_{C}^{\alpha+2\pi} \psi(x)\cos nx\,dx\right]$$

$$b_n = \frac{1}{\pi}\left[\int_{\alpha}^{C} \phi(x)\sin nx\,dx + \int_{C}^{\alpha+2\pi} \psi(x)\sin nx\,dx\right]$$

3. Change of Interval

Instead of 0 to 2π or $-\pi$ to $+\pi$, we come across -1 to $+1$ or $-C$ to $+C$ or 0 to $2l$ or 0 to $2C$:

$$f(x) = \frac{a_0}{2} + \sum_{n=1}^{\infty} a_n \cos\left(n\pi\frac{x}{l}\right) + \sum_{n=1}^{\infty} b_n \sin\left(n\pi\frac{x}{l}\right)$$

4. In Half Range Series of SINE

Expressed as sine or cosine terms to obtain Fourier series expansion of $f(x)$. If total period is -1 to $+1$ or 0 to 2, then half range is 0 to 1.

In half range series of since, the function $f(x)$ is:

$$f(x) = \sum_{n=1}^{\infty} b_n \sin\left(n\pi\frac{x}{l}\right)$$

$$b_n = \frac{2}{l}\int_{0}^{l} f(x)\sin\left(n\pi\frac{x}{l}\right) dx$$

where $a_0, a_1, a_2, \ldots, a_n = 0$

In case of cosine series:

$$f(x) = \frac{a_0}{2} + \sum_{n=1}^{\infty} a_n \cos\left(n\pi\frac{x}{l}\right)$$

where $a_0 = \frac{2}{l}\int_0^l f(x)\,dx$

$$a_n = \frac{2}{l}\int_0^n f(x)\cos\left(n\pi\frac{x}{l}\right)dx$$

Example 3.2 Find Fourier series of

$$f(x) = \pi - x;\ 0 < x < 1$$

$$= \pi(2 - x);\ 1 < x < 2$$

In the interval 0 to 2.

Solution:

$$f(x) = \frac{a_0}{2} + \sum_{n=1}^{\infty} a_n \cos nx + \sum_{n=1}^{\infty} b_n \sin nx$$

$$f(x) = \frac{a_0}{2} + \sum_{n=1}^{\infty} a_n \cos\left(n\pi\frac{x}{C}\right) + \sum_{n=1}^{\infty} b_n \sin\left(n\pi\frac{x}{C}\right)$$

where $a_0 = \frac{1}{C}\int_0^{2C} f(x)dx$

$$a_0 = \frac{1}{C}\int_0^{2C} f(x)dx$$

$$a_n = \frac{1}{C}\int_0^{2C} f(x)\cos\left(n\pi\frac{x}{C}\right)dx$$

Here, $C = 1$

$$b_n = \frac{1}{C}\int_0^{2C} f(x)\sin\left(n\pi\frac{x}{C}\right)dx$$

$$a_0 = \int_0^1 \pi\, x\, dx + \int_1^2 \pi(2 - x)dx$$

$$= \left.\frac{\pi x^2}{2}\right|_0^1 + \left.\left(2\pi x - \frac{\pi x^2}{2}\right)\right|_1^2$$

$$= \frac{\pi}{2} + 4\pi - 2\pi - 2\pi + \frac{\pi}{2} = \pi$$

$$a_n = \int_0^1 \pi x \cos n\pi x\, dx + \int_1^2 \pi(2 - x)\cos n\pi x\, dx$$

First term is integrated using integral of $udv = uv$ – integral of vdu.

Here $u = x$, $du = dx$, $dv = \cos n\pi\, dx$;

$$v = \frac{\sin n\pi\, x}{n\pi}$$

$$= \pi\left[\left.\frac{x \sin n\pi\, x}{n\pi}\right|_0^1 - \int_0^1 \frac{\sin n\pi x}{n\pi} dx\right] = 0$$

$$-\left.\frac{\pi \cos n\pi\, x}{n^2\pi^2}\right|_0^1 = \frac{(\cos n\pi - 1)}{n^2\pi}; \quad 2^{\text{nd}} \text{ term,}$$

$$\int_1^2 \pi(2-x) \cos n\pi\; xdx = \pi\int_1^2 (2 \cos n\pi\, xdx - \int_1^2 x \cos n\pi\, xdx$$

$$= \pi\left[\left.\frac{2 \sin n\pi\, x}{n\pi}\right|_1^2 - \left.\left(\frac{x \sin n\pi x}{n\pi}\right)\right|_1^2 + \left.\frac{\cos n\pi\, x}{n\pi}\right|_1^2\right]$$

$$a_n = \frac{1}{\pi n^2}\left[\cos n\pi - 1 + \cos n\pi - \cos 2n\pi\right]$$

$$a_1 = -\frac{4}{1^2\,\pi}$$

$$a_2 = 0$$

$$a_3 = -\frac{4}{3^2}\pi$$

$$b_n = \int_0^1 \pi\, x \sin n\pi\, x\, dx + \int_1^2 \pi(2-x)\sin n\pi\, xdx$$

First term is:

$$= \pi\left[\left.\frac{-x \cos n\pi\, x}{n\pi}\right|_0^1 + \int_0^1 \frac{\cos n\pi\, xdx}{n\pi}\right]$$

$$= \pi\left[\frac{-\cos n\pi}{n\pi} + \left.\frac{\sin n\pi x}{n^2\pi^2}\right|_0^1\right]$$

$$= \frac{-\cos n\pi}{n}.$$

Second term is:

$$\pi\int_1^2 2 \sin n\pi\, xdx - \int_1^2 x \sin n\pi\, xdx$$

$$= \pi\left[\frac{-2\cos n\pi\, x}{n\pi}\Bigg|_1^2 + \frac{x\cos n\pi\, x}{n\pi}\Bigg|_1^2 - \frac{\sin n\pi\, x^2}{n^2\pi^2}\Bigg|_1^2\right]$$

$$= \pi\left[\frac{-2\cos 2n\pi}{n\pi} + \frac{2\cos n\pi}{n\pi} + \frac{2\cos 2n\pi}{n\pi} - \frac{\cos n\pi}{n\pi}\right]$$

$$= \frac{\cos n\pi}{n} - \frac{\cos n\pi}{n} = 0$$

$$f(x) = \frac{\pi}{2} - \frac{4}{\pi}\left[\frac{\cos \pi x}{1^2} + \frac{\cos 3\pi x}{3^2} + \frac{\cos 5\pi x}{5^2} + \ldots\right]$$

Example 3.3 A thin rod of length l with its ends A and B are maintained at 0ºC and 100ºC respectively until steady conditions are reached. If B is suddenly reduced to 0ºC and maintained at 0ºC, find the temperature T at a distance x from A at any time t.

Solution: Heat conduction equation is:

$$\frac{du}{dt} = C^2\frac{d^2u}{dx^2} \qquad \ldots(1)$$

At steady state,

$$\frac{du}{dt} = 0 = \frac{d^2u}{dx^2}$$

$$u = ax + b$$

u at $x = 0$ is 0;

So, $b = 0$; $u = 100$ at $x = l$; So $u = 100\ \frac{x}{l}$

Initial condition is:

$$u(x.0) = 100\frac{x}{l};$$

Boundary conditions are $u(0, t) = 0$ for all t values;

Solution to Eq. (1) is of the form:

$$u(x, t) = (c_1 \cos Px + c_2 \sin Px)\exp(-C^2p^2t)$$

$$u(0, t) = 0 = C_1 \exp(-C^2p^2t)$$

So, $C_1 = 0$

$$u(l, t) = C_2 \sin pl \times \exp(-C^2p^2t) = 0$$

$$\sin pl = 0;$$

So,

$$pl = n\pi$$

$$p = \frac{n\pi}{l}$$

where n is integer. $u(x, t) = b_n \sin \frac{n\pi x}{l} \times \exp\left(-\frac{C^2 n^2 \pi^2 t}{l^2}\right)$

where $b_n = C^2$

$u(x, t)$ depends on different values of n from $n = 1$ to ∞.

$$u(x, t) = \sum_{n=1}^{\infty} b_n \sin\left(\frac{n\pi x}{l}\right) \times \left(-\frac{C^2 n^2 \pi^2 t}{l^2}\right)$$

At $t = 0$;

$$u(x, 0) = \sum_{n=1}^{\infty} b_n \sin \frac{n\pi x}{l}$$

$$u(x, 0) = 100\frac{x}{l} = \sum_{n=1}^{\infty} b_n \sin \frac{n\pi x}{l}$$

B_n is evaluated by Fourier series expansion.

$$B_n = \frac{2}{l}\int_0^l \left(\frac{100x}{l}\right) \sin\left(\frac{n\pi x}{l}\right)$$

$$= \left(\frac{200}{l^2}\right)\frac{\left[x\left(-\cos\frac{n\pi x l}{l}\right) - 1\right]}{\left(\frac{n\pi}{l}\right)} - \left.\frac{\sin\frac{n\pi x}{l}}{\left(\frac{n\pi}{l}\right)^2}\right|_0^l$$

$$= \frac{200}{l^2}\left(\frac{-l^2 \cos n\pi}{n\pi}\right) = \frac{200(-1)^{n+1}}{(n\pi)}$$

$$u(x, t) = \left(\frac{200}{\pi}\right)\sum_{n=1}^{\infty}\left[\frac{(-1)^{n+1}}{n}\right]\sin\left(\frac{n\pi x}{l}\right)\exp\left(-\frac{C^2 n^2 \pi^2 t}{l^2}\right)$$

where $C^2 = \frac{K}{(Cp\,\rho)}$ is thermal diffusivity.

Example 3.4 The ends A and B of a copper rod of 4 cm long have their temperatures kept at 0ºC and 80ºC respectively, until steady state conditions prevail. The temperature of end B is reduced suddenly to 40ºC and kept so, while that of end A is kept at 0ºC. Find the temperature $u(x, t)$ as a function of length (l) and time (t), the time elapsed after sudden change of temperature at end B.

Solution:

$$u(x, t) = ut(x, t) + us(x)$$

Given

$$us(x) = \frac{40x}{l}$$

$$us(0) = 0$$

$$us(l) = 40$$

Given $$u(0, t) = 0$$

$$u(l, t) = 40$$

$$ut(x, t) = u(x, t) - us(x)$$

$$ut(0, t) = u(0, t) - us(0) = 0 - 0 = 0$$

$$ut(l, t) = u(l, t) - us(l) = 40 - 40 = 0$$

$$ut(x, 0) = u(x, 0) - us(x) = \frac{80x}{l} - \frac{40x}{l} = \frac{40x}{l}$$

$$u(x, t) = (C_1 \cos px + C_2 \sin px) \exp\left(\frac{-C^2 p^2 t}{l^2}\right)$$

$$ut(0, t) = 0 = C_1 \exp\left(\frac{-C^2 p^2 t}{l^2}\right)$$

So, $C_1 = 0$

$$ut(l, t) = 0 = C_2 \sin pl \exp\left(\frac{-C^2 p^2 t}{l^2}\right)$$

So, $\sin pl = n$p $$p = \frac{n\pi}{l};$$

$$ut(x, t) = \sum C_n \sin\left(\frac{n\pi x}{l}\right) \exp\left(\frac{-C^2 n^2 \pi^2 t}{l^2}\right)$$

$$ut(x, 0) = \frac{40x}{l} = \sum C_n \sin\left(\frac{n\pi x}{l}\right)$$

$$C_n = \frac{2}{l}\int_0^l \left(\frac{40x}{l}\right) \sin\left(\frac{n\pi x}{l}\right) dx$$

$$= \left(\frac{80}{l^2}\right)\int_0^l x \sin\left(\frac{n\pi x}{l}\right) dx$$

$$= \left(\frac{80}{l^2}\right)\left[-\left(\frac{lx}{n\pi}\right) \times \cos\left(\frac{n\pi x}{l}\right)\Big|_0^l + \int_0^l \frac{l}{n+1}\right] \cos\left(\frac{n\pi x}{l}\right) dx$$

$$C_n = -\frac{80}{n\pi \cos(n\pi)}$$

$$C_1 = -\frac{80}{\pi}\left(-\frac{1}{1}\right) = \frac{80}{\pi}$$

$$C_2 = -\frac{80}{\pi} \times \frac{1}{2}$$

$$ut(x, t) = ut(x, t) + us(x)$$

where $us(x) = \frac{40x}{l}$, $l = 40$

$$u(x, t) = x + \frac{80}{\pi}\sum_{n=1}^{\infty}\frac{(-1)^{n+1}}{n}\sin\left(\frac{n\pi x}{40}\right)\exp\left(\frac{-C^2 n^2 \pi^2 t}{1600}\right) \quad \textbf{Ans.}$$

Example 3.5 If the above rod is of copper with specific gravity of 8.89, thermal conductivity of (K) = 0.84 Cals/(cm.°C.sec) and specific heat = 0.092 Cal/(gm.°C). Find the temperature at a point 10 cm. From end A, 5 minutes after the change.

Solution:

$$u(x, t) = u(10{,}300)$$

$$= 10 + \left(\frac{80}{\pi}\right)\sin\left(\frac{10\pi}{40}\right)\exp\left(-\pi^2 C^2 \times \frac{300}{1600}\right)$$

$$-\left(\frac{80}{2\pi}\right) \times \left(\frac{20\pi}{40}\right) \times \exp\left(-4\pi^2 C^2 \times \frac{300}{1600}\right)$$

$$C^2 = \frac{K}{C_p\rho} = \frac{0.84}{(0.092 \times 8.89)} = 1.03$$

$$u(x, t) = 10 + \left(\frac{80}{\pi}\right)\left[0.70\exp(-0.0318) - 0.35 + 0.12 - \text{small term}\right]$$

$$= 12.7°\text{C} \quad \textbf{Ans.}$$

Example 3.6 A slab of porous material is to be dried from both faces at a constant rate of w lb/(hr. sq.ft). The initial moisture content of the slab is uniform at Co pcf(pounds/cubic feet). The movement of moisture in the slab is in x-direction only and the rate of movement is proportional to the concentration gradient of the moisture in the slab, $\frac{dc}{dx}$. If slab is of thickness, l determine the concentration of the moisture within the slab as a function of time and distance. Show that as drying time increases the concentration distribution becomes parabolic.

Solution:

$$\frac{d^2C}{dx^2} = \frac{1}{D}\frac{dC}{dt} \quad \ldots(1)$$

$$F = -\frac{DdC}{dx} \text{ at } x = 0 \text{ is } -w$$

$$F = -\left.\frac{DdC}{dx}\right|_{x=l} \text{ is } w$$

$$C = X(x)\,T(t) \quad \ldots(2)$$

$$X''T = \frac{T'X}{D}$$

$$\frac{X''}{X} = \frac{T'}{TD} = a \text{ (constant)} = -p^2 \text{ (say)} \qquad \ldots(3)$$

$$X'' + p^2X = 0$$

$$X = C_1 \cos px + C_2 \sin px \qquad \ldots(4)$$

$$T' + P^2TD = 0$$

$$T = C^3 \exp(-p^2Dt) \qquad \ldots(5)$$

$$C(x, t) = (C_{11} \cos px + C_{12} \sin px) \exp(-p^2tD) \qquad \ldots(6)$$

Using boundary conditions:

$$-\frac{D\,dC}{dx} = -w \text{ at } x = 0 \text{ gives:}$$

$$-D(-C_{11}\, p \sin px + C_{12}\, P \cos px) \exp(-p^2Dt) = -w \qquad \ldots(7)$$

$$C(0, 0) = C_{11} = C_0$$

$$-D\frac{dC}{dx} \text{ at } x = 1 = -D(C_{12}\, p \cos pl) \exp(-p^2\, Dt) = 0 \qquad \ldots(8)$$

$$-D\left(\frac{dC}{dx}\right) \text{ at } x = 0 \text{ is } -D(C_{12}\, p \exp(-p^2Dt) = 0 \qquad \ldots(9)$$

From Eqs. (8) and (9)

$$= \cos pl = -1$$

$$pl = 2n\pi$$

$$p = \frac{2n\pi}{l}$$

$$C(x, t) = C_{12} \sin\left(\frac{2n\pi x}{l}\right) \exp\left(\frac{-4n^2\pi^2 Dt}{l^2}\right)$$

$$= \sum C_{12} \sin\left(\frac{m\pi x}{l}\right) e^{-m^2\pi^2 Dt/l^2} \qquad (m = 2n)$$

$$C(x, 0) = \sum C_{12} \sin\frac{m\pi x}{l} = C_0$$

$$C_{12} = \frac{2}{l}\int_0^l C_0 \sin\left(\frac{m\pi x}{l}\right) dx = \left(\frac{2C_0}{m\pi}\right)\left[1-(-1)^m\right]$$

$$C(x, t) = \left(\frac{2C_0}{\pi}\right)\left[\left(\frac{1}{m}\right)\left(\frac{\sin m^2\pi x}{l}\right)\exp\left(\frac{-m^2\pi^2 Dt}{l^2} + \ldots\right)\right] \quad m = 1, 3, 7, 9, \ldots$$

Example 3.7 Isotherm migration method

The unsteady state heat conduction equation in one dimension for temperature dependent thermal conductivity is

$$\rho C\frac{dT}{dt} = \frac{d}{dx}\left(K\frac{dT}{dx}\right) \quad \ldots(1)$$

The gradient on the left hand side can be written as

$$\frac{dT}{dt} = -\frac{dT}{dx}\times\frac{dx}{dt} \quad \ldots(2)$$

The partial derivative on the right hand side of Eq. (1) can be written as:

$$\frac{d}{dx} = \frac{d}{dt}\times\frac{dT}{dx} \quad \ldots(3)$$

Repeated use of Eq. (3) and substitution of Eq. (3) in Eq. (1), gives

$$\frac{dx}{dt} = -\left(\frac{dx}{dt}\right)^{-2}\left(\frac{d^2x}{dT^2}\right)-\left(\frac{1}{\rho C}\right)\left(\frac{dx}{dT}\right)^{-1}\times\frac{dK}{dt} \quad \ldots(4)$$

Equation (4) is the new equation of heat condition where the location of the isotherm, $x(T, t)$ is the dependent variable, this is the place of Eq. (1) where the temperature $T(x, t)$ was the dependent variable. Heat balance on interface gives

$$K\frac{dT}{dx} \text{ at } x_i = LP\frac{dx_f}{dt} \quad \ldots(5)$$

This can be written as:

$$K\left(\frac{dx}{dT}\right)^{-1} \text{ at } x_i = LP\frac{dx_i}{dt} \quad \ldots(6)$$

Now we have to solve the problem using Eqs. (4) and (6). The initial temperature is T_f and surface temperature is T_s. The temperature at any location in slab varies between T_f and T_s. One can divide the range of temperature $(T_s - T_f)$ into a number of isotherms and trace their location as a function of time, thus creating a temperature–time grid. We will call T_f, the first isotherm and T_s the least isotherm with m isotherms at a constant temperature interval.

$$\Delta t = \frac{(T_s - T_f)}{m} \quad \ldots(7)$$

One can fit a parabola through the third and second isotherms with zero slope for initial temperature. The location of the isotherm is

$$x_1 = \left(X_2 - \left[\frac{(T_2 - T_1)}{(T_s - T_f)}X_s\right]^{1/2}\right)\Bigg/\left(1 - \left[\frac{(T_2 - T_f)}{(T_s - T_f)}\right]^{1/2}\right) \quad \ldots(8)$$

Finite Difference Formulation

Equation (4) can be written as

$$X_i^{n+1} = X_i^n + \Delta t[4\alpha_i \left[\frac{(X_{i+1}^n + X_{i-1}^n - 2X^n)}{(X_{i-1}^n - X_{i+1}^n)^2}\right] - \frac{1}{(\rho_i C_i)}\left[\frac{(K_{i-1} - K_{i+1})}{(X_{i-1}^n - X_{i+1}^n)}\right] \quad \ldots(9)$$

where i represents the ith Isotherm and n^{th} time increment. Equation (6) can be written as

$$X_i^{n+1} = \frac{X_i^n + \Delta t \cdot \Delta T}{(L\rho_i)}\left\{\frac{K_i}{(X_2^n - X_1^n)}\right\} \quad \ldots(10)$$

Solution: Distance, cms

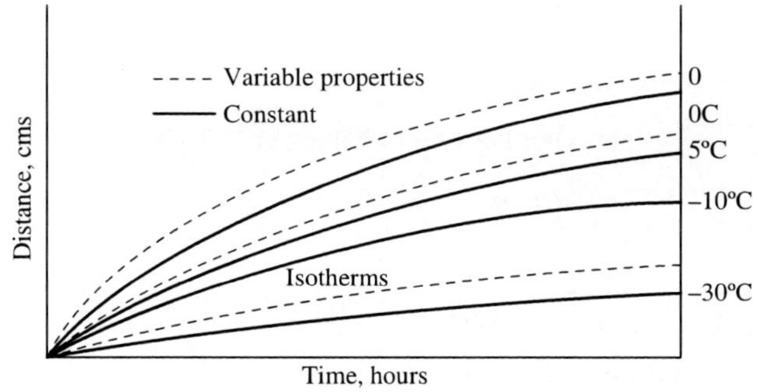

Fig. 3.8

Example 3.8 Using the method of separation variables, solve $\frac{du}{dx} = \frac{2du}{dt} + u$ where $u(x, 0)$ $= \exp(-3x)$.

Solution: Assuming (using separation variable method)

$$u(x, t) = X(x)\text{T}(t)$$

The given equation becomes

$$X'T = 2XT' + XT$$

or $$\frac{X' - X}{(2X)} = \frac{T'}{T} = K(\text{say})$$

$$X' - X - 2KX = 0$$

or $$\frac{X'}{X} = 2K + 1 \quad \ldots(1)$$

and $$\frac{T'}{T} = K \quad \ldots(2)$$

Solving Eq. (1) gives:

$$\log X = (1 + 2K)X + \log C$$

or $$X = C \exp[(T + 2K)x]$$

From Eq. (2), $$\log T = kt + \log c$$

or $$T = c' \exp(kt)$$

$$u(x, t) = XT = c'\,c \exp((1 + 2k)x) * (\exp kt) \qquad \text{...(3)}$$

Now, $$u(x, 0) = \exp(-3x)$$

$$CC' \exp[(1 + 2K)x] = \exp(-3K)$$

$$CC' = 1$$

$$1 + 2K = -3$$

i.e., $$K = -2$$

Substituting these values in Eq. (3) we get

$$u(x, t) = 1 \times \exp(-3x) \times \exp(-2t)$$

$$u = \exp(-3x - 2t)$$

Example 3.9 $$\frac{d^2y}{dt^2} = \frac{a^2d^2y}{dx^2} \qquad \text{...(1)}$$

In the given PDE whose solution is given as:

$$y(x, t) = (A \cos px + B \sin px)(C \cos pat + D \sin pat) \qquad \text{...(2)}$$

where A, B, C, D are constants. The given boundary conditions are:

$$y(0, t) = 0 \qquad \text{...(3)}$$

$$y(l, t) = 0 \qquad \text{...(4)}$$

$$y(x, 0) = f(x) \text{ at } t = 0 \qquad \text{...(5)}$$

$$\left(\frac{dy}{dt}\right) \text{ at } t = 0 \text{ is } 0 \qquad \text{...(6)}$$

Applying B.C. 3, we have

$$y(0, t) = A(C \cos pat + D \sin pat) = 0$$

Giving $A = 0$

By Eq. (2) $$y(x, t) = B \sin px(C \cos pat + D \sin pat) \qquad \text{...(7)}$$

Applying boundary conditions 4 to 7, we get

$$y(l, t) = B \sin pl(C \cos pat + D \sin pat) = 0$$

B cannot be zero, so $\sin pl = 0$; $p = \left(\frac{n\pi}{l}\right)$

Equation (7) reduces to:

$$y(x, t) = B\sin\left(\frac{n\pi x}{l}\right)\left[C\cos\frac{n\pi t}{l} + D\sin\frac{\pi t}{l}\right] \qquad \text{...(8)}$$

Applying B.C. 6, we get

$$\frac{dy}{dt} = B\sin\left(\frac{n\pi x}{l}\right)\left[-\frac{Cn\pi x}{l} \times \sin\left(n\pi\frac{at}{l}\right) + Dn\frac{\pi a}{l} \times \cos n\frac{\pi at}{l}\right]$$

At $t = 0$, $\quad \frac{dy}{dt}$ at $t = 0 = B\sin\frac{n\pi x}{l} \times D\sin\frac{n\pi a}{l}$

Thus, if $D = 0$, Eq. (8) is reduce to

$$y(x, t) = B\sin\left(\frac{n\pi x}{l}\right) \times C\cos\left(\frac{n\pi at}{l}\right)$$

$$y(x, t) = P\sin\left(\frac{n\pi x}{l}\right)\cos\left(\frac{n\pi at}{l}\right) \qquad \text{...(9)}$$

Applying boundary condition 5 to Eq. (9), we have

$$y(x, 0) = \sum Pn\sin\left(\frac{n\pi x}{l}\right)$$

$$Pn = \left(\frac{2}{l}\right)\int_0^l f(x)\sin\left(\frac{n\pi x}{l}\right)dx$$

$$y(x, t) = \sum Pn\sin\left(\frac{n\pi x}{l}\right) \times \cos\left(\frac{n\pi at}{l}\right) \quad \textbf{Ans.}$$

Example 3.10 Given $\frac{d^2u}{dt^2} = \frac{a^2d^2y}{dx^2} - 2b\frac{dy}{dt}$

Boundary conditions are

$$y(0, t) = 0; \qquad \text{...(1)}$$

$$y(l, t) = 0 \qquad \text{...(2)}$$

$$\frac{dy}{dt}(x, t) = 0 \text{ at } t = 0; \qquad \text{...(3)}$$

$$y(x, 0) = f(x) \qquad \text{...(4)}$$

Solution:

$$y(x, t) = X(x)\,T(t)$$

$$XT'' = a^2X''T - 2bXT'$$

$$\frac{T''}{T} = \frac{a^2X''}{X} - \frac{2bT'}{T}$$

$$\frac{X''}{X} = \frac{(T'' + 2bT')}{(a^2T)} = K = -P^2$$

$$X'' - KX = 0 \quad \ldots(5)$$

$$T'' + 2bT' - Ka^2T = 0 \quad \ldots(6)$$

Taking $K = -p^2$

$$X'' + p^2X = 0 \quad \ldots(7)$$

$$X = A_1 \cos px + A_2 \sin px \quad \ldots(8)$$

$$T'' + 2bT' + p^2a^2T = 0 \quad \ldots(9)$$

Roots are:

$$M = -b \pm (b^2 - a^2p^2)^{1/2}$$

$$= -b \pm (-(a^2p^2 - b^2)^{1/2};$$

Since b is small;

$$m = -b \pm i(a^2p^2 - b^2)^{1/2}$$

$$T = \text{Exp}\,(bt)\,[A_3 \cos (a^2p^2 - b^2)^{1/2}t + A_4 \sin (a^2p^2 - b^2)^{1/2}t] \quad \ldots(10)$$

$$y(x, t) = \text{Exp}\,(bt)(A_1 \cos px + A_2 \sin px)\,(A_3 \cos(a^2p^2 - b^2)^{1/2}t + A_4 \sin(a^2p^2 - b^2)^{1/2}\,t \quad \ldots(11)$$

Using B.C.1. Using condition –1, in Eq. (11) gives $A_1 = 0$;

Using B.C.2;

$$A \sin pl = 0$$

So,

$$p = \frac{n\pi}{l}$$

Using

$$\frac{dy}{dt} = 0 \text{ at } (x, t)$$

$$A_4 = \frac{b\,A_3}{(a^2p^2 - b^2)^{1/2}}$$

$$y(x, t) = \exp(-b \times t)\sum(A_{31} \cos Knt + A_{41} \sin knt) \sin\frac{n\pi x}{l}$$

$$y(x, 0) = \sum_{n=1}^{\infty}\left(A_{31} \sin\frac{n\pi x}{l}\right) = f(x) = A_3 n \sin\frac{n\pi x}{l}$$

$$A^{3n} = \int_0^l \frac{2}{l} f(x) \sin\left(\frac{n\pi x}{l}\right) dx$$

Example 3.11

$$\frac{du}{dt} = \frac{Kd^2u}{dx^2} \quad \ldots(1)$$

Solution. Assuming

$$x = C \exp(\lambda t + ux) \quad \ldots(2)$$

Subsituting Eq. (2) in Eq. (1) gives:

$$\lambda = K\mu^2 \quad \ldots(3)$$

$u = u_0 \sin nt$ as $x \to \infty$, the solution should be imaginary to satisfy the B.C.;

$$\lambda = ip \qquad ...(4)$$

$$\mu^2 = \frac{ip}{K}$$

$$\mu = \pm\left(\frac{ip}{K}\right)^{1/2}$$

$$\mu = \pm\left(\frac{ip}{K}\right)^{1/2} = \pm(i)^{1/2}\left(\cos\frac{\pi}{2} + i\sin\frac{\pi}{2}\right)^{1/2}$$

$$\mu = \pm\left(\frac{p}{K}\right)^{1/2}\left(\cos\frac{\pi}{4} + i\sin\frac{\pi}{4}\right)$$

$$= \pm\left(\frac{p}{K}\right)^{1/2}\left[\frac{(1+i)}{2^{1/2}}\right]$$

We have to take μ with its real part. Negative since $\mu \to 0$ as $x \to \infty$.

$$\mu = -\left(\frac{p}{K}\right)^{1/2}\left[\frac{(1+i)}{2^{1/2}}\right]$$

$$u(x, t) = C\exp\left[-pt - \left(\frac{p}{K}\right)^{1/2}\left(\frac{1+i}{2^{1/2}}\right)x\right]$$

or
$$u(x, t) = C\exp\left[-\left(\frac{p}{2K}\right)^{1/2} \times \sin\left[pt - \left(\frac{p}{2K}\right)^{1/2}x\right]\right]$$

or
$$u = C\exp\left[-\left(\frac{p}{2K}\right)^{1/2} \times \sin\left(pt - \left(\frac{p}{2k}\right)^{1/2}x\right)\right]$$

or
$$u = C\exp\left[-\left(\frac{p}{2K}\right)^{1/2} \times \left(\cos pt - \left(\frac{p}{2k}\right)^{1/2}x\right)\right]$$

are possible solutions. As $u \to \infty$

$$u(x, t) = 0$$

Example 3.12 Solve $\dfrac{d^2u}{dx^2} + \dfrac{d^2u}{dy^2} = 0$ ...(1)

(steady state heat conduction equation)

$$u = u(x, y)$$

Equation (1) becomes $[u = X(x)\,Y(y)]$

$$X''Y + Y''X = 0$$

$$\frac{X''}{X} = -\frac{Y''}{Y} = +K \qquad \text{...(2)}$$

$$X'' - KX = 0 \qquad \text{...(3a)}$$

$$Y'' + KY = 0 \qquad \text{...(3b)}$$

Case 1: For $K = p^2$

$$Y = C_3 \cos py + C_4 \sin py \qquad \text{...(4)}$$

$$X = C_1 e^{px} + C_2 e^{-px}$$

$$u(x, y) = C_1 e^{px} + C_2 e^{-px}) (C_3 \cos py + C_4 \sin py) \qquad \text{...(5)}$$

Case 2: For $K = -p^2$

$$X'' + p^2 X = 0$$

$$X = (C_5 \cos px + C_6 \sin px)$$

$$Y'' - p^2 Y = 0$$

$$Y = C_7 e^{py} + C_8 e^{-py}$$

$$u(x, y) = (C_5 \cos px + C_6 \sin px) (C_7 e^{py} + C_8 e^{-py}) \qquad \text{...(6)}$$

Case 3: For $K = 0$

$$X'' = 0$$

$$X = C_9 x + C_{10}$$

$$Y'' = 0$$

$$Y = C_{11} y + C_{12}$$

$$u(x, y) = X(x) Y(y) = (C_9 x + C_{10}) (C_{11} y + C_{12}) \qquad \text{...(7)}$$

The suitable solution is selected based on the boundary conditions.

Example 3.13 Laplace equation in polar coordinates. The coordinates are: r, θ, z.

$$\frac{r^2 d^2u}{dr^2} + \frac{r\, du}{dr} + \frac{d^2u}{d\theta^2} + \frac{d^2u}{dz^2} = 0$$

Leaving $\frac{d^2u}{dz^2}$: $\quad \frac{r^2 d^2u}{dr^2} + \frac{r\, du}{dr} + \frac{d^2u}{d\theta^2} = 0 \qquad \text{...(1)}$

$$u = R(r)\,\phi(\theta) = R\phi$$

$$\frac{du}{dr} = \phi \frac{dR}{dr}$$

$$\frac{du}{d\theta} = R\frac{d\phi}{d\theta}$$

$$\frac{d^2u}{d\theta^2} = R\frac{d^2\phi}{d\theta^2}$$

$$r^2\phi\frac{d^2R}{dr^2} + r\phi\frac{dR}{dr} + R\frac{d^2\phi}{d\theta^2} = 0$$

$$\phi(r^2R'' + rR') + R\phi'' = 0$$

$$\frac{r^2R'' + rR'}{R} = -\frac{\phi''}{\phi} = K$$

$$\frac{r^2R'' + rR'}{R} = k$$

$$r^2R'' + rR' - kR = 0 \qquad \ldots(2)$$

$$-\frac{\phi''}{\phi} = k$$

$$\phi'' + k\phi = 0 \qquad \ldots(3)$$

k has three possible solutions ($k = p^2$; $-p^2$; 0).

Case 1: For $k = p^2$,

$$r^2R'' + rR' - p^2R = 0$$

$$\frac{r^2d^2R}{dr^2} + r\frac{dR}{dr} - p^2R = 0 \qquad \ldots(1)$$

$$(D^2r^2 + Dr - p^2)R = 0$$

$$D^2r^2 + Dr - p^2 = 0$$

Using $r = e^t$; $\frac{dr}{dt} = e^t$; $\qquad dr = e^t\,dt$

$$\frac{dt}{dr} = \frac{1}{r}$$

$$\frac{dR}{dr} = \frac{dR}{dt}\times\frac{dt}{dr} = -\frac{1}{r}\frac{dR}{dt}$$

$$= -\frac{1}{r^2}\frac{dR}{dt} + \frac{1}{r}\frac{d}{dr}\left(\frac{dR}{dt}\right)$$

$$r\frac{dR}{dt} = \frac{dt}{dr}$$

$$\frac{d^2R}{dr^2} = -\frac{d}{dr}\left(\frac{dR}{r\,dt}\right) = -\frac{1}{r^2}\frac{dR}{dt} + \frac{1}{r}\frac{d}{dr}\left(\frac{dR}{dt}\right)$$

$$= -\frac{1}{r^2}\frac{dR}{dt} - \frac{1}{r}\frac{d}{dt}\left(\frac{dR}{dt}\times\frac{dt}{dr}\right)$$

$$= -\frac{1}{r^2}\frac{dR}{dt} + \frac{1}{r^2}\frac{d^2R}{dt}$$

$$\frac{r^2d^2R}{dr^2} = -\frac{dR}{dt} + \frac{d^2R}{dt^2}$$

Equation 1 becomes:

$$-\frac{dR}{dt} + \frac{d^2R}{dt^2} + \frac{dR}{dt} - p^2R = 0$$

$$\frac{d^2R}{dt^2} - p^2R = 0 \qquad \ldots(2)$$

$$R = C_1e^{pt} + C_2\,e^{-pt}$$

$$= C_1\,e^{p\ln r} + C_2\,e^{-p\ln r}$$

$$R = C_1r^p + C_2\,r^{-p} \qquad \ldots(3)$$

For $K = +p^2$: $\phi'' + p^2\phi = 0;\ (\phi'' + K\phi) = 0$

$$\phi = C_3\cos p\theta + C_4\sin p\theta \qquad \ldots(4)$$

$$u(r, \theta) = \phi(\theta)R(r) = (C_1r^p + C_2r^{-p})(C_3\cos p\theta + C_4\sin p\theta) \qquad \ldots(5)$$

Case 2: For $K = -p^2$

$$r^2R'' + rR' + p^2R = 0$$

Substituting $r = e^z$

$$\ln r = z$$

$$\frac{dR}{dr} = \left(\frac{dR}{dz}\right)\times\frac{dz}{dr} = -\frac{1}{r} - \frac{dR}{dz}$$

$$r\frac{dR}{dr} = \frac{dR}{dz}$$

$$\frac{d^2R}{dr^2} = -\frac{1}{r^2}\frac{dR}{dz} + \frac{1}{r^2}\frac{d^2R}{dz^2}$$

Substituting in the above equation,

$$-\frac{dR}{dz} + \frac{d^2R}{dz^2} + \frac{dR}{dz} + p^2R = 0$$

Giving: $$\frac{d^2R}{dz^2} + p^2R = 0$$

$$R = C_5 \cos pz + C_6 \sin pz$$

The solution is:

$$R = C_5 \cos p(\ln r) + C_6 \sin p(\ln r)$$

$$\phi'' - p^2\phi = 0$$

$$\phi = C_7\, e^{p\theta} + C_8\, e^{-p\theta}$$

$$u = R\phi = (C_5 \cos p(\ln r) + C_6 \sin p(\ln r))$$

$$(C_7\, e^{p\theta} + C_8\, e^{-p\theta}) \quad \ldots(6)$$

Case 3: For $K = 0$

$$\frac{r^2 d^2R}{dr^2} + r\frac{dR}{dr} = 0$$

$$r\frac{d^2R}{dr^2} + \frac{dR}{dr} = 0$$

Using substitution, $t = 1\text{n}\, r$

i.e., $$r = e^t;$$

$$-\frac{dR}{dt} + \frac{d^2R}{dt^2} + \frac{dR}{dt} = 0$$

$$\frac{d^2R}{dt^2} = 0$$

$$R = C_9 t + C_{10}$$

$$R = C_9 \ln r + C_{10}$$

$$\phi'' + K\phi = 0$$

$$\phi'' = 0$$

$$\phi = C_{11}\, \theta + C_{12}$$

$$u = (C_9 \ln r + C_{10})\,(C_{11}\theta + C_{12}) \quad \ldots(7)$$

The possible solutions are:

$$u(r, \theta) = (C_1\, r^p + C_2\, r^{-p})(C_3 \cos p\theta + C_4 \sin p\theta) \quad \ldots(8)$$

$$u(r, \theta) = C_5 \cos p(\ln r) + C_6 \sin p(\ln r)\,(C_7\, e^{p\theta} + C_8\, e^{-p\theta}) \quad \ldots(9)$$

$$u(r, \theta) = (C_9 \ln r + C_{10})\,(C_{11}\, \theta + C_{12}) \quad \ldots(10)$$

Suitable solution is selected based on the boundary conditions.

Example 3.14 A square plate has its faces and the edge $y = 0$ insulated. Its edges at $x = 0$ and $x = \pi$ are at a temperature of 0°C and its $y = \pi$ at a temperature of $F(x)$. Show that: $u(x, y) = \frac{2}{\pi}\sum \frac{\cos hy}{\cosh n\pi} \times \text{ sine } nx \int_0^{\pi} F(x) \sin x \, dx$

Solution: Let $u(x, y)$ be the temperature at any point $p(x, y)$, so it satisfies,

$$\frac{d^2u}{dx^2} + \frac{d^2u}{dy^2} = 0 \qquad \text{...(1)}$$

The boundary conditions are

$$u(0, y) = 0 \qquad \text{...(2)}$$

$$u(\pi, y) = 0 \qquad \text{...(3)}$$

$$\left(\frac{du}{dy}\right)_{y=0} = 0 \qquad \text{...(4)}$$

Since the edge $y = 0$ is insulated then, so no heat flows at all points on $y = 0$. Perpendicular to this edge then $\frac{du}{dy}$ at $y = 0$ at all points on $y = 0$.

$$u(x, \pi) = F(x); 0 < x < \pi \qquad \text{...(5)}$$

The solutions are

$$u(x, y) = (C_1 \cos nx + C_2 \sin nx) \times (C_3 \cosh py + C_4 \sinh py)$$

Using (1), $u(0, y) = 0$ gives

$$u(x, y) = C_1 (C_3 \cosh ny + C_4 \sinh ny) \text{ gives } C_1 = 0$$

Using $\frac{du}{dy}$ at $y = 0$ gives

$$C_2 \sin nx(C_3 n \sinh ny + C_4 \cosh ny) = 0$$

Gives: $C_2 \sin nx \times C_4 = 0$, giving $C_4 = 0$

$$u(x, y) = C_2 \sin nx \, C_3 \cosh ny$$

$$= \sum_{n=1}^{\infty} A_n \sin nx \cosh ny$$

Using Eq. (4):

$$u(x, y) = \sum_{n=1}^{\infty} A_n \sin nx \cosh ny$$

We now express $F(x)$ as a half range series with interval 0 to π.

$$A_n \cosh n\pi = -\frac{2}{\pi}\int_0^{\pi} F(x) \sin nx \, dx$$

$$A_n = \frac{2}{\pi \cosh n\pi}\int_0^{\pi} F(x) \sin x \, dx$$

Example 3.15 Form the partial differential equation for $z = f_1(y + 2x) + f_2(y - 3x)$

Solution:

$$\frac{dz}{dx} = 2f_1'(y+2x) - 3f_2'(y+3x) \quad \ldots(1)$$

$$\frac{d^2z}{dx^2} = 4f_1'' + 9f_2'' \quad \ldots(2)$$

$$\frac{dz}{dy} = f_1^1 + f_2^1 \quad \ldots(3)$$

$$\frac{d^2z}{dy^2} = f_1'' + f_2'' \, ; \quad \frac{\partial^2 z}{\partial x \, \partial y} = 2f_1'' - 3f_2'' \quad \ldots(4)$$

From Eqs. (3) and (4)

$$\frac{d^2z}{dy^2} = \frac{1}{2}\left(\frac{d^2z}{dx\,dy} + 3\,f_2''\right) + f_2''$$

$$\frac{d^2z}{dy^2} = \frac{1}{2}\left(\frac{d^2z}{dx\,dy} + 3f_2''\right) + f_2''$$

$$\frac{d^2z}{dy^2} = \frac{d^2y}{2\,dx\,dy} + \frac{5}{2} - f_2'' \quad ..(5)$$

From Eqs. (1) and (5),

$$\frac{d^2z}{dx^2} - 4\frac{d^2z}{dy^2} = 5f_2''$$

From Eq. (5)

$$\frac{1}{2} - \frac{d^2y}{dy\,dx} + \frac{1}{2}\left(\frac{d^2z}{dx^2} - 4\frac{d^2z}{dy^2}\right) = \frac{d^2z}{dy^2}$$

$$2\frac{d^2z}{dy^2} = \frac{d^2y}{dy\,dx} + \frac{d^2z}{dx^2} - 4\frac{d^2z}{dy^2}$$

$$\frac{d^2z}{dx^2} + \frac{d^2y}{dy\,dx} - 6\frac{d^2z}{dy^2} = 0 \textbf{ Ans.}$$

Example 3.16 Solve $\dfrac{d^2u}{dy\,dx} = 4x \sin(3xy)$

Solution: Integrating with respect to x, we get,

$$\frac{du}{dy} = 4\int x \sin(3\,xy)\,dx + f(y)$$

$$= 4\left(-\frac{x\cos(3xy)}{3y} + \int \frac{\cos(3xy)}{3y}\right) + f(y)$$

$$\frac{du}{dy} = -\frac{4x}{3y}\cos(3xy) + \frac{4}{9y^2}\sin(3xy) + f(y)$$

Integrating with respect to y, we get

$$u = -\frac{4x}{3} - \int -\frac{1}{y}\cos(3xy)dy + \frac{4}{9}\int \frac{1}{y^2}\sin(3xy)\,dy$$

$$+\int y(fy)\,dy + g(x)$$

$$u = -\frac{4x}{3}\left(\frac{1}{y}\frac{\sin 3xy}{3x} + \int \frac{1}{y^2}\frac{\sin 3xy}{3x}dy\right)$$

$$+\frac{4}{9}\int -\frac{1}{y^2}\sin(3xy)\,dy + g(x) + V(y)$$

$$u = -\left(\frac{4}{9y}\right)\sin(3xy) + V(y) + g(x) \textbf{ Ans.}$$

Example 3.17 Solve the heat conduction equation: $\dfrac{d\theta}{dt} = K\dfrac{d^2\theta}{dx^2}$ with the boundary conditions:

$$\frac{d\theta}{dx} \text{ at } x = 0$$

and

$$\frac{d\theta}{dx} \text{ at } x = l = 0$$

$$\theta(x, 0) = \theta_0.$$

Solution: Heat conduction solution is

$$\theta(x, t) = (A\cos px + B\sin px)\,e^{-(k)^{1/2}p^2 t}$$

$$\frac{d\theta}{dx} = \left(-Ap\sin px + BP\cos px\right)e^{-(K)^{1/2}p^2 t} = 0 \text{ at } x = 0$$

Giving $B = 0$; $\quad BP\,e^{-(K)^{1/2}}p^2 t = 0$

$$\frac{d\theta}{dx} \text{ at } x = l = \left(-Ap\sin pl\right)e^{(K)^{1/2}p^2 t} = 0 \text{ (at } x = l)$$

Giving $\sin pl = 0$; $\quad pl = n\pi$

$$p = \frac{n\pi}{l}$$

$$\theta(x, t) = A\cos\frac{n\pi x}{l} e^{-(k)^{1/2}} n^2\pi^2 \frac{t}{l^2}$$

$$\theta(x, t) = \sum A_n \cos\left(\frac{n\pi x}{l}\right) e^{-(K)^{1/2}} n^2\pi^2 \frac{t}{l^2}$$

$$\theta(x, 0) = \theta_0 = A_0 + A_n \cos\frac{n\pi x}{l}$$

$$A_0 = \left(\frac{2}{l}\right)\int_0^l f(x)\,dx$$

$$= \frac{2}{l}\int_0^l \theta_0\,dx = \left(\frac{2}{l}\right)\theta_0 x\Big|_0^l = 2\,\theta_0 = A_0$$

$$A_n = \left(\frac{2}{l}\right)\int_0^l \theta_0 \cos\left(\frac{n\pi x}{l}\right)dx$$

$$= \left(\frac{2}{l}\right)\theta_0 \sin\frac{n\pi x}{l}\Big/\left(\frac{n\pi}{l}\right)\Big|_0^l$$

$$= 2\theta_0 \times \frac{0}{(n\pi)} = 0 = A_n$$

$$\theta(x, t) = \frac{A_0}{2} + \sum A_n \cos\frac{n\pi x}{2l}$$

$$= \theta_0 + \frac{4\theta_0}{\pi}\cos\frac{\pi x}{l} e^{-K(1/2)1^2\pi^2 t/l^2} - \frac{4\theta_0}{3}\cos\frac{3\pi x}{l} e^{-K^{\frac{1}{2}}3^2\pi^2 t/l^2} + \ldots$$

Example 3.18 Three-dimensional solution: x, y, z coordinates (Cartesian):

$$\frac{d^2u}{dx^2} + \frac{d^2u}{dy^2} + \frac{d^2u}{dz^2} = \left(\frac{l}{K}\right)\frac{du}{dt} \qquad \ldots(1)$$

where u is temperature, t is time and k is a constant.

Solution: Using separation variable method:

$$u(x, y, z, t) = X(x)Y(y)Z(z)T(t) \qquad \ldots(2)$$

Equation (1) becomes

$$X''YZT + XYZ''T + XY''ZT = \left(\frac{1}{k}\right)XYZT'$$

$$\frac{X''}{X} + \frac{Y''}{Y} + \frac{Z''}{Z} = \left(\frac{1}{k}\right)\frac{T'}{T} \quad \ldots(3)$$

$$\frac{X''}{X} = -n^2$$

$$\frac{Y''}{Y} = -m^2$$

$$\frac{Z''}{Z} = -l^2$$

$$\frac{T'}{(kT)} = -p^2 \quad \ldots(4)$$

$$X_n(x) = A_n \cos nx + B_n \sin nx$$

$$Y_n(y) = C_m \cos my + D_m \sin my$$

$$Z_l(z) = E_l \cos lz + F_l \sin lz$$

$$T_P(t) = G_l e^{-p^2 Kt}$$

$$= H_{mnl} e^{-(m^2 + n^2 + l^2)kt} \quad \ldots(5)$$

$$u_{mnl}(x, y, z, t) = H_{mnl}(A_n \cos nx + B_n \sin nx) \times (C_m \cos my$$

$$+ D_m \sin my)\,(E_l \cos lz + F_l \sin lz)\; e^{-(n^2 + m^2 + l^2)kt} \quad \ldots(6)$$

General solution of equation (1) is

$$u_{nml}(x, y, z, t) = \sum_{n=1}^{\infty}\sum_{m=1}^{\infty}\sum_{l=1}^{\infty} u_{mnl}(x, y, z, t)$$

Final solution is obtained using boundary conditions to evaluate the constants.

Example 3.19 Solve spherical coordinates equation:

$$\frac{d^2u}{dr^2} + \left(\frac{2}{r}\right)\frac{du}{dr} + \left(\frac{1}{r}\right)\frac{d^2u}{d\theta^2} + \cot\frac{\theta}{r^2}\frac{du}{d\theta} + \left(\frac{1}{r^2}\right)\sin^2\theta\frac{d^2u}{d\phi^2} = \left(\frac{1}{k}\right)\frac{du}{dt} \quad \ldots(1)$$

$$u(r, \theta, \phi, t) = R(r)\,\theta(\theta)\,\phi(\phi)\,T(t) \quad \ldots(2)$$

Solution: Equation (1) becomes using Eq. (2):

$$\frac{R''}{R} + \left(\frac{2}{r}\right)\frac{R'}{R} + \left(\frac{1}{r^2}\right)\frac{\theta'}{\theta} + \frac{\cot\theta\;\theta'}{x^2\;\;\theta} + \frac{1}{r^2\sin^2\theta}\frac{\phi''}{\phi} = \left(\frac{1}{k}\right)\frac{T''}{T} \quad \ldots(3)$$

Taking the constant of separation as $-\lambda^2$

From Eq. (3) $T' = -k\lambda^2 T$ $\quad \ldots(4)$

$$\frac{1}{R}\left(R'' + \frac{2}{r}R'\right) + \left(\frac{1}{r^2}\right)\left(\theta'' + \frac{\cot\theta}{\theta}\theta'\right) + \frac{1}{r^2\sin^2\theta}\frac{\phi''}{\phi} = -\lambda^2$$

$$\left[\frac{1}{R}\left(R'' + \frac{2}{r}R'\right) + \frac{1}{r_\theta^2}(\theta'' + \cot\theta\theta') + \lambda^2\right]r^2\sin^2\theta = -\frac{\phi''}{\phi}$$

Right hand side depends on ϕ and left hand side depends only on r and θ. So it can be true only if each side is equal to same constant, say m^2.

Thus, we have

$$\phi'' + m^2\phi = 0 \qquad \ldots(5)$$

and

$$\left(\frac{1}{R}\right)\left(r^2R'' + 2rR'\right) + \left(\frac{1}{\theta}\right)(\theta'' + \cot\theta\,\theta') + \lambda^2 r^2 = \frac{m^2}{\sin^2\theta}$$

$$= -\left(\frac{1}{R}\right)(r^2R'' + 2rR') - \lambda^2 r^2$$

Further r and θ are independent. So the above can only be true if each side is equal to the same constant, say $-n(n + 1)$.

Then we have

$$r^2R'' + 2rR' + [\lambda^2 r^2 - n(n + 1)]R = 0 \qquad \ldots(6)$$

and

$$\theta''\cot\theta\theta' + \left[n(n+1) - \left(\frac{m^2}{\sin^2\theta}\right)\right]\theta = 0$$

Now putting $\mu = \cos\theta$; the above equation becomes,

$$(1-\mu^2)\frac{d^2\theta}{du^2} - 2\mu\frac{d\theta}{du} + \left[n(n+1) - \frac{m^2}{1-\mu^2}\right]\theta = 0 \qquad \ldots(7)$$

which is associated Legendre equation.

The solution of Eqs. (4), (5), (6) and (7) are given by

$$T_\lambda(t) = A\lambda\, e^{-k\lambda^2 t}$$

$$\phi_m(\phi) = B_m \cos m\phi + C_n \sin m\phi$$

$$\theta_{mn}(\theta) = F_{mn}P_n^m(\cos\theta) + G_{mn}Q_n^m(\cos\theta)$$

Thus, the general solution is given by

$$u(r, \theta, \phi, t) = \sum_n\sum_m\sum_l T\lambda(t)\phi_m(\phi)\,R(r)\theta_m(\theta)$$

where $T\lambda(t)$, $\phi_m(\phi)$, $R(r(z))$, $\theta_{mn}(\theta)$ are defined earlier.

Example 3.20 Solving the cylinder equation

Given $$\frac{d^2u}{dr^2} + \left(\frac{2}{r}\right)\frac{du}{dr} = \left(\frac{1}{k}\right)\frac{du}{dt} \qquad \ldots(1)$$

Solution: $$u = R(r)\,T(t) \qquad \ldots(2)$$

Equation (1) becomes

$$R''T + \left(\frac{2}{r}\right)R'T = \left(\frac{1}{k}\right)RT'$$

$$\frac{R''}{R} + \left(\frac{2}{r}\right)\frac{R'}{R} = \left(\frac{1}{k}\right)\frac{T'}{T} = \lambda^2 \qquad \ldots(3)$$

$$T' + \lambda^2 kT = 0 \qquad \ldots(4)$$

$$T = C_1 e^{-\lambda^2 kt} \qquad \ldots(5)$$

$$\frac{r^2}{R}\left(\frac{d^2R}{dr^2} + \frac{2}{r}\frac{dR}{dr}\right) + \lambda^2 r^2 = 0 \qquad \ldots(6)$$

$n(n + 1)$ say

$$\frac{d^2R}{dr^2} + \frac{2}{r}\frac{dR}{dr} + \left(\frac{\lambda^2 - n(n+1)}{r^2}\right) = 0 \qquad \ldots(7)$$

Putting $R = \lambda r^{-1/2}\psi(r)$ in Eq. (7) gives

$$\frac{d^2\psi}{dr^2} + \frac{1}{r}\frac{d\psi}{dr} + \left[\frac{\lambda^2 - (n+1)^2}{r^2}\right]\psi = 0 \qquad \ldots(8)$$

$$\psi(r) = AJ_{n+1/2}(Ar) + BJ_{-(n+1/2)}(Ar) \qquad \ldots(9)$$

where A and B are constants.

$$\frac{R}{r} = (\lambda r)^{-1/2} A\left(J_{n+1/2}(\lambda r) + BJ_{-(n+1/2)}(\lambda r)\right) \qquad \ldots(10)$$

Thus, the solution of Eq. (1) is

$$u(r, t) = \sum_{n,\lambda} A_{n,\lambda}(\lambda r)^{-1/2} \times J_{n+1/2}(\lambda r)e^{-k\lambda^2 t} \qquad \ldots(11)$$

The constant $A_{n,\lambda}$, etc., are evaluated using the boundary conditions.

Example 3.21

$$\frac{d^2u}{dr^2} + \left(\frac{1}{r}\right)\frac{du}{dr} = \left(\frac{1}{k}\right)\frac{du}{dt} \qquad \ldots(1)$$

Solution: $$u(r, t) = R(r)\,T(t) \qquad \ldots(2)$$

Equation (1) becomes

$$\frac{1}{R}\left(\frac{d^2R}{dr^2}+\frac{1}{r}\frac{dR}{dr}\right) = \left(\frac{1}{KT}\right)\frac{dT}{dt} = -\lambda^2$$

$$\frac{dT}{dt} + k\lambda^2 T = 0 \qquad \ldots(3)$$

$$\frac{d^2R}{dr^2} + \left(\frac{1}{r}\right)\frac{dR}{dt} + \lambda^2 R = 0 \qquad \ldots(4)$$

Equation (3) is $\quad T(t) = C_1 e^{-\lambda^2 kt} \qquad \ldots(5)$

Putting $\lambda r = \mu$ in Eq. (4) gives

$$\frac{d^2R}{d\mu^2} + \left(\frac{1}{\mu}\right)\frac{dR}{dr} + R = 0 \qquad \ldots(6)$$

This is a Bessel equation of zero order and its solution is

$$R(r) = C_1 J_0(\lambda r) + C_2 J_0(\lambda r) \qquad \ldots(7)$$

Solution of Eq. (1) is

$$u(r, t) = \sum A_n J_0(\lambda r) e^{-\lambda^2 kt} \qquad \ldots(8)$$

Boundary conditions are used to evaluate A_n.

Example 3.22 Solution of $\dfrac{du}{dt} = C^2\left(\dfrac{d^2u}{dx^2}+\dfrac{d^2u}{dy^2}\right)$, where $C^2 = \dfrac{K}{(\rho C_p)}$ ...(1)

Solution:

$$u(x, y, t) = X(x)\, Y(y)\, T(t)$$

$$= XYT \text{ (say)}$$

Equation (1) becomes

$$\frac{1}{C^2T}\frac{dT}{dt} = \frac{1}{x}\frac{d^2x}{dx^2} + \frac{1}{Y}\frac{d^2Y}{dy^2} \qquad \ldots(2)$$

The equations are

(a)

$$\left(\frac{1}{X}\right)\frac{d^2X}{dx^2} = 0$$

$$\left(\frac{1}{Y}\right)\frac{d^2Y}{dy^2} = 0$$

$$\left(\frac{1}{c^2T}\right)\frac{dT}{dt} = 0$$

(b)
$$\left(\frac{1}{X}\right)\frac{d^2X}{dx^2} = p_1^2$$
$$\left(\frac{1}{Y}\right)\frac{d^2Y}{dy^2} = p_2^2$$
$$\left(\frac{1}{c^2T}\right)\frac{dT}{dt} = p^2$$

(c)
$$\left(\frac{1}{X}\right)\frac{d^2x}{dx^2} = -p_1^2$$
$$\left(\frac{1}{Y}\right)\frac{d^2Y}{dy^2} = -p_2^2$$
$$\left(\frac{1}{c^2T}\right)\frac{dT}{dt} = -p^2$$

where $p^2 = p_1^2 + p_2^2$

Based on the boundary conditions, A, B, C, (any one) is selected as a solution.

$$X = (A_1 \cos p_1x + B_1 \sin p_1x)$$
$$Y = (A_2 \cos p_2y + B_2 \sin p_2y)$$
$$T = A_3 e^{-c^2p^2t} \quad \ldots(3)$$
$$u(x, y, t) = (A_1 \cos p_1x + B_1 \sin p_1x)$$
$$(A'_2 \cos p_2y + B'_2 \sin p_2y) \times e^{-c^2}p^2t \quad \ldots(4)$$

where
$$A'_2 = A_2A_3$$
$$B'_2 = B_2 A_3$$

The constants A_1, B_1, p_1, A'_2, B'_2, p_2 and p are obtained by the boundary conditions.

Example 3.23 Solve

$$\frac{d^2u}{dr^2} + \left(\frac{2}{r}\right)\frac{du}{dr} + \left(\frac{1}{r^2\sin^2\theta}\right)\frac{d}{d\theta}\left(\sin\theta\frac{du}{d\theta}\right) = \left(\frac{1}{k}\right)\frac{du}{dt} \quad \ldots(1)$$

Solution: Let
$$u(r, \theta, t) = R(r)\, T(t)\, \phi(\theta) \quad \ldots(2)$$

Equation (1) becomes

$$\frac{1}{R}\left(\frac{d^2R}{dr^2} + \frac{2}{r}\frac{dR}{dr}\right) + \frac{1}{\phi r^2 \sin\theta}\frac{d}{d\theta}\left(\sin\theta\frac{d\phi}{d\theta}\right) = \left(\frac{1}{kT}\right)\frac{dT}{dt} \quad \ldots(3)$$
$$= -\lambda^2 \text{ say}$$

Giving possible solutions

$$\frac{dT}{dt} = -\lambda^2 kT$$

$$T(t) = c_1 e^{-\lambda^2 kt} \quad \text{...(4)}$$

$$\frac{r^2}{R} - \left(\frac{d^2R}{dr^2} + \frac{2}{r}\frac{dR}{dr}\right) + \lambda^2 r^2 = -\frac{1}{\phi \sin\theta}\frac{d}{d\theta}\left(\sin\frac{d\phi}{d\theta}\right)$$

$$\frac{d^2R}{dr^2} + \frac{2}{r}\frac{dR}{dr} + \left(\frac{\lambda^2 - n(n+1)}{r^2}\right)^{1/2} R = 0 \quad \text{...(5)}$$

and
$$-\left(\frac{1}{\phi}\sin\theta\right)\frac{d}{d\theta}\left(\sin\theta\frac{d\phi}{d\theta}\right) = n(n+1) \quad \text{...(6)}$$

Putting
$$R = (\lambda r)^{-1/2}\,\psi(r) \quad \text{...(7)}$$

We have

$$\frac{d^2\psi}{dr^2} + \left(\frac{1}{r}\right)\frac{d\psi}{dr} + \left(\lambda^2 - \frac{(n+1/2)^2}{r^2}\right)\psi = 0 \quad \text{...(8)}$$

In case $\frac{n+1}{2}$ is one, then zero nor an integer. We can write

$$\psi(r) = AJ_{n+1/2}(\lambda r) + BJ_{-(n+1/2)}(\lambda r) \quad \text{...(9)}$$

where A and B are constants.

$$R(r) = (\lambda r)^{-1/2}(AJ_{n+1/2}(\lambda r) + BJ_{-(n+1/2)}(\lambda r) \quad \text{...(10)}$$

Further substituting

$$\mu = \cos\theta \quad \text{...(11)}$$

We obtain
$$(1-\mu^2)\frac{d^2\phi}{du^2} - 2\mu\frac{d\phi}{d\mu} + n(n+1)\phi = 0 \quad \text{...(12)}$$

This is Legendre equation, whose solution is given by

$$\phi(0) = A' p_n(\mu) = A' p_n(\cos\theta)$$

where A' is constant. Thus, the solution of Eq. (1) can be put as

$$u(r, \theta, t) = \sum_{n,\lambda} A_{n,\lambda}(\lambda r)^{-1/2} J_{n+1/2}(\lambda r) P_n(\cos\theta) e^{-k\lambda^2 t} \quad \text{...(13)}$$

Using the boundary conditions, the constants $A_{n,\lambda}$; λ values are obtained.

EXERCISES

1. A metallic rod of 1 m length with insulated cylindrical surface is initially at 10ºC. Starting at $t = 0$, the ends are held at 0°C. Calculate the temperature distribution in the rod at a time $t > 0$. Using the equation $\frac{dT}{dt} = c^2 \frac{d^2T}{dx^2}$ where c^2 is a constant. x is the length measured from one end, t is time and T is the temperature.

2. For each of the following set of equations suggest an appropriate physical situation and solve the equations.

   (a) $\frac{dU}{dt} = a\frac{d^2U}{dx^2} - bU$ with boundary conditions

   $U = U'$ for $x = 0, t > 0$; $U = 0$ for $x \to \infty$ and $t > 0$.

   $U = 0$ for $x > 0$, and $t < 0$.

   (b) $\frac{dU}{dt} = a\frac{d^2U}{dx^2} - b\frac{dU}{dx}$ with boundary conditions

   $U = U^0$, for $x < 0$ and $t > 0$

   $U = \bar{U}$ for all x and $t > 0$

   $U = U^0 - c\frac{dU}{dx}$ for $x = 0, t > 0$

   where a, b, c and U^0 are constants.

   (c) $\frac{\partial^2 T}{\partial x^2} + \frac{\partial^2 T}{\partial y^2} + f(x, y) = 0$

3. Solve the following differential equations by the method of separation variables.

$$\frac{dT}{dt} = c^2 \frac{d^2T}{dx^2}$$

   where $T = f(x, t)$ and c^2 is constant. Given (i) $T(0, t) = 0$, (ii) $T(l, t) = 0$ (iii) $T(x, 0) = \frac{110x}{l}$.

4. Derive the equation for the temperature distribution $T(x, t)$ relating position and time in a semi-infinite solid $x \geq 0$, initially at temperature zero, when constant heat flux is maintained at the boundary $x = 0$. The differential equation is: $\frac{dT}{dt} = c^2 \frac{d^2T}{dx^2}$. Intial condition is: $T(x, 0) = 0$; Boundary condition: $-c^2 \frac{dT}{dx}$ at $x = 0$ = constant, for all t.

5. A semi-infinite solid bounded internally by the sphere $r = R$ is initially at uniform temperature of T_0. If the spherical surface at $r = R$ is suddenly brought and maintained at a temperature T_1, find temperature in the solid as a function of time and position.

6. The following differential equation often arises in steady state convective heat transport problems.

$$v_r \frac{dU}{dr} + \frac{V}{r}\left(\frac{dU}{d\theta}\right) = D\frac{d^2U}{dr^2}$$

with boundary conditions, $r = a, \theta = 0, U = U_0$; $r = b, \theta \geq 0, U = U_0$; $r = a, \theta > 0, U = 0$.

(a) Transform the equation to express U as a function of t and ϕ where $t = (8a\ w\ gD^2)^{1/2} \times \int_0^\theta \sin\theta\, d\theta$; $\phi = 2a^{-1}\, wg(r-a)^2 \sin^2\theta$; $v_r = \left(\frac{1}{r}\right)\frac{d\phi}{d\theta}$ and $v_\theta = \frac{d\phi}{dr}$.

(b) Considering b as infinity, find the solution of the differential equation derived in section (a) above. **Hint:** Similarly, transformation method can be employed.

7. Find the Fourier series for the function f defined by,

$$f(x) = \pi, -\pi \leq x \leq 0 = x, 0 \leq x \leq \pi$$

In the interval $-\pi \leq x \leq +\pi$

8. The following equation is obtained in process engineering.

$-GC_p\left(\frac{dt}{dx} + \frac{1}{v}\frac{dt}{d\theta}\right) = hA(t - t') = df\frac{dt'}{d\theta}$ where t, t' are dependent variables. x and θ are independent variables and the remaining are constants.

(a) Replace t, t', x and θ in the equation by the new variables, and

$$T = \frac{t - t_\theta}{t'_\theta - t_\theta}$$

$$T' = \frac{t' - t_\theta}{t'_\theta - t_\theta}$$

$$Z = \frac{h}{c_p}\frac{Ax}{G}$$

$$b = \frac{Ah}{df}\left(\theta - \frac{x}{V}\right)$$

9. Apply the method of partial differentiation to change $\frac{d^2V}{dx^2}$ from Cartesian coordinates to cylindrical coordinates of the following expression $\frac{d^2V}{dx^2} + \frac{d^2V}{dy^2} + \frac{d^2V}{dz^2} = 0$ where $V = f(x, y, z)$ in Cartesian and r, θ, z in cylindrical coordinate system.

10. Derive the unsteady state heat conduction equation for a cylindrical solid.

11. Convert $\frac{d^2u}{dx^2} + \frac{d^2u}{dy^2} + \frac{d^2u}{dz^2} = 0$ from Cartesian to spherical coordinates.

12. A metallic rod of 10 cm length with insulated cylindrical surface is initially at 100°C. Starting at $t = 0$, the ends are held at 25°C. Calculate the temperature distribution in the rod at time $t > 0$ using the equation

$$\frac{dT}{dt} = \frac{d^2T}{dx^2} C^2$$

where c^2 is a constant, x is length measured from any one end, t is time, and T is the temperature.

13. A thin metal rod of 100 cm length has its ends $x = 0$ and $x = 100$ cm are kept at 0°C. Initially, half the bar is at 60°C, while the other half is at 40°C. Assume the diffusivity, $C^2 = 0.16$ in CGS units, and that of the surface of the bar is insulated, find the relation relating the temperature of the bar with distance and time. The steady state conduction equation in one dimension in the insulated thin metal bar is

$$\frac{dT}{dt} = C^2 \frac{d^2T}{dx^2}$$

where $T = F(x, t)$

14. Solve $\frac{d^2T}{dy^2} = \left(\frac{cp\mu}{k}\right) y \frac{dT}{dx}$ with boundary conditions:

$T = T_1$ at $x = 0$, $y > 0$; $T = T_1$ where $y \to \infty$; $x > 0$; $T = T_0$ at $y = 0$, $y > 0$.

15. A rod of length l with insulated sides has uniform temperature, u_0 initially. Its ends are cooled to 0°C and are kept at that temperature. Find the temperature function $u(x, t)$.

❑❑❑

4

Numerical Methods

4.1 NUMERICAL INTEGRATION

4.1.1 Trapezoidal Rule

If the base of Fig. 4.1 having length $A - B$ is divided into n equal parts and $n + 1$ corresponding ordinates are $y_1, y_2, \ldots, y_n, y_{n+1}$. Then the area is given by: for three coordinates (y_1, y_2, y_3):

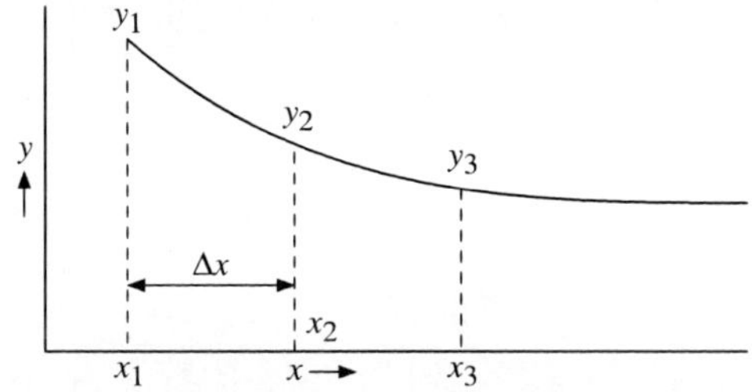

Fig. 4.1 Trapezoidal rule

$$I = \int_{x_1}^{x_3} y\,dx = x\frac{(y_1 + y_2)}{2} + x\frac{(y_2 + y_3)}{2}$$

$$= x\left(\frac{y_1}{2} + y_2 + \frac{y_3}{2}\right).$$

Extending this for n equal parts:

$$I = \int_{x_1}^{x_n} y\,dx$$

$$= x\left(\frac{y_1}{2} + y_2 + y_3 + y_4 + \ldots + y_n + \frac{y_n + 1}{2}\right)$$

Example 4.1 The top product leaving a batch distillation column is being continuously reprocessed. Using trapezoidal rule, find the area (Amount).

X	0	1	2	3	4	5	6	7
Y	2	6	18	38	66	102	146	198

Solution:

$$I = 1\left(\frac{2}{2} + 6 + 18 + 38 + 66 + 102 + 146 + \frac{198}{2}\right)$$

$$= 476 \textbf{ Ans.}$$

4.1.2 Simpson's Rule

The length A to B (x ordinate) is divided into n equal parts. The width of a part is h. Three points are accounted as a parabola for calculation. Parabola equation is:

$$y = ax^2 + bx + c$$

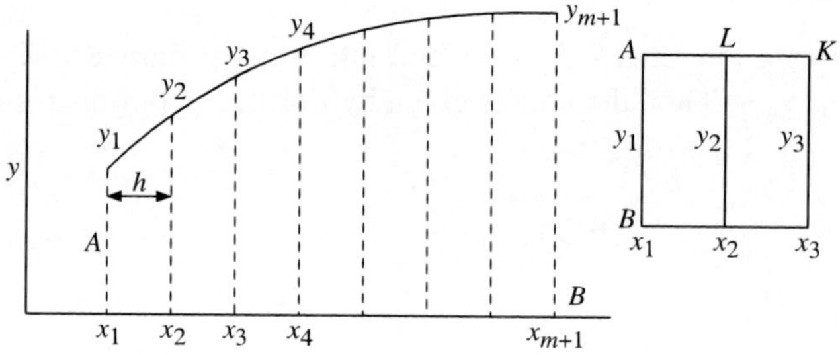

Fig. 4.2

$$\text{Area, } Ax_1x_3K = \int_{x_1}^{x_3} y\,dx = \int (ax^2 + bx + c)dx$$

$$= \left.\frac{ax^3}{3} + \frac{bx^2}{2} + cx\right|_{-h}^{h} = \frac{2ah^3}{3} + 2ch$$

$$= \frac{h}{3}(2ah^2 + 6c).$$

The values of a and c are replaced by y_1, y_2 and y_3.

$$y = ax^2 + bx + c;$$

$$y_1(x_1 = -h) = ah^2 - bh + c$$

$$y_2(x = 0) = c$$

$$y_3(x = h) = ah^2 + bh + c$$

$$y_1 + 4y_2 + y_3 = 2ah^2 + 6c$$

$$\text{So area} = \frac{h}{3}(y_1 + 4y_2 + y_3)$$

Next two strips areas are

$$(y_3, y_4, y_5) = \frac{h}{3}(y_3 + 4y_4 + y_5)$$

$$(y_5, y_6, y_7) = \frac{h}{3}(y_5 + 4y_6 + y_7)$$

If three points are not there, last two points are accounted by trapezoidal rule.

Example 4.2 The top product leaving a batch distillation column is being continuously reprocessed so that it is difficult to measure directly the amount produced. However, the flow rate is continuously measured and samples are taken at hourly intervals giving the following results.

Time, hr.	0	1	2	3	4	5	6	7	8
More volatile Component %	90	92	91	91	89	87	84	80	75
Flow rate, kg/min.	60	62	63	62	62	60	59	57	57

(a) What is the total amount produced during the period of 8 hours?

(b) What proportion of this total is the more volatile component?

Solution: (a) $f_1 = 60;\ f_2 = 62;\ f_3 = 63;\ f_4 = 62;\ f_5 = 62;\ f_6 = 60;\ f_7 = 59;\ f_8 = 57;\ f_9 = 57$

$$\text{Area of } f_1, f_2, f_3 = \frac{h}{3}(f_1 + 4f_2 + f_3) = \frac{1}{3}(60 + 4 \times 62 + 63) = \frac{371}{3}$$

$$\text{Area of } f_3, f_4, f_5 = \frac{h}{3}(f_3 + 4f_4 + f5) = (63 + 4 \times 62 + 62) = \frac{373}{3}$$

$$\text{Area of } f_5, f_6, f_7 = \frac{h}{3}(f_5 + 4 \times f_6 + f_7) = (62 + 4 \times 60 + 59) = \frac{361}{3}$$

$$\text{Area of } f_7, f_8, f_9 = \frac{h}{3}(59 + 4 \times 57 + 57) = \frac{344}{3}$$

Adding all gives 483 (per minute). **Ans.** = 28,980 per hour

(b) More volatile component

$$v_1 = 60 \times 0.90 = 54$$

$$v_2 = 62 \times 0.90 = 57.04$$

$$v_3 = 63 \times 0.91 = 57.33$$

$$v_4 = 0.91 \times 62 = 56.42$$

$$v_5 = 62 \times 0.89 = 55.18$$

$$v_6 = 60 \times 0.87 = 52.2$$

$$v_7 = 0.84 \times 59 = 49.56$$

$$v_8 = 0.80 \times 57 = 45.6$$

$$v_9 = 0.75 \times 57 = 42.75$$

$$\text{Area by Simpson's rule} = \frac{1}{3}(54 + 4 \times 57.04 + 57.33)$$

$$+ \frac{1}{3}(57.33 + 4 \times 56.42 + 55.18)$$

$$+ \frac{1}{3}(55.18 + 4 \times 4 \times 52.2 + 49.56)$$

$$+ \frac{1}{3}(49.6 + 4 \times 45.61 + 42.75)$$

$$= 421.98$$

In hour units: 421.98 × 60 kg; percentage volatile component

$$= \frac{421.98 \times 60}{28980} = 87.61 \quad \textbf{Ans.}$$

4.1.3 Simpson's One-Third Rule for Cubic Equation

$$y = a_0 + a_1 x + a_2 x^2 + a_3 x^3$$

Let

$$H = 2h(x_2 - x_0)$$

$$f(x)dx = \int_{x_0}^{x_0 + H} y\, dx$$

$$= a_0(x_2 - x_0) + \frac{a_1}{2}[(x_0 + H)^2 - x_0^2]$$

$$+ \frac{a_2}{3}[(x_0 + H)^3 - x_0^3] + \frac{a_3}{4}[(x_0 + H)^4 - x_0^4]$$

$$= a_0 H + \frac{a_1}{2}(2x_0 H + H^2) + \frac{a_2}{3}(3x_0^2 H + 3x_0 H^2 + H^2)$$

$$+ \frac{a_3}{4}(4x_0^3 H + 6x_0^2 H^2 + 4x_0 H^3 + H^4)$$

Since the curve passes through the points, we have

$$y_0 = a_0 + a_1 x_0 + a_2 x_0^2 + a_3 x_0^3$$

$$y_1 = a_0 + a_1\left(x_0 + \frac{H}{2}\right) + a_2\left(x_0 + \frac{H}{2}\right)^2 + a_3\left(x_0 + \frac{H}{2}\right)^3$$

$$y_2 = a_0 + a_1(x_0 + H) + a_2(x_0 + H)^2 + a_3 \times (x_0 + H)^3$$

$$y_0 + 4y_1 + y_2 = 6a_0 + a_1(6x_0 + 3H) + a_2(6x_0^2 + 2H^2 + 6x_1 H)$$
$$+ \; a_3\left(6x_0^3 + \frac{3}{2}H^3 + 9x_0^2 H + 6x_0 H^2\right)$$
$$= \frac{6}{H}\left(a_0 H + \frac{a_1}{2}(2x_0 H + H^2) + \frac{a_2}{3}(3x_0^2 H + H^3 + 3x_0 H^2)\right.$$
$$\left. + \; \frac{a_3}{4}(4x_0^3 H + H^4 + 6x_0^2 H^2 + 4x_0 H^3)\right)$$
$$= \frac{6}{H}\int_{x_0}^{x_2} f(x)\,dx$$

$$f(x)dx = \frac{H}{6}(y_0 + 4y_1 + y_2) = \frac{h}{3}(y_0 + 4y_1 + y_2)$$

For entire interval

$$\int_a^b y\,dx = \frac{h}{3}(y_0 + 4y_1 + y_2) + \frac{h}{3}(y_2 + 4y_3 + y_4) + \frac{h}{3}(y_4 + 4y_5 + y_6) + \ldots\ldots$$

4.1.4 Obtain Formula

$$\int_{x_0}^{x_1} f(x)\,dx = (x_1 - x_0)f(x_0) + \frac{1}{2}(x - x_0)^2 f(x_0, x_1) + \frac{1}{6}(x - x_0)^3 f(x_0, x_1, x_2) + E$$

where $E = \int_{x_0}^{x_1} (x - x_0)(x - x_1)(x - x_2) f(x_0, x_1, x_2)\,dx$

Solution: $f(x_0, x_1, x_2, \ldots, x_k) = f(x_1, x_2, \ldots, x_k) - \dfrac{f(x_0, x_1, \ldots, x_{k-1})}{(x_k - x_0)}$

From this, consider up to 4 terms of the above equation,

$$f(x) = f(x_0) + (x - x_0)f(x_0, x) + (x - x_0)(x - x_1)f(x_0, x_1, x)$$
$$+ \; (x - x_0)(x - x_1)(x - x_2)f(x_0, x_1, x_2, x)$$

$$\int_{x_0}^{x_1} f(x)\,dx = \int_{x_0}^{x_1} f(x_0)\,dx + \int_{x_0}^{x_1} (x - x_0)f(x_0, x)\,dx + \int_{x_0}^{x_1} (x - x_0)(x - x_1)f(x_0, x_1, x)\,dx$$
$$+ \int_{x_0}^{x_1} (x - x_0)(x - x_1)(x - x_2)f(x_0, x_1, x_2, x)\,dx$$

$$= f(x_0)(x - x_0) + \frac{1}{2}(x_1 - x_0)^2 f(x_0, x_1)$$

$$+ \int_{x_0}^{x_1} (x^2 - xx_1 - xx_0 - x_0x_1) f(x_0, x_1, x)\, dx$$

$$+ \int_{x_0}^{x_1} (x^3 - x^2x_1 - x^2x_0 - xx_0x_1 - x^2x_2 + xx_1x_2$$

$$+ xx_0x_2 + x_0x_1x_2)\, f(x_0, x_1, x_2, x)\, dx$$

$$= f(x_0)(x - x_0) + \frac{1}{2}(x_1 - x_0)^2 f(x_0, x_1)$$

$$+ \left(x^3 - x^2\frac{x_1}{2} - x^2\frac{x_0}{2} - x_0x_1\right) f(x_0, x_1, x)\Big|_{x_0}^{x_1}$$

$$+ \left(\frac{x^4}{4} - \left(\frac{x^3}{3}\right)x_1 - \left(\frac{x^3}{3}\right)x_0 - \left(\frac{x^2}{2}\right)x_0x_1 - \left(\frac{x^3}{3}\right)x_2\right.$$

$$\left. + \left(\frac{x^2}{2}\right)x_1x_2 + \left(\frac{x^2}{2}\right)x_0x_2\right) f_0(x_0, x_1, x_2, x)\Big|_{x0}^{x_1}$$

First term: $(x_1 - x_0)f(x_0)$

Second term: $\frac{1}{2}(x_1 - x_0)^2 f(x_0, x_1)$

Third term: $\left(x_1^3 - \frac{x_1^3}{2} - \left(\frac{x_1^2}{2}\right)x_0 - x_0\, x_1\right) f(x_0, x_1, x) - \left(x_0^3 - x_0^2\frac{x_1}{2} - \frac{x_0^3}{2} - x_0x_1\right) f(x_0, x_1, x)$

If $f(x_0, x_1, x_1) = f(x_0, x_1, x_2) = f(x_0, x_1, x_0)$ as x_2 is in between x_0 and x_1.

$$x_1^3 - \frac{x_1^3}{2} - \frac{x_1^2x_0}{2} - x_0\, x_1 - x_0^3 + \frac{x_0^2x_1}{2} + \frac{x_0^3}{2} + x_0x_1 \times (x_1 - x_0)^3 = x_1^3 - x_0^3 - 3x_1x_0(x_1 - x_0)$$

So, $\frac{x_1^3}{2} - \frac{x_1^2x_0}{2} + \frac{x_0^2x_1}{2} - \frac{x_0^3}{2} = \frac{1}{6}(x_1 - x_0)^3$

Fourth term: $(x - x_0)(x - x_1)\,(x - x_2)f(x_0, x_1, x_2, x)dx$.

So, $\int_{x_0}^{x_1} f(x)\,dx = (x_1 - x_0)f(x_0) + \frac{1}{2}(x_1 - x_0)^2 f(x_0, x_1) + \frac{1}{6}(x_1 - x_0)^3 f(x_0, x_1, x_2) + E$

4.1.5 C-Computer Program of Trapezoidal Rule

```
/*Program of Trapezoidal rule:
#include<stdio.h>
#include<conio.h>
#include<math.h>
#define MAX 100
Void main(void)
{
Int n,I;
Signed float a,b,h,sum,ict;
Float func(float x);
Crscr();
Print("\n FINAL VALUE OF X ….");
Scanf ("%f",&b);
N=(b-a)/h;
Sum=(func(a)+func(b))/2.0;
For(i=1;i<n-1;i++);
{
Sum=sum+func(a+i*h);
}
Ict=sum*h;
Printf("Integral value from %f to %f when h=%f",a,b,h,ict);
}
/*End of Main progam */
Foat func(float x);
{
Float f;
F=1.0-exp(-x/2.0);
Retun;
```

```
}
Data:
A(1.0)
B(10.0) In Next line h(1.0)
```

4.1.6 C–Program for Simpson's Method

```
/*Simpsons Numerical Integration Method
#include<stdio.h>
#include<conio.h>
#include<math.h>
#define MAX 100
Void main (void)
{
Int n,m,I;
Signed float a,b,h,sum=0.0,ics,x,f1,f2,f3;
Float func(float x);
Print f("\n intial value of x …");
Scan f("%f", & a);
Print f("/n Final value of x….");
Scanf("%f", &b);
Print f("\n Enter the number of segments (Even number)…");
Scanf("%d", n);
h=(b-a)/n;
m=n/2;
x=a;
f1=func(x);
for(i=1; I, = m; i++);
{
f2=func(x+h);
f3=func(x+2*h);
```

```
sum=sum+f1+4*f2+f3;
f1=f3;
x=x+2*h;
}
ics=sum*h/3.0
Print("Integral value rm %f o%f when h=%f is %f",
a,b,h,ics);
}
/*End of main program*/
float func(float x);
{
float f;
f = 1.0-exp(-x/2.0);
return(f);
}
Data
a(0.0)
b(11.0)
n(11.0)
```

4.2 INTERPOLATIONS AND EXTRAPOLATIONS

4.2.1 Lagrangian Method

$y = a_1(x - x_2)(x - x_3)(x - x_4) + a_2(x - x_1)(x - x_3)(x - x_4) \ldots + a_3(x - x_1)(x - x_2)(x - x_4) \ldots + a_4(x - x_1)(x - x_2)(x - x_3) \ldots.$

where $a_1 = \dfrac{y_1}{[(x_1 - x_2)(x_1 - x_3)(x_1 - x_4)\ldots.]}$

$$a_2 = \frac{y_2}{[(x_2 - x_1)(x_2 - x_3)(x_2 - x_4)\ldots.]}$$

$$a_3 = \frac{y_3}{[(x_3 - x_1)(x_3 - x_2)(x_3 - x_4)\ldots.]}$$

$$a_4 = \frac{y_4}{[(x_4 - x_1)(x_4 - x_2)(x_4 - x_3)(x_4 - x_5)\ldots.]}$$

Example 4.3 Using Lagrangian interpolation formula, fit a polynomial to the following data.

X	0	1	3	4
Y	–12	0	6	12

Solution: Here,

$$a_1 = -\frac{12}{[(0-1)(0-3)(0-4)]} = 1$$

$$a_2 = \frac{0}{[(1-0)(1-3)(1-4)]} = 0$$

$$a_3 = \frac{6}{[(3-0)(3-1)(3-4)]} = -1$$

$$a_4 = \frac{12}{[(4-0)(4-1)(4-3)]} = 1$$

$$Y = 1(x-1)(x-3)(x-4) + 0 - 1(x-0)(x-1)(x-4) + 1(x-0)(x-1)(x-3)$$

$$= x^3 - 7x^2 + 18x - 12 \textbf{ Ans.}$$

4.2.2 Lagrangian Interpolation Formula

For four points: $x_0, y_0; x_1, y_1; x_2, y_2; x_3, y_3$.

$$Y_{\text{(interpolated value)}} = y_0 \frac{(x-x_1)(x-x_2)(x-x_3)}{[(x_0-x_1)(x_0-x_2)(x_0-x_3)]} + y_1 \frac{(x-x_0)(x-x_2)(x-x_3)}{[(x_1-x_0)(x_1-x_2)(x_1-x_3)]}$$

$$+ y_2 \frac{(x-x_0)(x-x_1)(x-x_3)}{[(x_2-x_0)(x_2-x_1)(x_2-x_3)]} + y_3 \frac{(x-x_0)(x-x_1)(x-x_2)}{[(x_3-x_0)(x_3-x_1)(x_3-x_2)]}$$

Example 4.4 The thermal conductivity (K) versus temperature of a material is given below. Calculate K at 200°F.

Temp. T, Deg. F.	154	189	222	284
K	0.0493	0.0447	0.0404	0.0357

Solution: Here, $K = y$ and x is temperature (200)

$$Y(200) = 0.0493 \times \frac{(200-189)(200-222)(200-284)}{[(154-189)(154-222)(154-284)]}$$

$$+ 0.0447 \times \frac{(200-154)(200-222)(200-284)}{[(189-154)(189-222)(189-284)]}$$

$$+ 0.0404 \times \frac{(200-154)(200-189)(200-284)}{[(222-154)(222-189)(222-284)]}$$

$$+ 0.0357 \times \frac{(200-154)(200-189)(200-222)}{[(284-154)(284-189)(284-222)]}$$

$$Y_{(200)} = 0.0493 \times -0.06570 + 0.0447 \times 0.7747 + 0.0404 \times 0.2723$$
$$+ 0.0357 \times -0.01454 = 0.041782 \textbf{ Ans.}$$

4.2.3 Newton's Interpolation Formula

$$f(x) = f(x_0) + (x - x_0)\,\frac{f_0}{h} + (x - x_0)(x - x_1)\,\frac{f_0}{(2h^2)}$$

$$+ (x - x_0) \times (x - x_1)(x - x_2)\,\frac{f_0}{(2 \times 3h^3)} + (x - x_0)(x - x_1)(x - x_2)(x - x_3)\,\frac{f_0}{(4!h^4)} + \ldots.$$

Example 4.5 Calculate the approximate values of $f(x) = \sin x$ for $x = 0.50(0.02)0.070$ and for $0.50(0.02)1.70$ by applying the approximate Newton's formula to the following data.

X	0.50	0.70	0.90	1.1	1.3	1.5	1.7
$f(x)$	0.47943	0.64422	0.78333	0.89121	0.96356	0.99149	0.99166

Solution:

x_0	0.5	$f_0 = 0.47943$						
			$\Delta f_0 = 0.16479$					
x_1	0.7	$f_1 = 0.64422$		$\Delta^2 f_0 = -0.02568$				
			$\Delta f_1 = 0.13911$		$\Delta^3 f_0 = 0.00585$			
x_2	0.90	$f_2 = 0.78333$		$\Delta^2 f_1 = -0.03123$		$\Delta^4 f_0 = 0.00155$		
			$\Delta f_2 = 0.10788$		$\Delta^3 f_1 = 0.0043$		$\Delta^5 f_0 = -0.0014$	
x_3	1.10	$f_3 = 0.89121$		$\Delta^2 f_2 = -0.03353$		$\Delta^4 f_1 = 0.00141$		$\Delta^6 f_0 = .0028$
			$\Delta f_3 = 0.07235$		$\Delta^3 f_2 = -0.00289$		$\Delta^5 f_1 = 0.0014$	
x_4	1.30	$f_4 = 0.96356$		$\Delta^2 f_3 = -0.03842$		$\Delta^4 f_2 = 0.00155$		
			$\Delta f_4 = 0.03393$		$\Delta^3 f_3 = -0.00134$			
x_5	1.50	$f_5 = 0.99749$		$\Delta^2 f_4 = -0.03976$				
			$\Delta f_5 = -0.00583$					
x_6	1.70	$f_6 = 0.99166$						

Here, $x = 0.52$ and $h = 0.20$

$$f(0.52) = 0.47943 + \left[\frac{(0.52 - 0.50)}{0.20}\right] \times 0.16479 + \left[\frac{(0.52 - 0.50)(0.52 - 0.70)}{(2 \times 0.2^2)}\right] \times -0.02568$$

$$+ \left[\frac{(0.52 - 0.50)(0.52 - 0.70)(0.52 - 0.90)}{(6 \times 0.2^3)}\right] \times -0.00585$$

$$+ \left[\frac{(0.52 - 0.50)(0.52 - 0.70)(0.52 - 0.90)(0.52 - 1.10)}{(24 \times 0.2^4)}\right] \times 0.00155$$

$$+ \left[\frac{(0.52 - 0.50)(0.52 - 0.70)(0.52 - 0.90)(0.52 - 1.10)(0.52 - 1.3)}{(120 \times 0.2^5)}\right] \times 0.0014$$

$$+\left[\frac{(0.52-0.50)(0.52-0.70)(0.52-0.90)(0.52-1.10)(0.52-1.30)(0.52-1.50)}{(720\times 0.2^6)}\right]\times 0.0028$$

$$= 0.47943 + 0.016949 - 0.001556 + 0.000166725 - 0.000032027$$

$$+ 0.00002256345 - 0.00000737 = 0.498105$$ **Ans.**

Example 4.6 Given:

x	1.1275	1.1503	1.1735	1.1973
$f(x)$	0.11971	0.13954	0.15932	0.17903

Find $f(x)$ for $x = 1.160$ to 1.170 with 0.0020 difference using second difference $f(a_0, a_1, a_2)$.

Solution:

$$f(a_1, a_2, a_3) = \frac{f(a_2, a_3) - f(a_1, a_2)}{(a_3 - a_1)}$$

$$f(a_0, a_1, a_2) = \frac{f(a_1, a_2) - f(a_0, a_1)}{(a_2 - a_0)}$$

$$f(a_0, a_1) = \frac{f(a_0) - f(a_1)}{(a_1 - a_0)}$$

$$f(a_1, a_2) = \frac{f(a_1) - f(a_2)}{(a_1 - a_2)}$$

$$f(x) = f(x_0) + (x - x_0)\, f(x_0, x_1) + (x - x_0)\,(x - x_1)\, f(x_0, x_1, x_2)$$

$a_0 = 1.1275$ $\quad f(a_0) = 11971$

$$f(a_0, a_1) = \frac{(0.11971 - 0.13954)}{(1.1275 - 1.1503)} = 0.52544$$

$a_1 = 1.1503$ $\quad f(a_1) = 0.13954$ $\quad$ 7.1098

-109.99265

$$f(a_1, a_3) = \frac{(0.13954 - 0.15932)}{(1.15030 - 1.1735)} = 0.85239$$

$a_2 = 1.1735$ $\quad f(a_2) = 0.15932$ $\quad$ -0.446695

$$f(a_2, a_3) = \frac{(0.15932 - 0.17903)}{(1.1735 - 1.1972)} = 0.83164$$

$a_3 = 1.1972$ $\quad f(a_3) = 0.17903$

$$f(a_0, a_1, a_2) = \frac{(0.8529 - 0.52544)}{(1.1735 - 1.1275)} = 7.1098$$

$$f(a_1, a_2, a_3) = \frac{(0.83164 - 0.85259)}{(1.1972 - 1.1503)} = -0.446695$$

$$f(a_0, a_1, a_2, a_3) = \frac{(-0.446695 - 7.1098)}{(1.1972 - 1.1275)} = -109.99265$$

$$f(1.160) = 0.13954 + (1.160 - 1.1503) \times 0.85259$$

$$+ (1.160 - 1.1503)(1.160 - 1.1735) \times -0.446695$$

$$= 0.14957$$ **Ans.**

4.2.4 Gauss Formula

$$f(x) = f(x_0) + (x - x_0) \times \frac{\delta f_{1/2}}{h} + (x - x_0)(x - x_1)\frac{\delta^2 f_0}{2h^2} + (x - x_0)(x - x_1)(x - x_{-1})\frac{\delta^3 f_{1/2}}{3!h^3}$$

$$+ (x - x_0)(x - x_1)(x - x_{-1})(x - x_2)(x - x_{-2})\frac{\delta^4 f_0}{4!h^4}$$

$$+ (x - x_0)(x - x_1)(x - x_{-1})(x - x_2)(x - x_{-2})\frac{\delta^5 f_0}{5!h^5} + \ldots$$

If we write $s = \dfrac{(x - x_0)}{h}$

$$f(s) = f_0 + s\delta f^{1/2} + s(s-1)\frac{\delta^2 f_0}{2!} + s(s^2-1)\frac{\delta^3 f_{1/2}}{3!} + s(s^2-1)(s-2)\frac{\delta^4 f_0}{4!}$$

$$+ s(s^2-1)(s^2-4)(s^2-9)\delta^5 f^{1/2} + \ldots + Es.$$

Example 4.7 Calculate the approximate value of $x = 1.0$ from the below given data, first by the use of Gauss forward formula launched from $x = 0.90$ and second by use of the backward formula launched from $x = 1.10$.

Solution:

x_{-2}	0.5	$f = 0.47943$						
			$\delta f_{-3/2} =$ 0.16479					
x_{-1}	0.7	$f = 0.64422$		$\delta^2 f_1 =$ 0.025633				
			$\delta f_{-1/2} =$ 0.0.13911		$\delta^3 f_{-1/2} =$ –0.00555			
x_0	0.9	$f_0 = 0.78333$		$\delta^2 f_0 =$ –0.031233		$\delta^4 f_0 =$ 0.00125		
			$\delta f_{1/2} =$ 0.10788		$\delta^3 f_{1/2} =$ –0.00434		$\delta^5 f_{1/2} =$ 0.00016	
x_1	1.1	$f = 0.89121$		$\delta^2 f_1 =$ –0.03553		$\delta^4 f_1 =$ 0.00141		$\delta^6 f_1 =$ –0.00002
			$\delta f_{3/2} =$ 0.07235		$\delta^3 f_{3/2} =$ –0.00289		$\delta^5 f_{3/2} =$ 0.0014	
x_2	1.3	$f = 0.96356$		$\delta^2 f_2 =$ –0.03842		$\delta^4 f_2 =$ 0.00155		
			$\delta f_{5/2} =$ 0.03393		$\delta^3 f_{5/2} =$ –0.00134			
x_3	1.5	$f = 0.99749$		$\delta^2 f_3 =$ –0.03976				
			$\delta f_{7/2} =$ –0.00538					
x_4	1.7	$f = 0.99166$						

If we write $s = \dfrac{(x - x_0)}{h} = \dfrac{(1.0 - 0.9)}{0.20} = \dfrac{0.10}{0.20} = 0.50$

$$f_s = f_0 + s\delta f^{1/2} + \left[\frac{s(s-1)}{2!}\right]\delta^2 f_0 + \left[\frac{s(s^2-1)}{3!}\right]\delta^3 f_{1/2}$$

$$+ \left[\frac{s(s^2-1)(s-2)}{4!}\right]\delta^4 f_0 + \left[\frac{s(s^2-1)(s^2-4)(s^2-9)}{7!}\right]\delta^5 f_{1/2} + \ldots. Es.$$

$f_0 = f(0.90) = 0.78333$. By forward formula:

$$f(1.0) = 0.78333 + 0.50 \times 0.10788 + \left(\frac{0.5 \times -0.5 \times -0.03123}{2}\right)$$

$$+ \left[\frac{0.5 \times (0.25 - 1) \times -0.0043}{6}\right] + \left[\frac{0.5(0.25 - 1)(0.5 - 2) \times 0.00125}{24}\right]$$

$$+ \left[\frac{0.5(0.25 - 1)(0.25 - 4)0.00016}{120}\right]$$

$$+ \frac{0.5(0.25 - 1)(0.25 - 4)(0.25 - 9) \times 0.00002}{7!}$$

$$= 0.84147$$

4.2.5 Backward Formula

x_{-3}	0.5						
		$\delta f_{-5/2} = 0.16479$					
x_{-2}	0.7		$\delta^2 f_{-2} = -0.02563$				
		$\delta f_{-3/2} = 0.13911$		$\delta^3 f_{-3/2} = -0.0053$			
X_{-1}	0.9		$\delta^2 f_{-1} = -0.03123$		$\delta^4 f_{-1} = 0.00125$		
		$\delta f_{-1/2} = 0.10788$		$\delta^3 f_{-1/2} = -0.00430$		$\delta^5 f_{-1/2} = 0.00016$	
x_0	1.1		$\delta^2 f_0 = -0.03553$		$\delta^4 f_0 = 0.00141$		$\delta^6 f_0 = -0.00002$
		$\delta f_{1/2} = 0.07235$		$\delta^3 f_{1/2} = -0.00289$		$\delta^5 f_{1/2} = 0.00014$	
x_1	1.3		$\delta^2 f_1 = -0.03842$		$\delta^4 f_1 = 0.00155$		
		$\delta f_{3/2} = 0.03393$		$\delta^3 f_{3/2} = -0.00134$			
x_2	1.5		$\delta^2 f_2 = -0.03976$				
		$\delta f_{5/2} = -0.000583$					
x_3	1.7						

$$s = \frac{(x - x_0)}{h} = \frac{(1.0 - 1.1)}{0.20} = -0.50$$

$$f_0 = 0.89121$$

$$f_s = f_0 + s\delta f_{1/2} + s(s-1)\frac{\delta^2 f_0}{2} + s(s^2 - 1)\frac{\delta^2 f_{1/2}}{3!} + s(s^2-1)(s-2)\frac{\delta^2 f_0}{4!}$$

$$+ s(s^2 - 1)(s^2 - 4)\frac{\delta^5 f_{1/2}}{5!} + s(s^2 - 1)(s^2 - 4)(s - 3)\frac{\delta^6 f_0}{6!}$$

$$= \frac{0.89121 + (-0.5) \times 0.07235 + (-0.5) \times (0.25 - 1) \times -0.03553}{2}$$

$$+ \frac{-0.5 \times (0.25 - 1.0) \times -0.00289}{6} + \frac{(-0.5)(0.25 - 1)(-2.5) \times 0.00141}{24}$$

$$+ \frac{(-0.5)(0.25 - 1)(0.25 - 4) \times 0.00014}{120}$$

$$+\frac{(-0.5)(0.25-1)(0.25-2)(-0.5-3)\times -0.00002}{720}$$

$$= 0.8481725 \textbf{ Ans.}$$

4.2.6 Newton's Forward Formula

The vapour pressure of ammonia vapour with temperature are given below. Calculate the vapour pressure of ammonia at 105 °F using Newton's forward difference formula and Newton's backward difference formula.

Temp. °F	psi(p)	Δp	$\Delta^2 p$	$\Delta^3 p$	$\Delta^4 p$	$\Delta^5 p$	$\Delta^6 p$	$\Delta^7 p$
70	128.8							
		24.2						
80	153.0		3.4					
		27.6		0.3				
90	180.6		3.7		–0.2			
		31.3		0.1		0.6		
100	211.9		3.8		0.4		–1.3	
		35.1		0.5		–0.7		25
110	247.0		4.3		–0.3		1.2	
		39.4		0.2		0.5		
120	286.4		4.5		0.2			
		43.9		0.4				
130	330.3		4.9					
		48.8						
140	379.1							

(a) Newton's forward formula

$$p = p(t_0 + mh).$$

For $\quad p_0 = 153.0$

at $\quad t_0 = 90\ °\text{F};\ t = 105\ °\text{F}$

$$h = 10\ °\text{F}$$

$$m = \frac{(105-90)}{10} = \frac{15}{10} = 1.5$$

$$p(t_0 + mh) = p(t_0) + (m_1)\nabla p_0 + (m_2)\nabla^2 p_0 + (m_3)\nabla^3 p_0$$

$$p(105) = 180.6 + 1.5 \times 31.3 + \frac{1.5 \times 0.5 \times 3.8}{2}$$

$$= 180.6 + 46.95 + 1.425 = 228.98$$

(b) Using backward differences

$$t_0 = 120\ °\text{F}$$

$$t = t_0 + mh$$

$$105 = 120 - 15$$

$$m = -\frac{15}{10} = -1.5$$

$$p(t_0 + mh) = p(t_0) + (m_1)\nabla p_0 + (m_2)\nabla^2 p_0 + (m_3)\nabla^3 p_0$$

$$p(105) = 286.4 + (-1.5 \times 39.4) + \frac{-1.5 \times -0.5 \times 4.3}{2}$$

$$= 286.4 - 59.1 + 1.161 = 228.91 \textbf{ Ans.}$$

4.2.7 Stirling's Formula

$$f_s = f_0 + \left(\frac{s}{2}\right)(\delta f_{1/2} + \delta f_{-1/2}) + \left[\frac{s}{(2 \times 2!)}\right][(s-1) + (s+1)]\delta^2 f_0$$

$$+ \left[\frac{s(s^2-1)}{(2 \times 3!)}\right](\delta^2 f_{1/2} + \delta^2 f_{-1/2}) + \left[\frac{s(s^2-1)}{(2 \times 4!)}\right] \times [(s-2) + (s+2)]\delta^4 f_0$$

$$+ \left[\frac{s(s^2-1)(s^2-4)}{(2 \times 5!)}\right](\delta^5 f_{1/2} + \delta^5 f_{-1/2}) + \left[\frac{s(s^2-1)(s^2-4)}{(2 \times 6!)}\right][(s+3) + (s-3)]\delta^6 f_0 + Es$$

Example 4.8 Using Stirling's formula, calculate the values of $f(x)$ for $x = 1.00$ to 1.20 (0.02) from the data of the previous problem.

Solution: $s = \dfrac{1.0 - 1.1}{0.2} = -0.5$

$$f(1{\cdot}0) = 0.89121 + \left(\frac{-0.5}{2}\right)(0.0723 + 0.10788) + \left(\frac{-0.5}{4}\right)[(-0.5-1) + (-0.5+1)](-0.03553)$$

$$+ (-0.5)\frac{(0.25-1)}{2 \times 6}(-0.00430 - 0.00289) + (-0.5)(0.25-1)(0.25-4)[2 \times (-0.5)]$$

$$\times \frac{0.00141}{2 \times 24} + (-0.5)(0.25-1)(0.25-4)\frac{(0.00016 + 0.00014)}{2 \times 120}$$

$$+ (-0.5)(0.25-1)(0.25-4)(2 \times -0.5) \times \frac{0.00002}{2 \times 720}$$

$$= 0.89121 - 0.0450575 - 0.00444125 - 0.0002247055 - 0.000011016$$

$$- 0.0000017632 - 0.000000193$$

$$= 0.841474 \textbf{ Ans.}$$

4.2.8 Everett's Formula

$$f_s = (1 - s)\, f_0 - s\frac{(s-1)(s-2)}{3!}\,\delta^2 f_0 - s(s-1)(s-2)(s-3)\,\delta^2\frac{f_0}{5!}$$

$$+ s f_1 + (s+1)\, s\,(s-1)\,\delta^2\frac{f_1}{3!} + (s+2)(s+1)\, s\,(s-1)(s-2)\,\delta^4\frac{f_1}{5!}$$

Example 4.9 Using Everett's formula obtain the approximate value of $f(x) = \sin x$ for $x = 1.00(0.02)$ 1.20 from the following table:

x^{-3}	0.5	$f(0.5) = 0.47943$					
				$\delta^2 f^{-2} = 0.16479$			
x^{-2}	0.7	$f(0.7) = 0.64422$					
			$\delta f^{-1} = 0.13911$			$\delta^2 f^{-2} = 0.02568$	
x^{-1}	0.9	$f(0.9) = 0.78333$		$\delta^2 f^{-1} = -0.03123$	$\delta^3 f^{-1} = 0.00555$		
			$\delta f_0 = 0.10788$		$\delta^3 f^0 = 0.00430$	$\delta^4 f^0 = 0.00125$	
x^0	1.1	$f(1.1) = 0.89121$		$\delta^2 f^0$ 0.03553	$\delta^4 f_1 = 0.00141$	$\delta^5 f^1 = 0.00016$	$\delta^6 f^2 = 0$
			$\delta f^1 = 0.0723$		$\delta^3 f^1 = -0.00289$	$\delta^5 f^2 = 0.00016$	
x^1	1.3	$f(1.3) = 0.96356$		$\delta^1 f^1 = -0.03842$	$\delta^4 f^2 = 0.00157$		
			$\delta f^2 = 0.03393$	$\delta^3 f^2 = 0.00132$			
x^2	1.5	$f(1.5) = 0.99749$		$\delta^2 f^2 = 0.03976$			
			$\delta f^3 = -0.03393$				
x^3	1.7	$f(1.7) = 0.99166$					

Solution: $f(1.0) = (1 + 0.5) \times 0.89121 - (-0.5)(-0.5-1)(-0.5-2.0) \times \dfrac{0.03553}{6} - (-0.5+1)$

$$(-0.5)(-0.5-2.0)(-0.5-3.0) \times \frac{0.00141}{120} + (-0.5)(0.96356) + (1-0.5)(-0.5)$$

$$(0.5-1) \times \frac{0.03842}{6} + (-0.5+2)(0.5+1)(0.5)(0.5-1)(0.5-2) \times \frac{0.00141}{120}$$

$$f(1.00) = 1.336815 - 0.011103125 + 0.00002570312 - 0.48178 - 0.00240125 - 0.0000181641$$

$$= 0.84167687$$ **Ans.**

4.2.9 Truncation Error

Show that the magnitude of the truncation error to linear interpolation of the error function, erf x = $\dfrac{2}{\sqrt{\pi}}\displaystyle\int_0^x e^{-t^2}\,dt$ between x_0 and x_1 cannot exceed $\dfrac{(x_1-x_0)^2}{(2\pi e)^{1/2}}$ and hence is smaller than $\dfrac{(x_1-x_0)^2}{8}$ $E(x)$ $= \dfrac{1}{(n+1)!} f^{(n+1)}(\xi)\, \pi(x)$

If $n = 1$, first order interpolation; $E(x) = \dfrac{1}{2} f''(\xi)(x - x_0)(x - x_1)$

$$f'(\text{erf } x) = \frac{2}{\sqrt{\pi}} e^{-t^2}$$

$$f''(\text{erf } x) = \frac{2}{\sqrt{\pi}} 2t^2 e^{-t^2}$$

$$E(x) = \frac{1}{2} \times \frac{2}{\sqrt{\pi}} 2t\, e^{-t^2} \quad (x - x_0)(x - x_1)(x - x_0)(x - x_1)$$

can be written as $\frac{1}{4}(x_1 - x_0)^2$.

The maximum value of the whole is $\dfrac{(x_1 - x_0)^2}{2(2\pi e)^{1/2}}$

$$(2\pi_e)^{1/2} < 4 = \frac{(x_1 - x_0)^2}{8} \quad \textbf{Ans.}$$

4.2.10 Relations

1. $\Delta = E\Delta$; **2.** $\nabla = E^{-1}\Delta$

3. $E^{-1/2}\,\Delta = E^{1/2}\,\Delta = \delta$; **4.** $\Delta\nabla = \nabla\Delta = \Delta - \nabla = \delta^2$

5. $\mu = \frac{1}{2}(\nabla + \Delta)$; **6.** $E^{\pm 1/2} = \mu \pm \frac{1}{2}$

7. $\mu^2 = 1 + \frac{1}{4}\delta^2$;

Solution:

1. To prove $E\nabla = \Delta;\ \Delta f(x) = f(x + h) - f(x)$

$$\nabla f(x) = f(x) - f(x - h)$$

$$\mu f(x) = \frac{1}{2}\left[f\left(x + \frac{h}{2}\right) - f\left(x - \frac{h}{2}\right)\right]$$

$$E f(x) = f(x + h)$$

$$Df(x) = f'(x)$$

$$J f(x) = \int_{x}^{x+h} f(t)dt$$

$$E^{\alpha} f(x) = f(x + \alpha h)$$

$$\Delta f(x) = f(x + h) - f(x)$$

$$Ef(x) = f(x + h)$$

$$E\nabla f(x) = f(x + h) - f(x - h + h) = f(x + h) - f(x) = \Delta f(x)$$

So, $E\nabla = \Delta$

2. To prove $\nabla = E^{-1}\Delta$

$$\Delta f(x) = f(x+h) - f(x)$$

$$\nabla f(x) = f(x) - f(x-h)$$

$$E^{-1}f(x) = f(x-h)$$

$$E^{-1}\Delta f(x) = E^{-1}[f(x-h) - f(x)] = f(x) - f(x-h) = \Delta$$

Hence, $\nabla = E^{-1}\Delta$

3. To prove $E^{1/2}\,\nabla = \delta$

$$\Delta f(x) = f(x+h) - f(x)$$

$$Ef(x) = f(x+h)$$

$$E^{-1/2}\,\Delta f(x) = f\left(x+h-\frac{h}{2}\right) - f\left(x-\frac{h}{2}\right) = f\left(x+\frac{h}{2}\right) - f\left(x-\frac{h}{2}\right) = \delta f(x)$$

Since $\delta f(x) = f\left(x+\frac{h}{2}\right) - f\left(x-\frac{h}{2}\right)$

Hence, $E^{-1/2}\Delta = \delta$

$$E^{1/2}\nabla f(x) = E^{1/2}\,[f(x) - f(x-h)] = f\left(x+\frac{h}{2}\right) - f\left(x-\frac{h}{2}\right) = \delta f(x)$$

So, $E^{1/2}\nabla = E^{-1/2}\Delta = \delta$

4. To prove $\Delta\nabla = \nabla\Delta$

$$\nabla f(x) = f(x) - (x-h)$$

$$\Delta\nabla f(x) = f(x+h) - f(x) - f(x-h+h) - f(x-h)$$

$$= f(x+h) - f(x) - f(x) + f(x-h)$$

$$= f(x+h) + f(x-h) - 2f(x)$$

$$\Delta\nabla f(x) = \nabla[f(x+h) - f(x)]$$

$$= f(x+h) - f(x+h-h) - [f(x) - f(x-h)]$$

$$= f(x+h) - f(x) + f(x-h) - f(x)$$

$$= f(x+h) + f(x-h) - 2f(x)$$

Hence, $\Delta\nabla = \nabla\Delta$

5. To prove $\Delta - \nabla = \delta^2 = \nabla\Delta = \Delta\nabla$

$$(\Delta - \nabla)f(x) = f(x+h) - f(x) - [f(x) - f(x-h)$$

$$= f(x+h) + f(x-h) - 2f(x)$$

$$\delta f(x) = f\left(x+\frac{h}{2}\right) - f\left(x-\frac{h}{2}\right)$$

$$\delta\,[\delta f(x)] = \delta\left[f\left(x+\frac{h}{2}\right) - f\left(x-\frac{h}{2}\right)\right]$$

$$= f\left(x+\frac{h}{2}\right) + \left(\frac{h}{2}\right) - f\left(x+\frac{h}{2}\right) - \left(\frac{h}{2}\right) - \left[f\left(x-\frac{h}{2}\right) + \frac{h}{2} - f\left(x-\frac{h}{2}\right) - \frac{h}{2}\right]$$

$$= f(x+h) + f(x-h) - 2f(x)$$

Hence, $\Delta\nabla = \nabla\Delta = \Delta - \nabla = \delta^2$

6. To prove $\mu\delta = \frac{1}{2}(\nabla + \Delta)$

$$\frac{1}{2}(\Delta + \nabla)f(x) = \frac{[f(x+h) - f(x) + f(x) - f(x-h)]}{2}$$

$$= \frac{[f(x+h) - f(x-h)]}{2}$$

$$\delta f(x) = f\left(x+\frac{h}{2}\right) - f\left(x-\frac{h}{2}\right)$$

$$\mu f(x) = \frac{1}{2}\left[f\left(x+\frac{h}{2}\right) + f\left(x-\frac{h}{2}\right)\right]$$

$$\mu\delta f(x) = \frac{1}{2}\left[f\left(x+\frac{h}{2}+\frac{h}{2}\right) + f\left(x-\frac{h}{2}-\frac{h}{2}\right)\right] - \frac{1}{2}\left[f\left(x-\frac{h}{2}\right) + \frac{h}{2} + f\left(x-\frac{h}{2}-\frac{h}{2}\right)\right]$$

$$= \frac{1}{2}\,[f(x+h) + f(x) - [f(x) - f(x-h)]$$

$$= [f(x+h) - f(x-h)]/2$$

Hence, $u\delta = \frac{1}{2}(\nabla + \Delta)$

7. To prove $E^{\pm 1/2} = u \pm \frac{1}{2}\delta$

$$E^{1/2} f(x) = f\left(x+\frac{h}{2}\right)$$

$$E^{-1/2} f(x) = f\left(x-\frac{h}{2}\right)$$

$$\mu f(x) = \frac{1}{2}\left[f\left(x+\frac{h}{2}\right) - f\left(x-\frac{h}{2}\right)\right]$$

$$\frac{1}{2}\delta f(x) = \frac{1}{2}\left[f\left(x+\frac{h}{2}\right) - f\left(x-\frac{h}{2}\right)\right]$$

$$uf(x) + \frac{1}{2}\delta f(x) = \frac{1}{2}f\left(x+\frac{h}{2}\right)+\frac{1}{2}f\left(x-\frac{h}{2}\right)+\frac{1}{2}f\left(x+\frac{h}{2}\right)-\frac{1}{2}f\left(x-\frac{h}{2}\right)$$

$$= f\left(x+\frac{h}{2}\right) = E^{1/2}f(x)$$

So, $$\mu + \frac{1}{2\delta} = E^{1/2}$$

$$\mu f(x) - \frac{1}{2}\delta f(x) = \frac{1}{2}f\left(x+\frac{h}{2}\right)+\frac{1}{2}f\left(x-\frac{h}{2}\right)-\frac{1}{2}f\left(x+\frac{h}{2}\right)+\frac{1}{2}f\left(x-\frac{h}{2}\right)$$

$$= E^{-1/2} f(x)$$

So, $$E^{\pm 1/2} = \mu + \frac{1}{2}\delta$$

8. To prove $$\mu^2 = 1 + \frac{\delta^2}{4}$$

$$\mu f(x) = \frac{1}{2}\left[f\left(x+\frac{h}{2}\right)+f\left(x-\frac{h}{2}\right)\right]$$

$$\mu\,\mu\, f(x) = \mu^2 f(x) = \frac{1}{4}\left[f\left(x+\frac{h}{2}+\frac{h}{2}\right)+\left(x+\frac{h}{2}-\frac{h}{2}\right)+f\left(x-\frac{h}{2}+\frac{h}{2}\right)+f\left(x-\frac{h}{2}-\frac{h}{2}\right)\right]$$

$$\delta f(x) = f\left(x+\frac{h}{2}\right)-f\left(x-\frac{h}{2}\right)$$

$$\delta\,\delta\, f(x) = \delta^2 f(x) = f\left(\frac{x+h}{2}+\frac{h}{2}\right)+f\left(\frac{x+h}{2}-\frac{h}{2}\right)-f\left(\frac{x-h}{2}+\frac{h}{2}\right)-f\left(\frac{x-h}{2}-\frac{h}{2}\right)$$

$$= f(x+h) - 2f(x) + f(x-h)$$

$$\left[1+\frac{1}{4}\delta^2\right] f(x) = f(x) + \frac{1}{4}\,[f(x+h) - 2f(x) + f(x+h)]$$

$$= \frac{1}{4}f(x+h)+\frac{1}{4}f(x-h)+\frac{f(x)}{2} = \mu^2 f(x)$$

So, $$1 + \frac{\delta^2}{4} = \mu^2$$

4.2.11 Relations of Operators

Example 4.10 Express each of the operators E, Δ, ∇, δ, μ and $u\delta$ in terms of hD.

Solution: Taylor's expansion is

$$p(x+h) = p(x) + h\,p'(x) + h^2\frac{p''(x)}{2!} + h^n\,p\frac{n(x)}{n!}$$

This can be written as: $Ep(x) = 1 + \left(\frac{hD}{1!} + \frac{h^2D^2}{2!} \dots + \frac{h^nD^n}{n!} \right) p(x)$

Since the series in the parentheses is the expansion of the function e^{hD} we deduce the relationship, $E = e^{hD}$

Since $\Delta = E - 1; \Delta^r = (E - 1)^r$

$$= (e^{hD} - 1)^r = \left(\frac{hD}{1!} + \frac{h^2D^2}{2!} + \dots\dots \right)^r$$

So, $\Delta = (e^{hD} - 1).$

Since $\nabla = 1 - E^{-1}, \nabla = 1 - e^{-hD}$

$$\delta = E^{1/2} - E^{-1/2} = e^{hD/2} - e^{-hD/2}$$

$$= 2\sinh \frac{hD}{2}$$

$$\mu = \frac{1}{2} [E^{1/2} + E^{-1/2}]$$

$$= \frac{1}{2} [e^{hD/2} + e^{-hD/2}]$$

$$\mu = \frac{\cosh hD}{2}$$

$$\mu\delta = 2 \frac{\sinh hD}{2} \frac{\cosh hD}{2}$$

4.2.12 C–Program for Forward Interpolation

```
/* /Forward Interpolation Method
# include <studio.h>
# include <conio.h>
float h, xval, x(20), fdiff(20) (20), f(20);
float sum 1 = 0, sum = 0, fx;
float hpj = 1, prod = 1, denom = 1;
int i, j, k, l, m, n, fact = 1;
void main (void);
{
Crscr ( );
Print ("Program for Forward Difference Table");
```

```
}
Printf ("\n Enter value of h,n, xval, … "};
Scanf (%f %d f", &h &n, &x val);
n = n - 1;
I = 0;
For (j = 0; j <= n; j++);
{
Printf ("Enter value of x(%d), f(%d), …", j, j);
Scanf ("%f %f; & x(j), &f(j));
f-diff (i) (j) = f( );
}
Printf ("\n");
J = 0;
For (j = 0; j< = n; j++);
{
Printf ("\n%d%df-diff (%d) = %0.5 f", i, j, f-diff (i) (j));
}
Printf ("\n");
I = 1;
m = n;
while (i< = n);
{
For (j = 0; j < m; j++);
}
f-diff (i) (j) = f-diff (i - 1) (j + 1) - f - diff (i - 1) (j);
printf("\n%d%ddf-diff(%d)(%d)  = %0.5f", i, j, i, j, f-diffdiff
(i) (j));
printf("\n");
m = m + 1;
i++;
}
printf ("\n");
getch ( );
```

```
}
/* End of main program
int factorial (int val);
{
float fval = 1;
for (i = 1, i< = val; i ++);
{
fval* = i;
}
return (fval);
}
```

Data:

```
h(0.10)
n(10)
x(j) values to be given horizontally
y(j) values to be given horizontally
```

4.3 ORDINARY DIFFERENTIAL EQUATIONS

4.3.1 Numerical Methods for Solving First Order Differential Equations

(a) First order differential equations are of the form:

$$\frac{dy}{dx} = f(x, y)$$

$$y = f(x)dx + \text{const.}$$

Using initial condition, $$y = y_0 + \left(\frac{dy}{dx}\right)$$

Starting at x_0, y_0 calculate $\frac{dy}{dx}$ at x_0, y_0.

$$y_0 = f(x_0, y_0)$$

Euler Method

$$y_{11} = y_0 + \left(\frac{dy}{dx}\right)_{x_0,\, y_0} \times (x_1 - y_0)$$

$$(x_1 - x_0) = y_0 + f(x_0, y_0)\,(x_1 - x_0)$$

$$y_{21} = y_{11} + \left(\frac{dy}{dx}\right) x_1, y_{11} (x_2 - x_1)$$

$$= y_{11} + f(x_1, y_{11}) (x_2 - x_1)$$

$$y_{31} = y_{21} + \left(\frac{dy}{dx}\right) x_2, y_{21} (x_3 - x_2)$$

$$= y_{21} + f(x_2, y_{21}) \times (x_3 - x_2)$$

Like this it is continued.

Other methods: *Modified Euler Method:* Average of slope is taken, $\left(\frac{dy}{dx}\right)$ avg.

$$\left(\frac{dy}{dx}\right) \text{avg.} = \frac{[f(x_0, y_0) + f(x_1, y_{11})]}{2}$$

$$y_{11} = y_0 + \left(\frac{dy}{dx}\right) x_0, y_0 \times (x_1 - x_0)$$

$$= y_0 + f(x_0, y_0) (x_1 - x_0)$$

$$y_{12} = y_0 + \left(\frac{dy}{dx}\right)_{1 \text{ avg.}} (x_1 - x_0);$$

$$y_{1k} = y_0 + \left(\frac{dy}{dx}\right)_{K-1 \text{ avg.}} \times (x_1 - x_0);$$

$$\left(\frac{dy}{dx}\right)_{K-1 \text{ avg.}} = \frac{[f(x_0, y_0) + f(x_{-1}, y_{K-1})]}{2}$$

Example 4.11 $\frac{dy}{dx} = x + 0.1\, y^2$, $x_0 = 1.8$, $y_0 = 0$, x is from 1.8 to 3.0 with $dx = 0.30$. Find y versus x and tabulate.

Solution:

$$y_{11} = y_0 + \left(\frac{dy}{dx}\right)_{x_0, y_0} \times (x_1 - x_0)$$

$$\left(\frac{dy}{dx}\right)_{x_0, y_0} = 1.80 + 0.1 \times 0 = 1.8$$

$$y_{11} = 0 + 1.8 \times 0.30 = 0.540$$

$$\left(\frac{dy}{dx}\right)_{x_1, y_{11}} = 2.1 + 0.1 \times 0.54 \times 0.54 = 2.1292$$

$$y_{21} = y_{11} + \frac{dy}{dx}_{x_1, y_{11}} \times (x_2 - x_1)$$

$$= 0.4 + 2.1292 \times 0.30 = 1.179$$

$$\left(\frac{dy}{dx}\right)x_2, y_{21} = 2.4 + 1.179 \times 1.179 \times 0.10 = 2.544$$

$$y_{31} = 1.179 + 2.544 \times 0.30 = 2.0422$$

$$\frac{dy}{dx}x_3, y_{31} = 2.70 + 0.1 \times 2.42 \times 2.42 = 3.10$$

$$y_{41} = 2.042 + 3.1 \times 0.30 = 2.972$$

x	1.8	2.1	2.4	2.7	3.0
y	0.0	0.54	1.179	2.042	2. 972

Example 4.12 Modified Euler method: $\left(\frac{dy}{dx}\right) = y - \frac{2x}{y}$; $y(0) = 1$; $x(0) = 0$ with $h = 0.10$. Find $y(0.2)$.

Solution:

$$y_{11} = y_0 + \frac{dy}{dx}_{(x_0, y_0)} \times (x_1 - x_0)$$

Here, $(x_1 - x_0) = h$

$$\left(\frac{dy}{dx}\right)_{x_0, y_0} = y_0 - \frac{2x_0}{y_0} = 1 - \frac{2 \times 0}{1} = 1$$

$$y_{11} = 1 + 1 \times 0.1 = 1.1$$

$$\left(\frac{dy}{dx}\right)_{x_1, y_{11}} = 1.1 - 2 \times \frac{0.10}{1.1} = 0.918$$

$$\left(\frac{dy}{dx}\right)_{\text{avg}} = \left[\frac{\left(\frac{dy}{dx}\right)_{x_0, y_0} + \left(\frac{dy}{dx}\right)_{x_1, y_{11}}}{2}\right]$$

$$= \left(\frac{1.0 + 0.918}{2}\right) = 0.959$$

$$y_{12} = y_0 + \left(\frac{dy}{dx}\right)_{\text{avg}} (x_1 - x_0)$$

$$= 1.0 + 0.959 \times 0.10 = 1.0959$$

$$\left(\frac{dy}{dx}\right)_{x_1, y_{12}} = 1.0959 - \left(\frac{2 \times 0.10}{1.0959}\right) = 0.907$$

$$\frac{dy}{dx}_{\text{avg}} = \frac{1.0 + 0.907}{2} = 0.9535$$

$$y_{13} = y_0 + \left(\frac{dy}{dx}\right)_{\text{avg}} \times (0.10)$$

$$= 1 + 0.9535 \times 0.10 = 1.09535$$

$$\left(\frac{dy}{dx}\right)_{x_1, y_{13}} = 1.0 - \frac{2 \times 0.1}{1.09535} = 0.9128$$

$$\left(\frac{dy}{dx}\right)_{\text{avg}} = \left(\frac{1.0 + 0.9128}{2}\right) = 0.9564$$

$$y_{14} = y_0 + \left(\frac{dy}{dx}\right)_{\text{avg}} \times 0.10 = 1.0 + 0.9564 \times 0.10$$

$$= 1.09564$$

For $x = 0.20$: $y = 1.09564$ from previous calculation $= y_1$

$$\left(\frac{dy}{dx}\right)_{x_1, y_1} = 1.0964 - \left(\frac{2 \times 0.10}{1.09564}\right) = 0.91314$$

$$y_{21} = y_1 + \left(\frac{dy}{dx}\right)_{x_1, y_1} \times (x_2 - x_1)$$

$$= 1.0964 + 0.91314 \times 0.10 = 1.186874$$

$$\left(\frac{dy}{dx}\right)_{x_2, y_{21}} = 1.186874 - \left(\frac{2 \times 0.20}{1.186874}\right) = 0.84985$$

$$y_{22} = 1.0964 + \left(\frac{dy}{dx}\right)_{\text{avg}} \times 0.10 = 1.18379$$

$$\left(\frac{dy}{dx}\right)_{x_2, y_{22}} = 1.18379 - \left(\frac{2 \times 0.20}{1.18379}\right) = 0.845892$$

$$\left(\frac{dy}{dx}\right)_{x_2, y_{23}} = 1.18379 - \left(\frac{2 \times 0.20}{1.18350}\right) = 0.845892$$

$$\left(\frac{dy}{dx}\right)_{\text{avg}} = \left(\frac{0.91314 + 845635}{2}\right) = 0.87937$$

$$y_{24} = 1.0964 + 0.8793 \times 0.10 = 1.18358$$

x	0	0.10	0.20	0.20	0.20	0.20
y	1.0	1.09564	1.186374	1.18379	1.18359	1.18358

Based on accuracy, the calculations are continued.

4.4 MILNE METHOD

$$y_{n1} = y_{n-4} + \frac{4dx}{3} \times (2f_{n-1} - f_{n-2} + 2f_{n-3})$$

$$y_{n2} = y_{n-2} + \frac{dx}{3} (f_n + 4f_{n-1} + f_{n-2})$$

The first four points are calculated by Euler method or modified Euler method, etc. Subsequent points by equations of y_{n1} and y_{n2} given above.

Example 4.13

$$\frac{dy}{dx} = x + 0.1\,y^2$$

$$\frac{dy}{dx} = f(x, y) = f$$

$$x_0 = 1.80$$

$$dx = 0.20$$

$$y_0 = 0.0$$

x_n	1.80	2.00	2.20	2.40
y_n	0.0	0.38	0.81	1.29
dx	0.20	0.20	0.20	0.20
$\frac{dy}{dx}$	1.80	2.01	2.07	2.57
$\frac{dy}{dx} \times dx$	0.36	0.40	0.41	0.51
y_{b1}	0.36	0.78	1.26	1.80
$\left(\frac{dy}{dx}\right)_{b_1}$	2.01	2.26	2.26	2.92
$\left(\frac{dy}{dx}\right)_{avg}$	1.91	2.13	2.41	2.75

For $n = 4$. From equations of y_{n1} and y_{n2};

Solution:

$$y_{41} = y_0 + \frac{4dx}{3}\,(2f_3 - f_2 + 2f_1)$$

Here,

$$f = \frac{dy}{dx} = f(x, y) = 0.1y^2 + x$$

$$x_0 = 1.8$$

$$y_0 = 0.0$$

$$dx = 0.20$$

$$f_0 = 0.1 \times 0 + 1.8 = 1.8$$

$$f_1 = 0.1 \times 0.38 \times 0.38 + 2.0 = 2.01$$

$$f_2 = 2.27$$

$$f_3 = 2.57$$

$$y_{41} = 0 + \frac{0.8}{3}(2 \times 2.7 - 2.27 + 2 \times 2.01) = 1.84$$

5th, point is $x = 2.60$

$$f_4 = \frac{dy}{dx} = 2.60 + 0.10 \times 1.84 \times 1.84 = 2.94$$

From Eq. of y_{n2}

$$y_{n2} = y_{n-2} + \frac{dx}{3}(f_n + 4f_{n-1} + f_{n-2})$$

For $n = 4$

$$y_{42} = y_2 + \frac{dx}{3}(f_4 + 4f_3 + f_2)$$

$$= 0.81 + \frac{0.20}{3}(2.94 + 4 \times 2.57 + 2.27) = 1.84$$

$$\frac{dy}{dx} = 2.6 + 0.1 \times 1.84 \times 1.84 = 2.94$$

Since the $f_{4's}$ are nearly equal the iteration can be stopped and can go to next point, i.e., y_5. Otherwise dx has to be reduced for closeness.

4.5 RUNGE–KUTTA METHOD

The method is $$y_{n+1} = \frac{y_{n+1}}{6}(K_0 + 2K_1 + 2K_2 + K_3)$$

where $K_0 = hF(x_n, y_n)$

$$K_1 = hF(\left(x_n + \frac{h}{2}, y_n + \frac{K_0}{2}\right)$$

$$K_2 = hF\left(x_n + \frac{h}{2}, y_n + \frac{K_1}{2}\right)$$

$$K_3 = hF(x_n + h, y_n + K_2)$$

The calculations are continued.

Example 4.14 $$\frac{dy}{dx} = 2x^3 - 2xy$$

$$y(0) = 0$$

$$F = 2x^3 - 2xy$$

$$h = 0.10$$

$$K_0 = 0.1\ F\ (x_0, y_0) = 0.10F(0, 0) = 0$$

$$K_1 = 0.1F\left(x_0 + \frac{h}{2}, y_0 + \frac{K_0}{2}\right)$$

$$= 0.1F\ (0 + 0.05, 0 + 0)$$

$$= 0.1\ F\ (0.05, 0) = 0.10(2x^3 - 2xy) = 0.000025$$

$$K_2 = hF\left(0 + 0.05; \frac{0.000025}{2}\right)$$

$$= 0.1F(0.05, 0.0000125) = 0.000024875$$

$$K_3 = 0.1F(0.10, 0.000024875) = 0.0001995$$

$$y_1 = y_0 + \frac{1}{6}(0.0 + 2 \times 0.000025 + 2 \times 0.000024875 + 0.0001995)$$

$$= 0.0000541$$

Here, $y_0 = 0$

So, $y_1 = 0.0 + 0.0000541 = 0.0000541$

Analytical solution is $y = e^{-x^2} - 1 + x^2$

At $x = 0,\ y = 0$; at $x = 0.1$, $y = 0.00005$

4.6 TAYLOR'S EXPANSION METHOD

The Taylor's expansion about the origin is

$$y(x) = y(0) + y'(0)x + y''(0)\frac{x^2}{2!} + y'''(0)\frac{x^3}{3!} + \ldots\ldots$$

At $x = 0$, $y'(0) = 0, y(0) = 0, y''(0) = 0, y'''(0) = 0 \ldots\ldots$

Example 4.15 Solve $y' + 2xy + 2x^3$, $y(0) = 0$ using Taylor's expansion series.

Solution:

$$y' = 2x^3 - 2xy$$

$$y'' = 6x^2 - 2xy' - 2y$$

$$y''' = 12x - 2xy'' - 4y'$$

$$y^{iv} = 12 - 2xy''' - 6y''$$

Analytical solution is $y(x) = \dfrac{12x^4}{24} = 0.5x^4$

At $x = 0.1$; $y(0.10) = 0.00005$;

At $x = -0.1$; $y(-0.1) = 0.00005$;

At $x = 0.20$; $y(0.20) = 0.0008$;

At $x = -0.2$; $y(-0.2) = 0.00079$

Example 4.16 The concentration C variation with length z in a 3 feet continuous mass transfer device is described by the equation: $dC/C = (-k/v)\ dz$ where $k = 50/\text{hr}$, $v = 5$ feet/hr, C is lb. moles/cubic ft., z = length in feet; $C = 1.0$ lb moles/cubic ft. at $z = 0$. It is proposed to represent the above with 10 staged units of 0.3 foot in length. Develop solutions to this problem, using finite difference method of solving an ordinary differential equation and a lumped parameter model employing a method of solution of simultaneous linear algebraic equations.

Solution: Finite difference method:

$$\frac{dC}{dz} = \frac{\Delta C}{\Delta z} = \frac{C_{n+1} - C_n}{z_{n+1} - z_n} = \frac{K}{v_z} C_{\text{av}}$$

$$C_{n+1} = \frac{K\Delta z}{v_z} C_{\text{av}} + C_n$$

$$C_{\text{av}} = \frac{C_{n+1} + C_n}{2} \quad \text{and solving for } C_{n+1} \text{ (for } n = 0, 1, 2, \ldots)$$

Lumped parameter method:

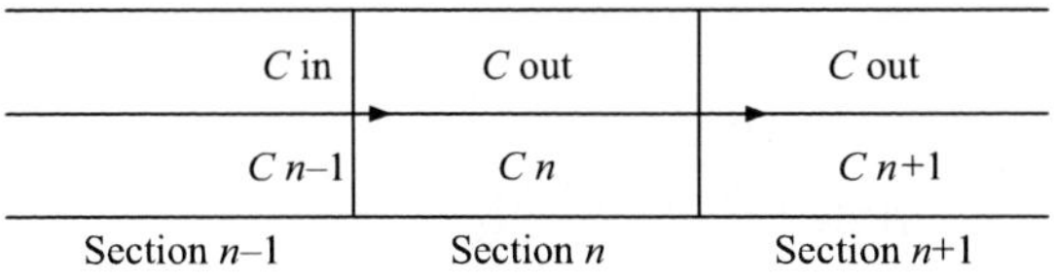

Fig. 4.3

Rearranging the above equation: $\dfrac{C_{\text{out}} - C_{\text{in}}}{\Delta z} = \dfrac{K}{v_z} C_{\text{out}}$

Example 4.17 The concentration of a pollutant in a flowing river is given by

$$\frac{dC}{dt} = -0.580\ C + 6 \times 10^{-5}$$

where $C_0 = 5.0 \times 10$ gm mole/cubic cm at $t = 0$ min.

Estimate C at $t = 1$ min and at $t = 5$. Use Runge–Kutta method.

Solution: $\dfrac{dy}{dx} = f(x, y)$, the solution takes the form:

$$y_{n+1} = y_n + \frac{h}{6}\ (D_1 + 2D_2 + 2D_3 + D_4)$$

where $D_1 = hf(x_n, y_n)$

$$D_2 = hf\left(x_n + \frac{h}{2}, y_n + \frac{D_1}{2}\right)$$

$$D_3 = f\left(x_n + \frac{h}{2}, y_n + \frac{D_2}{2}\right)$$

$D_4 = hf(x_n + h, y_n + D_3)$. h is increment in x.

For equation of form: $\frac{dC}{dt} = a + bC$ where at $t = 0$, $C = C_0$. The Runge–Kutta algorithm becomes for $t = h$; $C_1 = C_0 + \frac{h}{6}(D_1 + 2D_2 + 2D_3 + D_4)$.

4.6.1 C–Program for Euler's Method

```
/* Euler's method#
# include <stdio.h>
# include <conio.h>
# include <math.h>
# define MAX 100
Void main (void)
{
Int n, i;
Signed float x, y, xp, h, dy;
Float func(float x, foat y);
Clrscr ( );
Printf ("Enter Initial Values of X and Y …");
Scanf ("%f" %f, &x, &y);
Printf ("\n Enter the Value of X at which Y …");
Scanf (%f", &xp);
Printf ("\n Enter the Variation …");
Scanf (%f" &h);
n = (int) ((xp - x)/h + 0.5);
for (i = 1; i < = n; i++)
{
dy = h * func(x, y);
x = x + h;
y = y + dy;
printf(%d %f %f\n", i, x, y);
}
Printf ("value of Y at X = %f is %f"x, y);
}
```

```
float func(float x, float y)
{
foat f;
f = 2.0*y/x;
return (f);
}
Data:
x(0.0)       y(0.0)
xp(5.0)
h(0.20)
/* End of Main Program */
```

4.6.2 C–Computer Program for Milne Method

```
/* Milne's Method for First Order Differential Equation
# include <stdio.h>
# include <conio.h>
# include <math.h>
# define MAX 100
void main (void)
{
Int n, i;
Signed float x [MAX], y [MAX], xp, h, k1, k2, k3, k4, f2, f3, f4,
f5;
Float func(float, float);
Clrsc( );
Print("Enter the value of X at which Y …");
Scanf ("%f", &xp);
Printf("Enter the variation …");
Scanf ("%f", &h);
N = (int) ((xp - x) (1))/h + 0);
Printf("INITIAL VALUES\n\t X\t\t Y");
Printf (²\n\%f\t%f\n²x(1), y(1));
```

```
Print("\n values obtained by Runge-Kutta method \n");
For (i = 1; i < = 3; i++)
K1 = func(x(i), y(i));
K2 = func(x(i) + 0.5 * h, y(i) + k1 * h * 0.5);
K3 = func(x(i) + 0.5 * h, y(i) + k2 * h * 0.5h);
K4 = func(x(i) + h, y(i) + k3 * h);
x(i + 1) = x(i) + h;
y(i + 1) = y(i) + ((k1 + 2 * k2 + 2 * k3 + k4) * h/6.0);
printf(%d\t%f\t%f\n",  i, x(i + 1), y(i + 1));                  }
return (f)
}
printf("\n values obtained by Milne-Simpson method\n"); return(f)
for (i = 4; i < = n; i++)
{
f2 = func(x(i-2), y(i-2));
f3 = func(x(i-1), y(i-1));
f4 = func(x(i), y(i));
/* Predicted values of Y by Milne's Formula) */
y(i + 1) = y(i - 3) + (4.0 * h/3.0) * (2.0 * f2 - f3 + 2.0 * f4);
x(i + 1) = x(i) + h;
f5 = func(x(i + 1), y(i + 1));
/* Corrected value of y (By Simpson's Formula) */
y(i + 1) = y(i - 1) + (h/3.0) * (f3 + 4.0 * f4 + f5);
printf("%d\t%f\t%f\n", i, x(i + 1), y(i + 1));
printf("Value of y at x = %f is %f", xi), y(i));
}
/* End of Main Program */
float func(float x, float y)
{
float f;
f = -(y * y);
```

4.6.3 C–Program for Runge–Kutta Method

```
# include <stdio.h>
# include <conio.h>
# include <math.h>
# define MAX 100
void main (void)
{
Int n, i;
Signed float x, y, xp, h, k1, k2, k3, k4;
Float func(float, float);
clrscr( );
printf("\n Enter initial values of X and Y …");
scanf("%f", &xp);
printf("\n Enter the variation …");
scanf("%f" &h);
n = (int) ((xp - x)/h + 0.5);
for (i = 1; i < = n, i++)
k1 = func(x, y);
k2 = func(x + 0.5 * h, y + k1 * 0.5 * h);
k3 = func(x + 0.5 * h, y + k2 * 0.5 * h);
k4 = func(x + h, y + k3 * h);
x = x + h;
y = y + ((k1 + 2 * k2 + 2 * k3 + k4) * h/6.0);
printf("%d\t%f\t%f\t%f\n", i, x, y);
}
printf("Value of y at x = %f is %f," x, y);
}
/* END OF MAIN PROGRAM */
float func(float x, float y)
{
float f;
f - x * x + y * y;
return (f);
}
DATA:
```

```
x(0.0)  y(0.0)
xp (11.0)
h (0.50)
```

4.7 SECOND ORDER DIFFERENTIAL EQUATIONS

Numerical method for second order differential equation

For solving differential equations like:

$$y'' = f(x, y, y') \qquad \ldots(1)$$

$$y(x_0) = y_0$$

$$y'(x_0) = y'_0$$

Using Taylor's series

$$y(x + h) = y(x) + h\,y'(x) + \frac{h^2}{2}y''(x)$$

For derivatives also using Taylor's series:

$$y'(x + h) = y'(x) + h\,y''(x) + h^2\,\frac{y'''(x)}{2!} + \ldots$$

Neglecting higher order terms,

$$y(x + h) = y(x) + h\,y'(x) + h^2\,\frac{y''(x)}{2}$$

$$y'(x + h) = y'(x) + h\,y''(x)$$

The procedure is

Step 1: $y_0'' = f(x_0, y_0, y_0')$ is computed from the given equation.

Step 2: $y_1 = y_0 + h\,y_0' + h^2\,\frac{y''}{2}$ is computed, which approximates $y(x_1) = y(x + h)$

Step 3: $y_1' = y_0' + h\,y_0''$ which is needed in the next step calculation.

Step 4: Compute $y_1'' = f(x_1, y_1, y_1')$ from given equation.

Step 5: $y_2 = y_1 + h\,y_1' + h^2\,\frac{y''}{2}$,which corresponds to $y(x_2) = y(x_0 + 2h)$.

Step 6: $y_2' = y_1' + h\,y_1''$. In this way is continued until $y_n'' = f(x_n, y_{n'}, y_n')$ is computed from given equation 1.

Then $y_{n+1} = y_n + h\,y_n' + h^2\,y_n''/2$ which is an approximation for $y(x_{n+1})$ and furthermore $y_{n+1'} = y_n' + h\,y_n''$ which is an approximation for $y_{n+1'}$ needed in the next stage.

The method is an approximation of the curve by portions of a parabolas.

Example 4.18 Solve the following differential equation using numerical method subject to the conditions that when $x = 0$, $y = 2$ and $dy/dx = -1$. Calculate the values of y for $x = 0$ to 0.60 with the increment of 0.2 for x and tabulate y versus x.

$$\frac{d^2y}{dx^2} + \frac{2dy}{dx} + 2y = \sin x;\ x \text{ is in radians for } x.$$

Solution:

$$\frac{d^2y}{dx^2} = \sin x - 2y - \frac{2dy}{dx}$$

$$x = 0$$

$$y = 2$$

For $n = 0, x = 0,$

$$y_0'' = 0 - 2 \times 2 - 2 \times -1 = -2$$

$$y_1 = y_0 + h\, y_0' + h^2 \frac{y_0''}{2}$$

$$h = 0.2$$

$$y_1 = 2 + 0.2 \times -1 + 0.2 \times 0.2 \times -\frac{2}{2} = 1.76$$

At $x_1 = 0.2$

$y_1 = 1.76$ **Ans.**

$$y'' = y_0 + h\, y_0'' = -1 + 0.2 \times -2 = -1.4$$

$$y_1'' = \sin 0.2 - 2 \times 1.76 - 2 \times -1 = -0.55213$$

$$y_2 = y_1 + h\, y_1' + h^2 \frac{y''}{2}$$

$$= 1.76 + 0.2 \times -1.4 + 0.2 \times 0.2 \times \frac{-0.5213}{2} = 1.4696$$

$$y_2' = y_1' + 0.2 \times y_1'' = 1.4 + 0.2 \times -0.5213 = -1.5043$$

$$y_2'' = \sin(0.40) - 2 \times 1.4696 - 2 \times -1.5043 = 0.4583$$

$$y_3 = y_2 + h\, y_2' + \frac{h^2\, y_2''}{2}$$

$$= 1.4696 + 0.2 \times -1.043 + 0.2 \times 0.20 \times \frac{0.4588}{2} = 1.1779$$

x	0	0.20	0.40	0.60
y	2.0	1.76	1.4696	1.1779

4.7.1 Runge–Kutta–Nystrom Method

For solving second order differential equation (Fourth order method).

$$A_n = \frac{h}{2} f(x_n, y_n, y'_n)$$

$$B_n = \frac{h}{2} f\left(x_n + \frac{h}{2}, y_n + B_n, y'_n + A_n\right)$$

where $B_n = \frac{h}{2}\left(y'_n + \frac{A_n}{2}\right)$

$$C_n = \left(\frac{h}{2}\right) f\left(x_n + \frac{h}{2}, y_n + b_n, y'_n + B_n\right)$$

$$D_n = \frac{h}{2} f(x_n + h, y_n + \delta_n, y'_{n+2} + C_n)$$

where $\delta_n = h(y_n' + C_n)$

Then $y_{n+1} = y_n + h\,(y_n + K_n)$ where $K_n = \frac{1}{3}\,(A_n + B_n + C_n)$ which is an approximation for $y(x_{n+1})$.

$y'_{n+1'} = y_n' + K_n^*$ where, $K_n^* = \frac{1}{3}\,(A_{n+2} + B_{n+2} + C_n + D_n)$ which is an approximation for $y'_{(xn+1)}$ needed in the next step.

4.8 PARTIAL DIFFERENTIAL EQUATIONS

(a) Solving one-dimensional unsteady state heat conduction equation. This comes under parabolic type.

$$\frac{du}{dt} = \frac{C^2 d^2 u}{dx^2} \qquad \text{...(1)}$$

where $u = u(x, t)$, temperature;

t = is time,

x = position,

C^2 = constant;

$$u(x, t) = u(ih, jh) = u(i, j) \qquad \text{...(2)}$$

$\frac{du}{dt}$ is written as

$$\frac{u_i, j+1 - u_i, j}{k} \qquad \text{...(3)}$$

$$\frac{d^2u}{dx^2} = \frac{\left[\left(\frac{du}{dx}\right)x - \left(\frac{du}{dx}\right)x - dx\right]}{h}$$

$$\frac{du}{dx}x = \frac{u_{i,\,j+1} - u_{i,\,j}}{h}$$

$$\left(\frac{du}{dx}\right)_x = \frac{\left(u_{i-1,\,j+1} - u_{i,\,j}\right)}{h}$$

$$\frac{d^2u}{dx^2} = \frac{u_{i-1,\,j-2}u_{i,\,j} + u_{i,\,j+1}}{h^2} \quad \text{...(4)}$$

Substituting Eqs. (4) and (3) in Eq. (1) gives:

$$\frac{u_{i,\,j+1} - u_{i,\,j}}{k} = \left(\frac{C^2}{h^2}\right)(u_{i-1,\,j} - 2u_{i,\,j} + u_{i+1,\,j}) \quad \text{...(5)}$$

$$u_{i,\,j+1} = a\,u_{i-1,\,j} + (1 - 2a)\,u_{i,\,j} + au_{\,i+1,\,j} \quad \text{...(6)}$$

where $a = \frac{kC^2}{h^2}$. Equation (6) works only for $0 < a < \frac{1}{2}$. For $a > \frac{1}{2}$, solutions are different.

Example of calculation: For $a = \frac{1}{2}$ $(h = 1, k = \frac{1}{8}, C^2 = 4)$; $u(0, t) = u(5, t) = 0$; $u(x, 0) = 5x - x^2$, $u_{0,\,i} = 0$, and $u_{5,\,j} = 0$ for all values of j. The entries in the first and last columns are zeroes. $u(x, 0) = 5x - x^2$; $u_{i,\,0} = 5i - i^2 = 0, 4, 6, 6, 4, 0$ (for $i = 0$ to 5); $u_{0,\,0} = 0$, $u_{1,\,0} = 4$, $u_{2,\,0} = 6$, $u_{3,\,0} = 6$, $u_{4,\,0} = 4$, $u_{5,\,0} = 0$.

Putting $j = 0$, we have

$$u_{i,\,j+1} = au_{i-1,\,j} + au_{i+1,\,j} \quad \text{...(7)}$$

$$u_{i,\,1} = a_{u_{i-1}},0 + a_{u_{i+1}},0 = \frac{1}{2}(u_{i-1,\,0} + u_{i+1,\,0}) \quad \text{...(8)}$$

Giving $i = 1$ to 4 in Eq. (8) gives

$$u_{1,\,1} = \frac{1}{2}(u_{0,\,0} + u_{2,\,0}) = \frac{1}{2}(0 + 6) = 3$$

$$u_{2,\,1} = \frac{1}{2}(u_{1,\,0} + u_{2,\,0}) = \frac{1}{2}(4 + 6) = 5$$

$$u_{3,\,1} = \frac{1}{2}(u_{2,\,0} + u_{3,\,0}) = \frac{1}{2}(6 + 6) = 6$$

$$u_{4,\,1} = \frac{1}{2}(u_{3,\,0} + u_{4,\,0}) = \frac{1}{2}(6 + 4) = 5$$

$$u_{5,\,1} = \frac{1}{2}(u_{4,\,0} + u_{5,\,0}) = \frac{1}{2}(4 + 0) = 2$$

These are the entries in second row.

$j\backslash i$	0	1	2	3	4	5
0	0	4	6	6	4	0
1	0	3	5	6	5	0
2	0	2.5	4.5	5.5	3	0
3	0					0
4	0					0
5	0					0

Using $j = 1$ in Eq. (7) gives

$$u_{i,2} = \frac{1}{2}(u_{i-1,1} + u_{i+1,1})$$

$$u_{1,2} = \frac{1}{2}(u_{0,1} + u_{2,1}) = \frac{1}{2}(0 + 5) = 2.5$$

$$u_{2,2} = \frac{1}{2}(u_{1,1} + u_{3,1}) = \frac{1}{2}(3 + 6) = 4.5$$

$$u_{3,2} = \frac{1}{2}(u_{2,1} + u_{4,1}) = \frac{1}{2}(5 + 6) = 5.5$$

$$u_{4,2} = \frac{1}{2}(u_{3,1} + u_{5,1}) = \frac{1}{2}(6 + 0) = 3.0$$

Next for $j = 3$; $j = 4$; $j = 5$ in Eq. (7) (varying i value), the entire table is calculated.

Example 4.19 For a is $< \frac{1}{2}$; Eq. (6) becomes

$$u_{i,j+1} = a \times u_{i-1},j + (1 - 2a)\, u_{i,j} + au_{i+1,j}$$

For $a = \frac{1}{8}$;

$$u_{i,j+1} = \left[\frac{1}{8}(u_{i-1,j})\right] + \left[u_{i,j} \times \frac{3}{4}\right] + \left[u_{i+1}, \frac{j}{8}\right]$$

$$u_{i,j+1} = \frac{1}{8}(u_{i-1,j} + 6\, u_{i,j} + u_{i+1,j}) \qquad \text{... (9)}$$

Equation (9) gives the table of values by varying i and j (as integers).

(b) Two-dimensional equation. Solving

$$\frac{d^2u}{dx^2} + \frac{d^2u}{dy^2} = 0 \qquad \text{...(1)}$$

by numerical method.

Solution: $\frac{d^2u}{dx^2}$ in numerical form is $\frac{1}{h^2}\,(u_{i-1,j} - 2\,u_{i,j} + u_{i,j+1})$ ...(2)

$\frac{d^2u}{dy^2}$ in numerical form is $\frac{1}{h^2}\,(u_{i-1,j-1} - 2\,u_{i,j} + u_{i,j+1})$ ...(3)

Adding both and equating to zero gives

$$u\,i,j = \frac{1}{4}\,[u_{i-1,j} + u_{i+1,j} + u_{i,j+1} + u_{i,j-1}] \quad ...(4)$$

Sometimes a formula similar to 4 is used, given by

$$u_{i,j} = \frac{1}{4}\,[u_{i-1,j+1} + u_{i+1,j-1} + u_{i+1,j+1} + u_{i-1,j-1}] \quad ...(5)$$

Equation (4) is the standard five-point formula. Equation (5) is the diagonal five-point formula, Eq. (4) shows that the value of $u\ i, j$ at interior mesh point is the average of the values at function neighbouring points to the left, right, above and below. Equation (5) uses the average of the four diagonal points.

b's are the boundary conditions.

* b 1, 5	* b 2, 5	* b 3, 5	* b 4, 5	* b 5, 5
* b 1, 4	* u 2, 4	* u 3, 4	* u 4, 4	* b 5, 4
* b 1, 3	* u 2, 3	* u 3, 3	* u 4, 3	* b 5, 3
* b 1, 2	* u 2, 2	* u 3, 2	* u 4, 2	* b 5, 2
* b 1, 1	* b 2, 1	* b 3, 1	* b 4, 1	* b 5, 1

By Equation

$$u_{3,3} = \frac{1}{4}\,[b_{5,5} + b_{5,1} + b_{1,1} + b_{1,5}] \text{ (diagonal)}$$

$$u_{2,4} = \frac{1}{4}\,[b_{1,5} + u_{3,3} + b_{3,5} + b_{1,3}] \text{ (diagonal)}$$

$$u_{4,4} = \frac{1}{4}\,[b_{3,5} + b_{5,3} + b_{5,5} + u_{3,3}] \text{ (diagonal)}$$

$$u_{4,2} = \frac{1}{4}\,[u_{3,3} + b_{5,1} + b_{3,1} + b_{5,3}] \text{ (diagonal)}$$

$$u_{2,2} = \frac{1}{4}\,[b_{1,3} + b_{3,1} + u_{3,3} + b_{1,1}] \text{ (diagonal)}$$

Other points are calculated from sides.

$$u_{2,3} = \frac{1}{4}[b_{1,3} + u_{3,3} + u_{2,4} + u_{2,2}] \text{ (sides)}$$

$$u_{3,4} = \frac{1}{4}[u_{2,4} + u_{4,4} + u_{3,5} + u_{3,3}] \text{ (sides)}$$

$$u_{4,3} = \frac{1}{4}[u_{3,3} + b_{5,3} + u_{4,4} + u_{4,2}] \text{ (sides)}$$

$$u_{3,2} = \frac{1}{4}[u_{2,2} + u_{4,2} + u_{3,3} + u_{3,1}] \text{ (sides)}$$

Equations (4) or (5) are used depending on the values available. Having found all the u values, their accuracy is improved by iteration of the Eq. (4).

$$u_i^{n+1} = \frac{1}{4}(u_{i-1,j}^{n+1} + u_{i,j}^n + u_{i,j-1}^n) \quad \ldots(6)$$

This is repeated until previous value and present value do not differ much.

Example 4.20 Solve $\frac{d^2u}{dx^2} + \frac{d^2u}{dy^2} = 0$ for the mesh shown below. Let $u_1, u_2, \ldots u_9$ be the interior values of u;

C

	0	500	1000	500	0	
	1000	u_1(1125)	u_2	u_3	1000	
	2000	u_4	u_5(1500)	u_6	2000	B
A	1000	u_7	u_8	u_9	1000	
	0	500	1000	500	0	

D

Since the boundary values are symmetrical about AB and CD, $u_7 = u_1$; $u_8 = u_2$; $u_9 = u_3$; $u_3 = u_1$; $u_6 = u_4$; $u_9 = u_7$

Solution: The values to be calculated are: u_1, u_2, u_4 and u_5. Using standard formula,

$$u_5 = \frac{1}{4}[u_2 + u_8 + u_4 + u_6]$$

$$= \frac{1}{4}[1000 + 1000 + 2000 + 2000] = 1500$$

$$u_1 = \frac{1}{4}[0 + u_5 + 2000 + 1000] = 1125 \text{ (diagonal)}$$

$$u_5 = 1500$$

$$u_4 = \frac{1}{4}[u_1 + u_7 + 2000 + 1500\ (u_5)]$$

$u_1 = 1125,\ u_7 = u_1 = 1125$ (standard formula)

$$u_4 = \frac{1}{4}[1125 + 1125 + 2000 + 1500] = 1438 \text{ (standard formula)}$$

$$u_2 = \frac{1}{4}[1125 + 1125 + 1000 + 1500]$$

$= 1187.5$ (standard formula)

Starting values are $u_1 = 1125$

$u_2 = 1188(1187.5)$

$u_4 = 1438$

$u_5 = 1500$

Iteration procedure is:

For $n + 1$:

$$u_{1,\,n+1} = \frac{1}{4}[1000 + u_{2,\,n} + 500 + u_{4,\,n}]$$

$$u_{2,\,n+1} = \frac{1}{4}[u_{1,\,n} + u_{1\,+n+1} + 1000 + u_{5,\,n}]$$

$$u_{4,\,n+1} = \frac{1}{4}[2000 + u_{5,\,n} + u_{1,n+1} + u_{1,\,n}]$$

$$u_{5,\,n+1} = \frac{1}{4}[u_{4,\,n+1} + u_{4,\,n} + u_{2,\,n+1} + u_{2,\,n}]$$

$n = 0$ (first iteration)

$$u_{1,\,1} = \frac{1}{4}[1000 + u_{2,\,0} + 500 + u_{4,\,0}]$$

$$= \frac{1}{4}[1000 + 1188 + 500 + 1438] = 1032$$

$$u_{2,\,1} = \frac{1}{4}[u_{1,\,0} + u_{1,\,1} + 1000 + u_{5,\,0}]$$

$$= \frac{1}{4}[1125 + 1032 + 1000 + 1500] = 1164$$

$$u_{4,\,1} = \frac{1}{4}[2000 + u_{5,\,0} + u_{1,\,1} + u_{1,\,0}]$$

$$= \frac{1}{4}[2000 + 1500 + 1032 + 1125] = 1414$$

$$u_{5,\,1} = \frac{1}{4}\,[u_{4,\,1} + u_{4,\,0} + u_{2,\,1} + u_{2,\,0}]$$

$$= \frac{1}{4}\,(1414 + 1438 + 1164 + 1188) = 1301$$

Second trial (iteration) is $n = 1$. The trials are continued ($n = 0, 1, 2, 3, \ldots$) until the u values do not differ much.

(Reference: Higher Engineering Mathematics, B.S. Grewal, Khanna Publishers, Delhi)

Two-dimensional heat conduction equation in unsteady state:

$$\frac{dT}{dt} = C^2\left(\frac{d^2T}{dx^2} + \frac{d^2T}{dy^2}\right) \quad \ldots 0$$

$$T(x, y, t) = X(x)\,Y(y)\,\Theta(t)$$

The infinite rectangular bar can be formed from two infinite plates of thickness $2l_1$ and $2l_2$ respectively. For the two infinite plates, the differential equations governing temperature distribution are given by

For plate-1 of thickness $2\,l_1$:

$$\frac{dT_1}{dt} = C^2\,\frac{d^2T_1}{dx^2} \quad \ldots(1)$$

For plate-2 of thickness $2l_2$:

$$\frac{dT_2}{dt} = C^2\,\frac{d^2T}{dy^2} \quad \ldots(2)$$

The solution of Eqs. (1) and (2) are of form

$$T_1 = T_1(x, t) \quad \ldots(3)$$

and

$$T_2 = T_2\,(y, t) \quad \ldots(4)$$

$$T(x, y, t) = T_1(x, t),\, T_2\,(y, t) \quad \ldots(5)$$

From Eq. (5), we get

$$\frac{d^2T}{dx^2} = \frac{T_2 d^2T_1}{dx^2} \quad \ldots(6)$$

$$\frac{d^2T}{dy^2} = \frac{T_1 d^2T_2}{dy^2} \quad \ldots(7)$$

$$\frac{dT}{dt} = \frac{T_1 dT_2}{dt} + \frac{T_2 dT_1}{dt} \quad \ldots(8)$$

From Eqs. (2), (3), (4) and (8), we get

$$\frac{dT}{dt} = \frac{T_1 C^2 d^2T_2}{dy^2} + \frac{T_2 C^2 d^2T_1}{dx^2} \quad \ldots(9)$$

Substituting Eqs. (6), (7) and (9) in 0, we get

$$\frac{T_2C^2d^2T_1}{dx^2}+\frac{T_1C^2d^2T_2}{dy^2} = C^2\left(\frac{T_2d^2T_1}{dx^2}+\frac{T_1d^2T_2}{dy^2}\right) \quad \ldots(10)$$

The temperature distribution in a rectangular bar is given by the product of the temperature distribution in two infinite plates as

$$\left[\left(\frac{T-T_\infty}{T_i-T_\infty}\right)\right]\text{bar} = \left[\left(\frac{T-T_\infty}{T_1-T_\infty}\right)\right]\text{ plate 1} \times \left[\left(\frac{T-T_\infty}{T_i-T_\infty}\right)\right]\text{ plate 2} \quad \ldots(11)$$

The heat transfer from a two-dimensional body can be obtained by superimposing the heat transfer solutions from one-dimensional body. The heat transfer of a two-dimensional body which is obtained as an intersection of two one-dimensional bodies 1 and 2, given by :

$$\left(\frac{Q}{Q_i}\right)\text{2-}D\text{ body} = \left(\frac{Q}{Q_i}\right)_1 + \left(\frac{Q}{Q_i}\right)_2\left[1-\left(\frac{Q}{Q_i}\right)_1\right]$$

Example 4.21 A large rectangular iron bar of cross section 10 cm by 6 cm which is initially at a uniform temperature of 630 °C is exposed to an environment at 30 °C with a convection heat transfer coefficient of 500 W/m^2. Calculate the temperature at the centre of the bar at 100 seconds after it is exposed to the environment.

Solution: For iron K = 73 W/(m.K) and $C^2 = 2.034 \times 10^{-5}$ m^2/s. It is given that T_1 = 630 °C, T_∞ = 30 °C; t = 100 s and h = 500 W/(m^2.K). The rectangular iron body can be considered a two-dimensional body obtained as an intersection of (1) an infinite flat plate of thickness ($2l_1$) 10 cm, and (2) an infinite flat plate of thickness ($2l_2$) 6 cm. That is for plate 1, l_1 is 5 cm and for plate 2, l_2 = 3 cm

For plate 1:
$$F(0)_1 = \frac{C^2t}{l_1^2} = 2.034\times10^{-5}\times\frac{100}{(0.05)^2} = 0.8136$$

$$(B_i)^1 = h\times\frac{l_1}{K} = 500\times\frac{0.05}{73} = 0.3425$$

$$\left(\frac{1}{B_i}\right)^1 = 2.92$$

From the graph of B_i and F_0,

$$\left[\left(\frac{T_0-T_\infty}{T_i-T_\infty}\right)\right] = 0.85 \text{ at } F_0 = 0.8136 \text{ and } \frac{1}{B_i} = 2.92$$

For plate 2:
$$(F_0)^2 = C^2\times\frac{t}{(l_2)^2} = 2.034\times10^{-5}\times\frac{100}{(0.03)^2} = 2.26$$

$$(B_i)^2 = h\times\frac{l_2}{K} = 500\times\frac{0.03}{73} = 0.2055$$

$$\left(\frac{1}{B_i}\right)^2 = 4.87$$

From graph of $\frac{1}{B_i}$ and F_0, $\left[\frac{T_0 - T_\infty}{T_i - T_\infty}\right]^2 = 0.68$.

So, $\frac{T_0 - T_\infty}{T_i - T_\infty}$ plate-1 $\frac{T_0 - T_\infty}{T_i - T_\infty}$ plate-2 $= 0.85 \times 0.68 = 0.578$

So, $$\left[\left(\frac{T_0 - T_\infty}{T_i - T_\infty}\right)\right] = 0.578$$

Here, $T_\infty = 30$, $T_i = 630$, giving $T_0 = 376.8\,°C$ **Ans.**

Two-dimensional unsteady state heat conduction (numerical method):

$$\frac{dT}{dt} = C^2\left(\frac{d^2T}{dx^2} + \frac{d^2T}{dy^2}\right) \quad ...(1)$$

Central estimate for $\frac{dT}{dt}$ is: $\frac{(T^{p+1}m, n - T^{p-1}m, n)}{(2dt)}$ ...(2)

Forward $$\frac{dT}{dt} = \frac{\left(T^{p+1}_{m,n} - T^{p}_{m,n}\right)}{dt} \quad ...(3)$$

Backward $$\frac{dT}{dt} = \frac{\left(T^{p}_{m,n} - T^{p-1}_{m,n}\right)}{dt} \quad ...(4)$$

$$\frac{d^2T}{dx^2} = \frac{(T^p_{m+1,n} + T^p_{m-1,n} - 2T^p_{m,n})}{(dx)^2} \quad ...(5)$$

$$\frac{d^2T}{dy^2} = \frac{(T^p_{m,n+1} + T^p_{m,n-1} - 2T^p_{m,n})}{(dy)^2} \quad ...(6)$$

$$C^2\left\{\frac{[T^p_{m+1,n} + T^p_{m-1,n} - 2T^p_{m,n}]}{(dx)^2} + \frac{[T^p_{m,n+1} + T^p_{m,n-1} - 2T^p_{m,n}]}{(dy)^2}\right\} \quad ...(7)$$

Temperature at node m, n at time increment $p + 1$ is calculated, by knowing temperatures at nodes ($m, n; m + 1, n; m - 1, n; m, n + 1; m, n - 1$ at time increment p. If we take

$$dx = dy \quad ...(8)$$

Eq. (7) is $$T^{p+1}_{m,n} = \frac{c^2 dt}{(dx)^2}\left[(T^p_{m+1,n} + T^p_{m-1,n} + T^p_{m,n+1} + T^p_{m,n-1} + 1 - 4C^2\frac{dt}{(dx)^2}\right]T^p_{m,n} \quad ...(9)$$

$$T_{m,n}^{p+1} = (F_0)_\Delta(T_{m+1,n}^p + T_{m-1,n}^p + T_{m-1,n+1}^p + T_{m,n-1}^p) + (1-4(F_0)_\Delta)T_{m,n}^p \quad ...(10)$$

$F_0 = \dfrac{C^2 dt}{(dx)^2}$ is network Fourier number. We can write,

$$T_m^{p+1} = (F_0)_\Delta(T_{m+1}^p + T_{m-1}^p)[1-2(F_0)_\Delta]T_m^p \quad ...(11)$$

$$T_{m,n,l}^{p+1} = F_0\left[T_{m+1,n,l}^p + T_{m-1,n,l}^p + T_{m,n+1,l}^p + T_{m,n-1,l}^p + T_{m,n,1}^p + T_{m,n,l-1}^p\right]$$
$$+ \left[1-6(F_0)_\Delta\right]T_{m,n,l}^{p+1} \quad ...(12)$$

The values of F_0 should be chosen in such a way, $(F_0)_\Delta < \dfrac{1}{2}$ for one-dimensional body, $(F_0)_\Delta < \dfrac{1}{4}$ for a two-dimensional body and $(F_0)_\Delta < \dfrac{1}{6}$ for three-dimensional body. If we choose $(F_0)_\Delta = \dfrac{1}{4}$ i.e., two-dimensional, then Eq. (9) becomes

$$T_{m,n}^{p+1} = \frac{(T_{m+1,n}^p + T_{m-1,n}^p + T_{m,n-1}^p + T_{m,n+1}^p)}{4} \quad ...(13)$$

$$(F_0)_\Delta = \frac{1}{4} = \frac{C^2 dt}{(dx)^2}$$

or $$dt = \frac{(dx)^2}{4C^2} \quad ...(14)$$

Example 4.22 A nickel steel plate ($C^2 = 0.5 \times 10^{-5}$ m²/s) of 30 cm thick is initially at 30 °C. Suddenly, one face of the wall is raised to 670 °C, while the other face is insulated. Estimate the temperature distribution that exists in the plate 25 minutes after the surface temperature is raised.

Solution: $dx = 5$ cm; $(F_0)_\Delta = \dfrac{1}{2} = \dfrac{C^2 dT}{(dx)^2} = 0.5 \times 10^{-5} \times \dfrac{dt}{(0.05)^2}$ gives $dt = 250$ sec

Number of time increments $= 25 \times \dfrac{60}{250} = 6$.

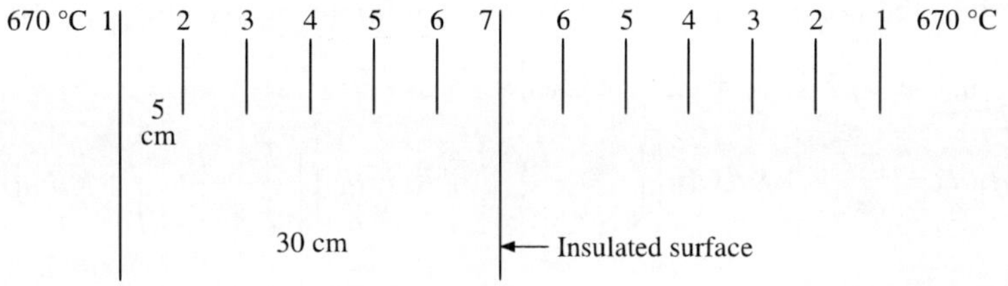

Fig. 4.4

For one-dimensional heat flow: We have

$$T_m^{p+1} = (F_0)_\Delta(T_{m+1}^p + T_{m-1}^p) + [1-2(F_0)_\Delta]T_m^p$$

Choosing $(F_0)_\Delta = \dfrac{1}{2}$, the above equation becomes,

$$T_m^{p+1} = \frac{(T_{m+1}^p + T_{m-1}^p)}{2}$$

At time $t = 0$, the surface 1 is at 670 °C and the rest of the plate (all nodes 2 to 7) is at 30 °C. The calculations are summarized in the Table below.

Temperature in °C at nodes

Increment	1	2	3	4	5	6	7	8
0	670	30	30	30	30	30	30	30
1	670	350	30	30	30	30	30	30
2	670	350	190	30	30	30	30	30
3	670	430	190	30	30	30	30	30
4	670	430	270	70	30	30	30	30
5	670	470	70	70	50	50	30	30
6	670	470	320	110	50	50	30	30

4.8.1 Three-Dimensional Heat Conduction Equation in Steady State

$$\frac{d^2T}{dx^2} + \frac{d^2T}{dy^2} + \frac{d^2T}{dz^2} = 0$$

where T is a function of x, y, z. The solution is of the form

$$\left(\frac{1}{X}\right)\frac{d^2T}{dx^2} = -\left(\frac{1}{Y}\right)\frac{d^2T}{dy^2} = -\left(\frac{1}{Z}\right)\frac{d^2T}{dz^2}$$

which are solved subject to the given boundary conditions.

In a three-dimensional problem, there will be six neighbouring nodes of any internal node. Heat flow by conduction into the nodal point O from all the neighbouring six nodes is given by

$$Q_{1-0} + Q_{2-0} + Q_{3-0} + Q_{4-0} + Q_{5-0} + Q_{6-0} = 0$$

$$Kdy\,dx\,\frac{T_1 - T_0}{dz} + Kdx\,dz\left[\left(\frac{T_2 - T_0}{dy}\right)\right] + K\,dy\,dz\left[\left(\frac{T_3 - T_0}{dx}\right)\right] + Kdz\,dx\left[\left(\frac{T_4 - T_0}{dy}\right)\right]$$

$$+ K\,dx\,dy\left[\left(\frac{T_5 - T_0}{dz}\right)\right] + K\,dx\,dy\left[\left(\frac{T_6 - T_0}{dz}\right)\right] = 0$$

If $\quad dx = dy = dz = b,\ K(T_1 - T_0 + T_2 - T_0 + T_3 - T_0 + T_4 - T_0 + T_5 - T_0 + T_6 - T_0) = 0$

$$T_1 + T_2 + T_3 + T_4 + T_5 + T_6 - 6T_0 = 0$$

4.8.2 Three-Dimensional Numerical Formulation

Consider a three-dimensional block of 3 × 4 × 4 cm with front surface exposed to convection environment with $T = 10\,°C$ and $h = 500\ W/m^2\ °C$. The four sides are maintained at 100 °C and the black surface is insulated. We choose $dx = dy = dz = 1$ cm and set up the nodes as shown. The front surface has nodes 11, 12, 13, 14, 15, 16. The next dz nodes are 21, 22, 23, 24, 25, 26 and so on. We shall use the resistance formulation in the form of equation: $T_i \dfrac{\dfrac{\Sigma T_j}{R_{ij}}}{\dfrac{1}{\Sigma R_{ij}}}$ to set up the nodal equations.

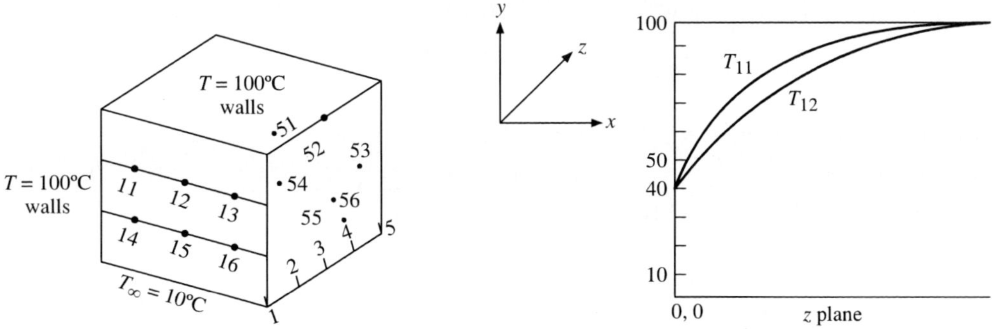

$\Delta x = \Delta y = Dz = 1$ cm

$K = 2.0$ W/(m°C)

Convection front $T_{11} = T_{12} = T_{13} = T_{14} = T_{15} = T_{16}$

Fig. 4.5

Solution: All of the interior nodes for z-planes 2, 3, 4 have resistances of $\dfrac{1}{R} = \dfrac{KA}{dx} = 2 \times \dfrac{(0.01)^2}{0.01}$ $= 0.02 = \dfrac{1}{R_{11-21}} = \dfrac{1}{R_{21-22}}$ etc.

The surface conduction resistances for surface z-plane are $\dfrac{1}{R_{11-12}} = K$ $\dfrac{A}{dx} = 2 \times 0.01 \times \dfrac{0.01}{(2 \times 0.01)} = \dfrac{1}{R_{11-14}}$ etc. The surface convection resistances are $\dfrac{1}{R_{11}} = hA =$ $500 \times (0.01)^2 = 0.05$. For surface nodes like 11 the $\sum \dfrac{1}{R_{ij}}$ term in the above equation of T_i becomes, $\sum \dfrac{1}{R_{11-j}} = 4 \times 0.01 + 0.02 + 0.05 = 0.11$.

While for the interior nodes, we have $\dfrac{1}{R_{21-j}} = 6 \times 0.02 = 0.12$. For the insulated brick surface nodes $\dfrac{1}{R_{51-j}} = 4 \times 0.01 + 0.02 = 0.06$.

There are 30 nodes total 6 in each plane. We could write equation for all of them but prefer to take advantage of symmetry of the problem.

Thus, $T_{11} = T_{13} = T_{14} = T_{16}$ and $T_{12} = T_{15}$, etc. We may then write the surface model equations as $T_{11} = [0.05\ T_\infty + 0.02\ T_{21} + 0.01\ (100 + 100 + T_{14} + T_{12})]/0.11$.

$$T_{12} = \frac{[0.05T_\infty + 0.02T_{22} + 0.01(100 + T_{11} + T_{15} + T_{13})]}{0.11}.$$

Inserting $T_\infty = 10$ and simplifying, we have

$$T_{11} = \frac{2.5 + 0.02T_{21} + 0.01\,T_{12}}{0.10}$$

$$T_{12} = \frac{1.5 + 0.02T_{22} + 0.02T_{11}}{0.10}$$

Following the same procedure, for the other Z-planes, we obtain

$$T_{21} = \frac{(200 + T_{11} + T_{31} + T_{22})}{5}$$

$$T_{22} = \frac{(100 + T_{12} + T_{32} + 2T_{11})}{5}$$

$$T_{31} = \frac{(200 + T_{31} + T_{51} + 2T_{42})}{5}$$

$$T_{32} = \frac{(100 + T_{22} + T_{42} + 2T_{31})}{5}$$

$$T_{41} = \frac{(200 + T_{31} + T_{51} + 2T_{42})}{5}$$

$$T_{42} = \frac{(100 + T_{32} + T_{52} + 2T_{41})}{5}$$

$$T_{51} = \frac{(2 + 0.02T_{41} + 0.01T_{52})}{0.05}$$

$$T_{52} = \frac{(1.0 + 0.02T_{42} + 0.02T_{51})}{0.05}$$

Solving the ten equations, gives

Z-plane	Node-1	Node-2
1	45.9	40.29
2	84.86	80.57
3	96.34	93.83
4	98.49	97.93
5	99.16	98.94

4.9 THREE-DIMENSIONAL HEAT CONDUCTION EQUATION IN UNSTEADY STATE

The solution for a three-dimensional block of sides: $2l_1$, $2l_2$, $2l_3$ can be expressed as a product of three infinite plates of: plate 1 of thickness $2l_1$, plate 2 of thickness $2l_2$ and plate 3 of thickness $2l_3$ as

$$\left[\left(\frac{T-T_\infty}{T_i-T_\infty}\right)\right]_{\text{block}} = \left[\left(\frac{T-T_\infty}{T_i-T_\infty}\right)\right]_{\text{Plate-1}} \times \left[\left(\frac{T-T_\infty}{T_i-T_\infty}\right)\right]_{\text{Plate-2}} \times \left[\left(\frac{T-T_\infty}{T_i-T_\infty}\right)\right]_{\text{Plate-3}}$$

$$\left(\frac{Q}{Q_i}\right) 3D_{\text{body}} = \left(\frac{Q}{Q_i}\right)_1 + \left(\frac{Q}{Q_i}\right)_2 \left[1-\left(\frac{Q}{Q_i}\right)_1\right] + \left(\frac{Q}{Q_i}\right)_3 \times \left[1-\left(\frac{Q}{Q_i}\right)_1\right] \times \left[1-\left(\frac{Q}{Q_i}\right)_2\right]$$

Charts are available for:

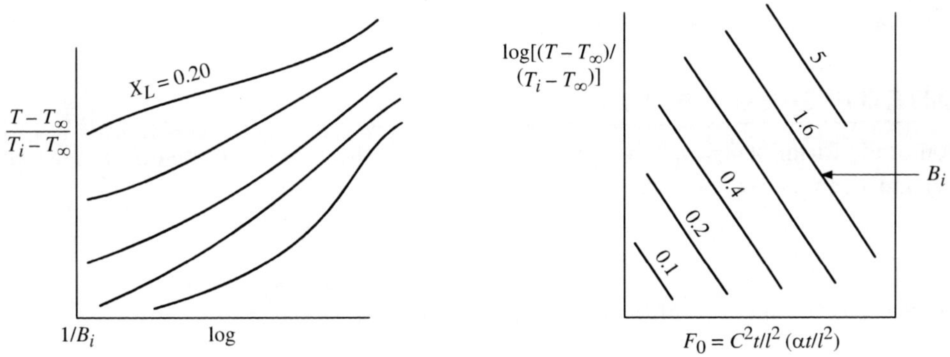

Fig. 4.6

(Ref.: Heat Transfer, Y.V.C. Rao, University Press, 2001.)

4.10 THREE-DIMENSIONAL SOLUTION OF TRANSIENT HEAT CONDUCTION

$$\frac{d^2T}{dx^2} + \frac{d^2T}{dy^2} + \frac{d^2T}{dz^2} = \left(\frac{1}{C^2}\right)\frac{dT}{dt} \qquad \text{...(1)}$$

$$T = f_1(x, t)\, f_2(y, t)\, f_3(z, t) \qquad \text{...(2)}$$

Substituting Eq. (2) in Eq. (1) gives

$$\frac{f_2 f_3 d^2 f_1}{dx^2} + \frac{f_3 f_1 d^2 f_2}{dy^2} + \frac{f_1 f_2 d^2 f_3}{dz^2} = \left(\frac{1}{C^2}\right) \frac{f_2 f_3 df_1}{dt} + \frac{f_3 f_1 df_2}{dt} + \frac{f_1 f_2 df_3}{dt}$$

Rearranging

$$f_2 f_3\left(\frac{d^2 f_1}{dx^2} - \frac{1}{C^2}\frac{df_1}{dt}\right) + f_3 f_1\left(\frac{d^2 f_2}{dy^2} - \frac{1}{C^2}\frac{df_2}{dt}\right) + f_1 f_2\left(\frac{d^2 f_3}{dz^2} - \frac{1}{C^2}\frac{df_3}{dt}\right) = 0 \qquad \text{...(3)}$$

Equation (3) is satisfied only if: $\dfrac{d^2 f_1}{dx^2} - \dfrac{1}{C^2}\dfrac{df_1}{dt} = 0$

$$\frac{d^2 f_2}{dy^2} - \frac{1}{C^2}\frac{df_2}{dt} = 0$$

$$\frac{d^2 f_3}{dz^2} - \frac{1}{C^2}\frac{df_3}{dt} = 0$$

These are three one-dimensional forms of Fourier equation. The solution for three-dimensional can be expressed as the product of the solutions of one-dimensional problem.

$$\frac{\theta_c}{\theta_i} = \frac{\theta_c}{\theta_i} A \times \frac{\theta_c}{\theta_i} B \frac{\theta_c}{\theta_i} C = \frac{T_c - T_\infty}{T_i - T_\infty}$$

from which T_c at any time can be evaluated.

4.11 FINITE DIFFERENCE EQUATIONS

An equation of the form: $\phi(x) = a_0 y_0 + a_1 y_1 + a_2 y_2 + a_3 y_3 \ldots$ + any n is a finite difference equation and $y = f(x)$ and x can take values of:

$$x_0, x_0 + h, x_0 + 2h, x_0 + 3h, \ldots x_0 + nh. \qquad \text{...(1)}$$

$$a_0, a_1, a_2, \ldots \qquad \text{...(2)}$$

and can be constants or functions of the independent variable x. $\phi(x)$ can be a constant or a function of x.

Equation (1) can be written as

$$\phi(x) = a_0 f(x_0) + a_1 f(x_1) + a_2 f(x_2) + \ldots..$$

$$\phi(x) = a_0 f(x_0) + a_1 f(x_0 + h) + a_2 f(x_0 + 2h) + \ldots + \text{an} f(x_0 + nh)$$

or

$$(a_0 + a_1 E + a_2 E^2 + a_3 E^3 + \ldots.. + a_n E^n) f(x_0) = \phi(x) \qquad \ldots(3)$$

$E = 1 + \Delta$. So Eq. (3) becomes

$$p_n \Delta^n + P_{n-1} \Delta^{n-1} + \ldots + p_1 \Delta + p_0) f(x_0) = \phi(x) \qquad \ldots(4)$$

P values are obtained from a values.

Example 4.23 Number of stages in an absorption column:

G_{N+1} moles of inert gas (dry gas) containing Y_{N+1} kg moles of solute per kg mole of dry gas is fed into the bottom of an absorption column. The solute is fed into the bottom of the absorption plate column. The solute is to be stripped from the gas by a liquid solvent of L_0 kg moles per hour of solute-free oil, sent from the top of the absorber. The composition of entering solvent is X_0 kg moles of solute/kg mole of inert solvent. The solute composition of exit gas is Y_1 kg moles of solute/ kg mole of inert gas. Calculate the number of stages (ideal) using finite difference method. Use $Y_m = K_m X_m$ equlibrium relation (Fig. 4.7).

Solution: Component (solute) balance over plate m gives. At steady state: Input of solute = Output of solute.

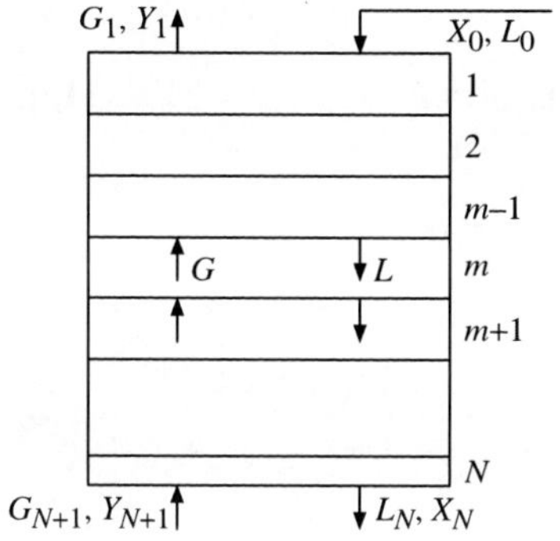

Fig. 4.7

$$L_sX_{m-1} + G_s Y_{m+1} = G_sY_m + L_sX_m \quad \text{...(1)}$$

Rearranging

$$L_s(X_{m-1} - X_m) = G_s(Y_m - Y_{m+1})$$

or
$$L_s(X_m - X_{m-1}) = G_s(Y_{m+1} - Y_m) \quad \text{...(1a)}$$

We have
$$y_m G_m = Y_m G_s$$

and
$$x_m L_m = X_m L_s \quad \text{...(2)}$$

$$Y_m = \frac{y_m G_m}{G_s} = K_m X_m \frac{L_s}{L_m} \quad \text{...(3)}$$

$$= \left(K_m x_m \frac{G_m}{G_s}\right)$$

$$Y_m = \frac{K_m X_m L_s G_m}{L_m G_s}$$

Rearranging
$$X_m = \frac{G_s}{L_s}\left(\frac{L_m}{K_m G_m}\right) Y_m \quad \text{...(4)}$$

Substituting Eq. (3) in Eq. (1) gives (eliminating X_m and X_{m-1})

$$Y_m = Y_{m+1} + \left(\frac{L_{m-1}}{K_{m-1}\, G_{m-1}}\right)\frac{Y_{m-1}}{[1 + (L_m/K_m G_m)]} \quad \text{...(5)}$$

$\frac{L_{m-1}}{K_m G_{m-1}}$ is termed absorption factor = A_{m-1} and $\frac{L_m}{K_m\, G_m} = A_m$

Equation (5) becomes in A terms as

$$Y_m = \frac{Y_{m+1} + A_{m-1} \times Y_{m-1}}{1 + A_m} \quad \text{...(6)}$$

Taking mean absorption factor in the column as A, Eq. (6) becomes

$$Y_{m+1} - (1 + A)\, Y_m + A\, Y_{m-1} = 0 \qquad \text{...(7)}$$

Using finite difference relations, $Y_{m+1} = E^2\, Y_m$, $Y_m = EY_{m-1}$. Eq. (7) becomes: $E^2 Y_{m+1} - E(1 + A)\, Y_{m-1} + A\, Y_{m-1} = 0$

Factoring: $(E - 1)\,(E - A)\, Y_{m-1} = 0$ ...(8)

The solution is

$$Y_{m+1} = C_1 + C_2\, A^{m-1} \qquad \text{...(9)}$$

or

$$Y_m = C_1 + C_2\, A^m \qquad \text{...(9}a\text{)}$$

where C_1 and C_2 are constants

$$Y_0 = KX_0 \qquad \text{...(10)}$$

Since A is assumed constant, Eq. (9) is

$$Y_0 = KX_0 = C_1 + C_2 \qquad \text{...(10}a\text{)}$$

Similarly,

$$Y_{N+1} = C_1 + C_2\, A^{N+1} \qquad \text{...(11)}$$

So,

$$Y_{N+1} - Y_0 = C_2(A^{N+1} - 1) \qquad \text{...(12)}$$

$$Y_{N+1} - Y_1 = C_2\,(A^{N+1} - A) \qquad \text{...(13)}$$

$$\left[\left(\frac{A^{N+1} - A}{A^{N+1} - 1}\right)\right] = \left[\left(\frac{Y_{N+1} - Y_1}{Y_{N+1} - Y_0}\right)\right]$$

This is *Kremser–Brown equation* used for calculating the number of stages. This is valid at low compositions. Here, N is the number of stages, $A = (L/m\ G)$ and Y is gas composition in ratio and x is liquid composition in mole fraction of solute in liquid, X is mole ratio of solute in liquid.

Example 4.24 A benzene-toluene feed containing 50 mole % benzene is sent continuously to a distillation column. The top product contains 98 mole % benzene and the bottom product contains 2 mole % benzene. The feed enters the column at its boiling point and the relative volatility of benzene to touene is 2.5. The reflux ratio is 3.0. Estimate the number of theoretical stages.

Solution: Assuming 100 kg moles of feed, D is moles of distillate and B is moles of bottom product. By material balance

$$100 = D + B \qquad \text{...(1)}$$

$$100 \times 0.50 = D \times 0.98 + B \times 0.02 \qquad \text{...(2)}$$

Solving Eqs. (1) and (2) gives:

$$D = 50 \text{ moles}$$

and

$$B = 50 \text{ moles}$$

A mass balance between the bottom of the column and some plate n in the stripping section is

$$Lx_{n+1} = Gy_n + Bx_B \qquad \text{...(3)}$$

where L is the number of moles liquid flow rate and G is the molar vapour flow rate up the column. Taking relative volatility as constant (α), the equilibrium relationship is:

$$y_n = \frac{x_n}{1+(\alpha-1)x_n} \qquad \text{...(4)}$$

Substituting Eq. (4) in Eq. (3) gives

$$Lx_{n+1} = \left[\frac{\alpha G x_n}{1+(\alpha-1)x_n}\right] + Bx_B \qquad \text{...(5)}$$

Rearranging $$x_{n+1} \times x_n + \frac{x_{n+1}}{\alpha-1} - \frac{\alpha G+(\alpha-1)\,Bx_B}{L(\alpha-1)}x_D - \frac{Bx_B}{L(\alpha-1)} = 0 \qquad \text{...(6)}$$

$$L = F + RD = (100 + 3 \times 50) = 250 \text{ kg moles/hour}$$

Let $$A = \frac{1}{(\alpha-1)} = \frac{1}{1.5} = 0.667$$

$$G = D;\ B = \alpha G + \frac{(\alpha-1)\,Bx_B}{L(\alpha-1)}$$

$$= \frac{2.5 \times 4 \times 50 + 1.5 \times 50 \times 0.02}{250 \times 1.5} = 1.337$$

$$C = \frac{Bx}{(\alpha-1)} = \frac{0 \times 0.02}{250 \times 1.5} = 0.00267$$

Equation (6) becomes

$$x_{n+1} \times x_n + Ax_{n+1} - Bx_n - C = 0 \qquad \text{...(7)}$$

Equation (7) is Ricatti equation.

The solution of Eq. (7) is obtained by the procedure:

$$z_n = \frac{1}{x_n - \delta} \qquad \text{...(8)}$$

δ is obtained from

$$\delta^2 + (A - B)\,\delta - C = 0$$

$$A - B = 0.667 - 1.337 = -0.667$$

$$\delta^2 - 0.667\,\delta - 0.00267 = 0$$

δ has roots as:

$$\delta = \frac{[0.667 \pm (0.667^2 + 4 \times 1 \times 0.00267)^{1/2}]}{2}$$

δ has values of 0.5059; –0.00395

The solution of Eq. (7) is

$$z_n = \frac{1}{x_n - \delta} = K\left[\frac{A+\delta}{B-\delta}\right]^n - \frac{1}{A+B+2\delta} \quad \text{...(9)}$$

Taking $\delta = -0.00395$, using the boundary condition at $n = 0$, $x_n = 0.02$ gives K value.

$$\frac{1}{0.02+0.00395} - \frac{1}{0.6779} = K\left[\frac{0.0667-0.00395}{1.337+0.00395}\right]^n$$

$$\frac{1}{0.02395} - \frac{1}{0.6779} = K\left(\frac{0.66305}{1.34095}\right)^n$$

$$40.275 = K(0.4945)^n$$

When $n = 0$, $K = 40.275$. The value of n for $x_n = 0.60$ is

$$\frac{1}{0.60 + 0.00395} = 40.275(0.4945)^n$$

$$0.4945^n = 0.0411$$

$$n = 4.53$$

Leaving reboiler as one theoretical stage

$n = 4.53 - 1 = 3.53$ **Ans.**

4.12 OTHER ONES

Eigenvalues by iteration. Let matrix $A = a_{jk}$. Taking a real vector X_0 (not equal to zero), with n components and compute vectors:

$$X_1 = AX_0$$

$$X_2 = AX_1; \ldots;$$

$$X_n = AX_{n-1}$$

Denoting X_{n-1} and X_n by y; $y = AX$; compute $m_0 = X^T X$; $m_1 = X^T y$. The quotient $q = \frac{m_1}{m_0}$ is an approximation for one of the eigenvalues of A (in general, for that eigenvalue which has the greatest absolute value among all eigenvalues of A).

Example 4.25 Let

$$A = \begin{pmatrix} 1 & 2 & 3 \\ 2 & 4 & 6 \\ 3 & 6 & 1 \end{pmatrix}$$

$$X_0 = \begin{pmatrix} 1 \\ 0 \\ 0 \end{pmatrix}$$

$$X_1 = AX_0 = \begin{pmatrix} 1 \\ 2 \\ 3 \end{pmatrix}$$

$$X_2 = AX_1 = \begin{pmatrix} 14 \\ 28 \\ 18 \end{pmatrix}$$

$$X_3 = AX_2 = \begin{pmatrix} 124 \\ 248 \\ 228 \end{pmatrix}$$

$$X_4 = AX_3 = \begin{pmatrix} 1304 \\ 2608 \\ 2088 \end{pmatrix}$$

Taking $X = X_3$; $Y = X_4$. We have

$$m_0 = -X^T X = \begin{pmatrix} 124 \\ 248 \\ 228 \end{pmatrix} \begin{pmatrix} 124 & 248 & 228 \end{pmatrix} = 128864$$

$$m_1 = X^T Y = \begin{pmatrix} 124 & 248 & 228 \end{pmatrix} \begin{pmatrix} 1304 \\ 2608 \\ 2088 \end{pmatrix} = 1284554$$

$$q = \frac{m_1}{m_0} = \frac{1284554}{128864} = 9.968$$

We can show that $\lambda = 10$ is an eigenvalue of A matrix.

Taking $X = X_2$ and $Y = X_3$, we obtain:

$$m_0 = 1304$$

$$m_1 = 12784$$

$$q = \frac{m_1}{m_0} = \frac{12784}{1304} = 9.80$$

4.12.1 C-Program for Newton-Raphson Method

```
# Newton-Raphson Method
# include <stdio.h>
# include <conio.h>
# include <math.h>
# define MAX (100)
float func(float x);
float func(float y);
int count = 1;
signed float xn, x0, fx, dfx, froot, root = 1, eps = 0.0001;
void main (void)
{
clr scr;
printf ("Enter the starting value of x0 …");
scanf ("%f", & x0);
while (count j = 0);
{
Fx = func(x0);
dfx = d func(x0);
xn = x0 - (fx/dfx);
count++;
if(fabs ((xn - x0)/xn) < = eps)
{
Root = xn;
froot=func(root);
printf("\n root = 0.4 f", root);
printf("\n No. of iterations = %d", count);
printf("\n f(%0.4f) = %f", root, func(root));
count = 0;
}
else
{
x0 = xn;
```

```
count ++
}
}
}
/* END OF MAIN PROGRAM
float func(float x);
{
float f;
f = (x * x * x) - 2.5 * (x * x) - 2.46 * x + 3.96;
return (f);
}
}
float defunct (float y);
{
float f;
f = 3 * y * y - 2.5 * y + 2.46;
return (f);
}
```

4.12.2 C-Program for Gauss Elimination Method

```
/* Simple Gauss Elimination Method
# include stdio.h
# include conio.h
# include math.h
# define MAX ARRAY 100
# define MAX IT 100
# define EPS 0.0001
float a (MAX ARRAY) (MAX ARRAY), b(MAX ARRAY), float Pivot, fact,
float pivot, fact, sum;
int n, i, j, k, count;
void main (void)
(
clr scr ( );
```

```
printf ("n Enter the Size of the System");
scanf ("%d", & n);
for (i = 1 = n; 1 ++)
printf ("ENTER COEFFICIENTS of a (%d) (%d) …", i);
scanf (% f, a(i)(j));
for(i = 1; i = n; i++)
(
printf ("ENTER COEFFICIENTS of b(%(d) …", i);
scanf (%f, & b(i));
)
count = 1;
(
for (k = 1; k = n - 1; k++)
(
pivot = a(k) (k);
if (pivot EPS)
(
count = 0;
return;
)
count  = 1;
for (i = k + 1; i = n, 1++)
(
fact = a(i) (k)/pivot;
for (j = k + 1; j = n; j++)
(
a(i) (j) = a(i) (j) - fact * a(k) (j);
b(i) = b(i) - fact * b(k);
)
)
x(n) = b(n)/a(n) (n);
for (k = n - 1; k = 1; k - 1)
(
sum = 0, 0;
```

```
(
sum = sum + a(k) (j) * x(j);
(
x(k) = (b(k) - sum)/a(k) (k);
for (i = 1; i = n; i++)
(
count = 0;
printf("n solution of x(%d) = %f" i, x(i)));
)
)
/* END OF MAIN PROGRAM
DATA
n
a(1, j)
b(i)
```

NUMERICAL SOLUTION EXAMPLE

A numerical solution of axial dispersion and intraparticle conduction effect on heat transfer in fixed beds. S.C. Dingra and M.K. Sarkar, Indian Chemical Engineer Journal, Section-A, Vol. 39, No. 3, July–Sep. 1997.

The equations describing dispersion, convection, and intraparticle conduction effects are as follows.

For the fluid phase:

$$\frac{dT_g}{dt} = L_{ax}\frac{d^2T_g}{dx^2} - V\frac{dT_g}{dx} - \frac{3}{R} - \frac{K_p}{c_p\rho}\frac{(1-E)dT}{dr} \quad \ldots(1)$$

For the particle:

$$\frac{dT}{dt} = \alpha_p\left(\frac{d^2T}{dr^2} + \frac{2}{r}\frac{dT}{dr}\right) \quad \ldots(2)$$

Equations (1) and (2) are coupled by the heat balance at the gas-solid interface.

$$K_p\left(\frac{dT}{dr}\right)_{r=R} = h(T_g - T_s) \quad \ldots(3)$$

The initial and boundary conditions for the step input are

$$T = T_s - T_{s1} \text{ at } t = 0 \text{ and } R > r > 0 \quad \ldots(4)$$

$$VT_{gi} = VT_g - \alpha_{ax}\frac{dT_g}{dx} \text{ at } x = 0 \text{ and } t \geq 0 \quad \ldots(5)$$

$$\frac{dT_g}{dx} = 0 \text{ at } x = L$$

and $$t \geq 0 \quad \ldots(6)$$

$$\frac{dT}{dr} = 0 \text{ at } r = 0$$

and $$t \geq 0 \quad \ldots(7)$$

The assumptions made in the above equations are: dispersed plug flow, constant thermal properties, no radial heat transfer, and no heat transfer from particle to particle by conduction. The equations in dimensionless form are:

$$\frac{dF_g}{dz} = \frac{\alpha_{ax}}{vL}\frac{d^2F_g}{dx^2}\frac{-3}{R}\frac{L}{V}\left(\frac{h}{C_p\rho}\right)\frac{1-\varepsilon}{\varepsilon}(F_g - F_s) \quad \ldots(8)$$

$$\frac{dF}{dz} = \frac{\alpha_p}{VR^2}R^2\frac{d^2F}{ds^2} + \left(\frac{2}{s}\right)\frac{dF}{ds} \quad \ldots(9)$$

$$\frac{KF}{hR}\left(\frac{dF}{dS}\right)_{S=1} = F_g - F_s \quad \ldots(10)$$

With the following initial and boundary conditions:

$$1 = \left(\frac{F_8 - \alpha_{ax}}{vL}\frac{dF_g}{dx}\right)_{x=0} \quad \text{for } z \geq 0 \quad \ldots(11)$$

$$F = F_s = 0 \text{ at } z = 0:$$

$$1 \geq s \geq 0 \text{ for } 1 \geq x \geq 0 \quad \ldots(12)$$

$$\left(\frac{dF_g}{dx}\right)_{x=1} = 0 \quad \ldots(13)$$

$$\left.\frac{dF}{dS}\right|_{s=0} = 0 \text{ for } z = 0 \quad \ldots(14)$$

where $F = \dfrac{T - T_{si}}{T_{gi} - T_{si}}$

$$F_s = \frac{T_g - T_{si}}{T_{gi} - T_{si}}$$

$$x = \frac{x}{L}$$

$$z = \frac{t}{L/V}$$

$$S = \frac{r}{R}$$

NUMERICAL SOLUTIONS

dX, dz, and dS are the increments in the lengths, time and radial dimensions respectively. The difference equation for the fluid phase obtained by approximating second order derivative term by Crank Nicholson technique and first order derivative by central difference is

$$uFg(n+1, i+1) + w\,F_g\,(n+1, i) + V\,F_g\,(n+1, i-1) + \frac{t}{2}\;F(n+1, i, k)$$

$$= -u\,F_g\,(n, i+1) + y\,F_g\,(n, i) - V\,F_g\,(n, i-1) - \left(\frac{1}{2}\right)\,F\,(n, i, k) \text{ for } i = 0, 1, \ldots \qquad \ldots(15)$$

where $$u = \frac{\alpha_{ag}}{2VL}\frac{dz}{(dx)^2} \times \frac{dz}{4dx}$$

$$W = \frac{L_{ag}}{2VL}\frac{dz}{(dX)^2} - \frac{3}{2R}\frac{L}{V} - \frac{u}{c_p p}\frac{1-\varepsilon}{\varepsilon}dz$$

$$t = \left(\frac{1}{R}\right)\left(\frac{L}{V}\right)\left(\frac{h}{c_p\rho}\right)\left(\frac{1-\varepsilon}{\varepsilon}\right)dz$$

$$V = \frac{\alpha_{ag}}{2VL}\frac{dz}{(dx)^2} + \frac{dz}{4dx}$$

$$Y = \frac{L_{ag}}{VL}\frac{dz}{(dx)^2} + \frac{3Lh}{2R\,V\,c_p p}\frac{1-\varepsilon}{\varepsilon}dz - 1$$

$$F(n, i, k) = F_g(n, i)$$

The fictitious temperatures appearing in Eq. (15) at $i = 0$ and $i = 1$ are eliminated by using the following finite difference approximation of the boundary Eqs. (11) and (13).

$$F_g(0, -1) + F_g\,(1, -1) - F_g\,(0, 1) + F_g(1, 1) + \frac{2VL\,dx}{\alpha_{ax}}\;(2F_g(0, 0) - F_g\,(1, 0)) \qquad \ldots(16)$$

$$F_g(n, i+1) = F_g(n, 1) \qquad \ldots(17)$$

The finite difference representation of the particle Eq. (9) obtained by approximating second order derivative by Crank–Nicholson and first order derivative term by central difference is

$$-\frac{M}{2}\left(1-\frac{1}{m}\right)\,F(n+1, i, m+1) + (1+M)\,F(n+1, i, m) + \frac{M}{2}\left(1-\frac{1}{m}\right)F(n+1, i, m-1)$$

$$= \frac{M}{2} - \left(1+\frac{1}{m}\right)\,F(n, 1, m+1)$$

$$+ (1-M)\,F(n, i, m) + \frac{M}{2}\left(1-\frac{1}{m}\right)\,F(n, i, m-1) \qquad \ldots(18)$$

where, $M = \dfrac{LLP}{VR^2}\dfrac{dz}{(dS \times dS)}$.

Equation (18) applies over the range $1 \le m \le K - 1$.

At the centre $m = 0$, the finite difference approximation of Eq. (9) obtained by using the condition of symmetry is

$$3MF(n + i, i, 1) + (1 + 3M)\, F(n + 1, i, 0) = 3MF\,(n, 1, 1) \qquad \ldots(19)$$

The hypothetical temperatures appearing in Eq. (18) when m is equal to K are eliminated by using the following finite difference approximation in Eq. (10).

$$\frac{KP}{2hR}\left[\frac{F(n+1, i, k+1) - F(n+1, i, K-1)}{2dS} + \frac{F(n, i, K+1) - F(n, i, K-1)}{2dS}\right]$$

$$= \frac{1}{2}\,(F_g(n + i, i) - F_g(n + 1, i) + F_g(n, i) - F_s(n, i)) \qquad \ldots(20)$$

At the surface $m = K$, the Eq. (18) becomes

$$-\frac{M}{K}\frac{dS}{p}hR\left(1+\frac{1}{K}\right)F_g\,(n+1, i) + \left[1 + M\left(\frac{dS}{KP}hR\left(1+\frac{1}{K}\right)\right)\right]$$

$$+ F(n + 1, i, K) - M\,F(n + 1, i, K - 1)$$

$$= \frac{M\,dS}{KP}hR\left(1+\frac{1}{K}\right) - F_g(n, i)$$

$$+\left[1 - M\left(1+\frac{dS}{KP}hR\left(1+\frac{1}{K}\right)\right) + M\,F(n, j, K-1 + M\,F(n, j, K-1)\right] \qquad \ldots(21)$$

The $(K + 2)$ algebraic Eqs. (15), (18), (19) and (21) describe the dispersion, convection and interparticle conduction model by finite difference approximation at the $(n + 1/2, 1)$ grid point.

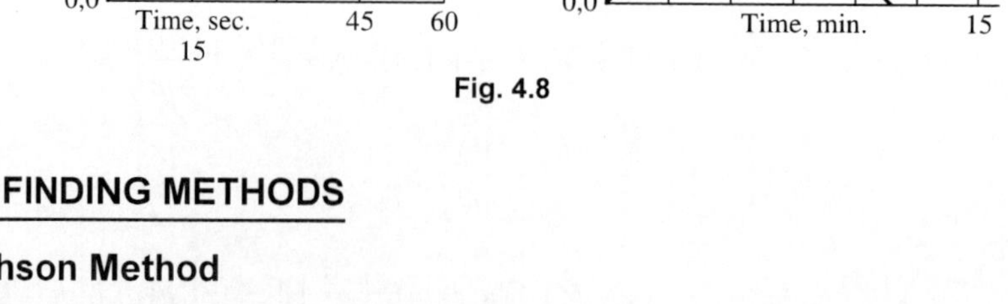

Fig. 4.8

4.13 ROOT FINDING METHODS

Newton*Raphson Method

This is based on: $f(x_i) = (x_1 - x_{I+1})\, f'(x_i)$

Rearranging: $x_{x+1} = x_i - \dfrac{f(x_i)}{f'(x_i)}$. This is termed Newton-Raphson formula.

Example: Estimate the root for: $f(x) = 3x^2 - x - 2$

$$f'(x) = 6x - 1$$

Trial 1: $i = 0$; $x_0 = 0.5$; $f(x_0) = 3.0 \times 0.25 - 0.5 - 2 = -1.75$

$f'(x_0) = 6 \times 0.5 - 1 = 2$

$$x_1 = x_0 - \frac{f(x_0)}{f'(x_0)} = 0.5 - \frac{1.75}{2} = -0.375\ (x_1)$$

Trial 2: $i = 1$; $x_1 = -0.375$; $f(x_1) = 3 \times -0.375^2 + 0.375 - 2 = 1.4844$

$f'(x_1) = 6 \times -0.375 - 1 = -2.250 - 1 = -3.25$

$$x_2 = -0.375 - \left(\frac{-1.4844}{-3.25}\right) = -0.375 - 0.4567 = -0.8317\ (x_2)$$

Trial 3: $i = 2$; $x_2 = -0.8317$; $f(x_2) = 3(-0.8317)^2 + 0.8317 - 2 = 0.9069$

$f'(x_2) = 6 \times (-0.8317) - 1 = -5.9902$

$$x_3 = -0.8317 - (0.969/-5.9902) = -0.6699$$

It reached to -0.667.

Based on starting point the other root 1.0 is reached.

This method fails for multiple roots. Modified Newton-Raphson method is used for multiple roots.

For the above example: Taking, $x_0 = 20$; $f(x_0) = 8$; $f'(x_0) = 11$; $x_1 = 1.27$

For $i = 1$; $x_1 = 1.27$; $f(x_1) = 1.569$; $f'(x_1) = 6.62$; $x_2 = 1.033$

For root is approaching 1.0.

For multiple roots the modified method is:

$$x_{i+1} = x_i - \frac{f(x_i)f'(x_i)}{f'(x)^2 - f(x)f''(x)}$$

Here also based on starting value different root is reached.

The Second Method

For some functions, to obtain the derivative gets very difficult or not easy to evaluate. For these functions, the derivative is approximated by backward finite divided difference.

$f(x_i) = \dfrac{f(x_{i-1}) - f(x_i)}{(x_{i-1} - x_i)}$. This approximation is used in the equation, giving:

$$x_{i+1} = x_i = \frac{f(x_i)(x_{i-1} - x_i)}{f(x_{i-1}) - f(x_i)}$$

The method requires two initial values of x.

Example: $f(x) = 3x^2 - x - 2$; $f'(x) = 6x - 1$

The roots are: $x = -2/3$; $x = 1.0$

Trial 1: $x_{-1} = 0.5$; $x_0 = 2.0$; $f(x_{-1}) = -1.75$; $f(x_0) = 8$

$$x_1 = x_0 - \frac{f(x_0)(x_{-1} - x_0)}{f(x_{-1}) - f(x_0)} = 2.0 - \left(\frac{8(0.5-2)}{-1.75-8}\right) = 1.77(x_1)$$

Trial 2: $x_0 = 2.0$; $f(x_0) = 8$ $x_1 = 1.77$; $f(x_1) = 5.629$

$$x_2 = x_1 - \frac{f(x_1)(x_0 - x_1)}{f(x_0) - f(x_1)} = 1.77 - \left(\frac{5.629(2.0-1.77)}{8-5.629}\right) = 1.224\ (x_2)$$

It is converging to 1.0. By changing initial value, we can reach the other root, –2/3.

Modified secant method: Instead of using two arbitrary values to estimate the derivative, an alternate method is:

$$f'(x_i) = \frac{f(x_i + \delta x_i) - f(x_i)}{\delta x_i}$$

$$x_{i+1} = x_i - \frac{\delta x_i \times f(x_i)}{f(x_i + \delta x_i) - f(x_i)}$$

Bisection Method

The method of working is:

1. Select lower x_1 and upper x_u for the root such that the function changes in sign. (say +ve to –ve and vice versa).
2. $x_f = (x_1 + x_u)/2$
3. The following decisions are taken to find where the root lies.
   (A) If $f(x_1) \times f(x_f) < 0$, then the root lies in the lower interval (x_1). Set $x_u = x_f$ and go to step 2.
   (B) If $f(x_1) \times f(x_f) > 0$, the root lies in the upper interval. Set $x_1 = x_f$ and go to step 2.
   (C) If $f(x_1) f(x_f) = 0$, the root equals to x_f. Terminate the calculation.

Example: $f(x) = 3x^2 - x - 2$

Trial 1: $x_u = 2.0$; $x_1 = 0.50$; $f(x_1) = -1.755$; $f(x_u) = 8.0$

2: $x_f = (0.50 + 2.0)/2 = 1.25$; $f(x_f) = 1.4375$

3. $f(f_u) \times f(x_1) = -1.755 \times 1.4375$ is less than zero. So $x_u = x_f$; $x_u = 1.25$

Trial 2: $x_1 = 0.50; x_u = 1.25; f(x_1) = -1.755; f(x_u) = 1.4375$

$$x_f = (0.50 + 1.25)/\ 2 = 0.875$$

$$f(x_f) = f(0.875) = -\ 0.578$$

Trial 3: $f(x_1) \times f(x_f) = -1.755 \times -\ 0.578$ is greater than zero.

So root lies in the upper interval (1.25).

$$x_u = 1.25; x_1 = x_f = 0.875; f(x_1) = -\ 0.578$$

$$x_f = (1.25 + 0.875)/2 = 1.064; f(x_f) = 0.3323$$

$$f(x_1)\,f(x_f) = -\ 0.578 \times 0.3323 = -\text{ve}$$

So root lies in lower interval, that is $x_1 = 0.875$; $x_u = 1.064$; $x_f = (0.875 + 1.064)/2 = 0.97$. So approaching 1.0 (final value). One more trial brings to 1.0.

False Position Method

The bisection method is inefficient sometimes. The magnitudes of $f(u_1)$ and $f(u_f)$ are not considered in bisection method. In false position, the slopes variation is also considered.

The equation is: $x_f = x_u - \dfrac{f(x_u)(x_1 - x_u)}{f(x_1) - f(x_u)}$

The value of x_f computed by the above equation, whichever of the two initial values, x_i or x_u. Yields a function value with same sign as $f(x_f)$. The process is repeated until the root is estimated with the desired error.

Example: $f(x) = 3x^2 - x - 2$; The roots are: 1 and –2/3.

Trial 1: $x_1 = 0.50; x_u = 2.0; f(x_1) = -1.755; f(x_u) = 8.0$

$$x_f = 2.0 - \frac{8.0(0.50 - 2.0)}{-1.755 - 8.0} = 2.0 - 1.23014 = 0.77\ (x_f)$$

$$f(x_f) = 3 \times 0.77^2 - 0.77 - 2 = -\ 0.9913$$

$$f(x_1)\,f(x_f) = -1.755 \times -\ 0.9913 = \text{Positive}$$

As an bisection method, the root lies in the upper interval.

$$x_u = 2.0; x_1 = 0.77; f(x_1) = -0.9913; f(x_u) = 8.0$$

$$x_f = 2.0 - \left(\frac{8.0(0.77 - 2.0)}{0.9913 - 8.0}\right) = 2.0 - 0.1337 = 1.863$$

Like this it is continued to reach final stage (1.0). To reach –2/3 the starting points are to be changed.

False position method fails for functions: $x^{10} - 2$ to be better than bisection method. Modified false position detects when one of the bounds is stuck. The function value at the stagnant bound divided is halved.

4.14 INTERPOLATION AND APPROXIMATION

Newton's polynomials and Lagrange polynomials. Spline interpolation bilinear interpolation.

Interpolation deals with the estimate the intermediate values between precise data points. The common method is, polynomial interpolation of form $f(x) = a_0 + a_1 x + a_2 x^2 + a_3 x^3 + ... + a_n x^n$.

Newton's Interpolating Polynomials

(a) Linear interpolation: $f(x) = a_0 + a_1 x$

$$\frac{f(x) - f(x_0)}{x - x_0} = \frac{f(x_1) - f(x_0)}{(x_1 - x_0)}$$

Rearranging: $f(x_1) = f(x_0) + \dfrac{f(x_1) - f(x_0)(x - x_0)}{(x_1 - x_0)}$

As the interval decreases, a continuous function is approximated by a straight line.

Example: Interpolating for $x = 1.5$, given at $x_1 = 2$ and at $x_0 = 1$. Function, is x^2.

$$f(1.5) = 1 + \frac{(4-1)(1.5-1)}{(2-1)} = 1 + \frac{3 \times 0.5}{1} = 2.5.$$

Actual value is $1.5^2 = 2.25$; Error is $= \dfrac{(2.5 - 2.25)}{2.5} \times 100 = 10\%$

A higher polynomial equation gives less error.

(b) Quadratic interpolation: If three data points are available then a second order polynomial equation can be used (parabola).

A convenient form is : $f_2(x) = b_0 + b_1(x + -x_0) + b_2(x - x_0)(x - x_1)$

Expanding, $f_2(x) = b_0 + b_1 x - b_1 x_0 + b_2 x_0 x_1 - b_2 \times x_0 - b_2 \times x_1 + b_2 x^2$

Collecting terms, $f_2(x) = a_0 + a_1 x + a_2 x^2$

where
$$a_0 = b_0 - b_1 x_0 + b_2 x_0 x_1$$
$$a_1 = b_1 - b_2 x_0 - b_2 x_1$$
$$a_2 = b_2$$
$$b_1 = f(x_0)$$

b_1 is obtained substituting $x = x_1$ in the starting equation,

$$b_1 = \frac{f(x_1) - f(x_0)}{(x_1 - x_0)}$$

b_2 is obtained using b_0, b_0, b_1 and $x = x_2$ in the staring equation.

$$b_2 = \frac{\frac{(f(x_2)-f(x_1))}{x_2-x_1} - \frac{(f(x_1)-f(x_0))}{x_1-x_0}}{(x_2-x_0)}$$

Example: $f(x) = x^2$

$$x_0 = 1;\ x_1 = 2;\ x_2 = 3$$

$$f(x_0) = 1;\ f(x_1) = 4;\ f(x_2) = 9$$

Use the polynomial to evaluate 2.5^2.

Solution: $b_0 = f(x_0) = 1$; $b_1 = (4 - 1)/1 = 3$; $b_2 = \frac{(9-4)}{1} - \frac{(4-1)}{1} = 2$

$$f_2(x) = b_0 + b_1(x - x_0) + b_2(x - x_0)(x - x_1)$$

$$= 1 + 3(x - 1) + 2(x - 1)(x - 2)$$

$$f_2(2.5) = 1 + 3(1.5) + 2(x - 1)(x - 2) = 7.0$$

Actual value is $2.5^2 = 6.25$; Error $= \frac{(6.25 - 7.0)}{6.25} = 11.6\%$

General form of Newton's Interpolating Polynomials

$$f_n(x) = b_0 + b_1(x - x_0) + b_2(x - x_0)(x - x_1) + \ldots + b_n(x - x_0)(x - x_1) \ldots (x - x_{n-1})$$

where $b_0 = f(x_0)$

$b_1 = f(x_1, x_0)$; $b_2 = f(x_2, x_1, x_0)$; ....

$b_n = f(x_n, x_{n-1}, \ldots\ x_1, x_0)$

The bracketed function evaluations are finite divided differences.

$$f(x_i, x_j) = \frac{f(x_i) - f(x_j)}{(x_i - x_j)}$$

$$f(x_i, x_j, x_k) = \frac{f(x_i, x_j) - f(x_j, x_k)}{(x_i - x_k)}$$

$$f(x_n, x_{n-1}, \ldots x_1, x_0) = \frac{f(x_n, x_{n-1}, \ldots\ x_1) - f(x_{n-1}, x_{n-2}, \ldots\ x_0)}{(x_n - x_0)}$$

$f_n(x) = f(x_0) + (x - x_0) f(x_1, x_0) + (x - x_0)(x - x_1) f(x_2, x_1, x_0) + \ldots + (x - x_0)(x - x_1) \ldots (x - x_{n-1})$ $f(x_n, x_{n-1}, \ldots x_0)$

Example: For function: $f(x) = x^2$

$x_i = 1\ (x_0)$ $f(x_i) = 1\ (f(x_0))$

$x_j = 2\ (x_1)$ $f(x_j) = 4\ (f(x_1))$

$x_k = 3\ (x_2) \qquad f(x_k) = 9\ (f(x_2))$

$x_1 = 4\ (x_3) \qquad f(x_1) = 16\ (f(x_3))$

$f(x_1, x_0) = (4 - 1)/1 = 3(b_1); f(x_2, x_1) = (9 - 4)/1 = 5; f(x_3, x_2) = (16 - 9)/1 = 7$

$b_2 = f(x_2, x_1, x_0) = (5 - 3)/(3 - 1) = 1$

$b_3 = f(x_3, x_2, x_1) = (7 - 5)/(4 - 2) = 1$

Third divided difference is: $f(x_3, x_2, x_1, x_0) = (1 - 1)/(4 - 1) = 0 = b_4$

$b_0 = f(x_0) = 1$

$f_3(x) = b_0 + b_1\ (x - x_0) + b_2\ (x - x_0)(x - x_1) + b_3\ (x - x_0)(x - x_1)(x - x_2)$

$f_3(x) = 1 + 3(x - 1) + 1(x - 1)(x - 2) + 1(x - 1)(x - 2)(x - 3) + 0(x - 1)(x - 2)(x - 3) \times (x - 4)$

$f(2.5) = 1 + 3(1.5) + 1(1.5)(0.5) + 1(1.5)(0.5)(-0.5) + 0 = 5.845$

Error = (6.25 – 5.845)/5.845 = 7% (less than 3 point one)

Error of Newton's interpolating polynomial is $(R_n) = f_{n+1}\ (x) - f_n\ (x)$

Lagrange Interpolating Polynomials

Represented as: $f_n(x_n) \qquad \sum_{n=0}^{n} \qquad L_i\ (x)\ f(x_i)$

where $L_i\ (x) = \prod_{\substack{j=0 \\ j \neq i}}^{n} \dfrac{(x - x_j)}{(x_i - x_j)}$ where π denotes the product of.

For example, expanding for $n = 1$:

$$f_1(x) = \frac{(x - x_1)}{(x_0 - x_1)} \quad f(x_0) + \frac{(x - x_0)}{(x_1 - x_0)} f(x_1)$$

For $n = 2$: $$f_2(x) = \frac{(x - x_1)(x - x_2)}{(x_0 - x_1)(x_0 - x_2)} \quad f(x_0) + \frac{(x - x_0)(x - x_2) f(x_1)}{(x_1 - x_0)(x_1 - x_2)} + \frac{(x - x_0)(x - x_1) f(x_2)}{(x_2 - x_0)(x_2 - x_1)} f(x_2)$$

Example: For $f(x) = x^2$

$x_0 = 1; f(x_0) = 1; f(x_1) = 2; f(x_1) = 4; x_2 = 3, f(x_3) = 9$

at $x = 2.5$

$$f(2.5) = f_1\ (2.5) = \frac{(2.5 - 2)}{(1 - 2)} \times 1 + \frac{(2.5 - 1)4}{(2 - 1)} = -0.5 + 6 = 5.5\ (6.25)$$

For $n = 3$

$$f_2(2.5) = \frac{(2.5 - 2)(-0.5)}{(-1)(-2)} \times 1 + \frac{(2.5 - 1)(2.5 - 3)}{(2 - 1)(2 - 3)} \times 4 + \frac{(2.5 - 1)(2.5 - 2)}{(3 - 1)(3 - 2)} \times 9$$

$$= 0.25/2 + 1 + (0.75 \times 9)/2 = 6.875\ (6.25)$$

Correct value is shown in bracket.

Spline Interpolation

The spline provides superior approximation of the behaviour of the functions that have local abrupt changes. The spline concept of the functions came from drafting technique of drawing smooth curve through a set of points.

Linear spones: $f(x) = f(x_0) + m_0\,(x - x_0)$ $\qquad x_0 \le x \le x_1$

$f(x) = f(x_1) + m_1\,(x - x_1)$ $\qquad x_1 \le x \le x_2$

.
.
.

$f(x) = f(x_{n-1}) + m_{n-1}\,(x - x_{n-1})$ $\qquad x_{n-1} \le x \le x_n$

where m is slope of straight line connecting the points.

$$m_1 = \frac{f(x_{i+1}) - f(x_i)}{(x_{i+1} - x_i)}$$

Spine examples are curve shapes.

Example: Given:

x	3.0	4.5	7.0	9.0
$f(x)$	2.5	1.0	2.5	0.5

Evaluate $f(x)$ at $x = 6$; $x = 6$ between 4.5 and 7.0.

***Solution*:** $m = (2.5 - 1.0)/(7.0 - 4.5) = 0.60$

$f(6) = f(4.5) + 0.60 \times (2.5 - 1.0) = 1.90$. It is coming between 1.0 and 2.5.

Quadratic Spines: This is a second order polynomial (parabola) fitting between given points. Each interval is represented by a quadratic equation between data points.

$$f_1(x) = a_1\,x^2 + b_1 x + c_1$$

For $n + 1$ points, there are n intervals and $3n$ unknown constants to evaluate. $3n$ equations or conditions are required to evaluate the unknowns.

The equations are:

(1) The functions values of the adjacent value polynomials must be equal at interior points (knots). This is in equation from is: $a_{i-1}\,x_{i-1}^2 + b_{i-1}\,x_{i-1} + c_{i-1} = f(x_{i-1})$

$$a_i\,x_{i-1}^2 + b_i\,x_{i-1} + c_i = f(x_{i-1})$$

for $i = 2$ to n. Since only interior points are used, each equation provides $(n - 1)$ conditions for a total of: $2 \times (n - 1) = 2n - 2$ conditions.

(2) The first function and last function pass through end points. This gives two conditions.

(3) The first derivatives at the interior knots must be equal.

First derivative of: $f(x) = ax^2 + 2bx + c$ used is $2xa + 2b$

so $2a_{i-1}x + b_{i-1} = 2a_i x_{i-1} + b_i$ for $i = 2$ to n.

This provides $(n - 1)$ conditions. Now total is: $2n - 2 + n - 1 + 2 = 3n - 1$.

We have $3n$ unknowns and one condition short. Taking this condition as 2nd derivative is zero at first point (i.e., $a_1 = 0$)

Example: Taking the data as:

x	3.0	4.5	7.0	9.0
$f(x)$	2.5	1.0	2.5	0.5

Using quadratic spine, find $f(6)$.

***Solution*:** We have 4 data points, so $n = 3$ intervals. $3n = 3 \times 3 = 9$ unknowns must be determined.

The (1) gives: $2n - 2$ equations, i.e.: $2 \times 3 - 2 = 4$

They are: $20.25\, a_1 + 4.5\, b_1 + c_1 = 1.0$

$$20.25\, a_2 + 4.5\, b_2 + c_2 = 1.0$$

$$49\, a_2 + 7\, b_2 + c_2 = 2.5$$

$$49\, a_3 + 7\, b_3 + c_3 = 2.5$$

The (3) gives two conditions, they are:

$$9a_1 + b_1 = 9a_2 + b_2$$

$$14a_2 + b_2 = 14a_3 + b_3$$

The (2) gives two conditions, they are:

$$9a_1 + 3b_1 + c_1 = 2.5;\ 81\, a_3 + 9b_3 + c_3 = 0.50$$

The (4) gives: $a_1 = 0$

These $3n - 1 = 3 \times 3 - 1 = 8$ equations are solved simultaneously to obtain $a_1, b_1, c_1 \ldots\ldots a_3, b_3, c_3$.

The values are: $a_1 = 0$; $b_1 = -1$; $c_1 = 5.5$; $a_2 = 0.64$; $b_2 = -6.76$; $c_2 = 18.46$; $a_3 = -1.6$; $b_3 = 24.6$; $c_3 = -91.3$

The equations are: $f_1(x) = -x + 5.5$ $\quad 3.0 \le x \le 4.5$

$$f_2(x) = 0.64x^2 - 6.76x + 18.46 \qquad 4.5 \le x \le 7.0$$

$$f_3(x) = -1.6x^2 + 24.6x - 91.3 \qquad 7.0 \le x \le 9.0$$

For $x = 6$; $f_2(x) = 0.64 \times 36 - 6.76 \times 6 + 18.46 = 0.94$ (near to correct value)

Cubic spine method works better than parabolic method.

Cubic Spines: This is a third order polynomial of form:

$$f(x) = a_1x^3 + b_ix^2 + c_ix + d_i$$

For $(n + 1)$ points, there are n intervals. So $4n$ ($n \times 4$ constants a, b, c, d) to be evaluated. $4n$ conditions are required to evaluate the unknowns.

These are:

(1) The function values are equal at interior knots $(2n - 2)$ conditions.

(2) The first and last function must pass through end points (2 conditions).

(3) The first derivatives at the interior points must be equal $(n - 1)$ conditions.

(4) The second derivatives at the interior knots are zero (2 conditions).

The above conditions give a total of $4n$ equations.

The cubic equation for each interval is:

$$f_i(x) = \frac{f_i''(x_{i-1})(x_i - x)^3}{6(x_i - x_{i-1})} + \frac{f_i''(x_i)(x_i - x_{i-1})^3}{6(x_i - x_{i-1})} + \frac{f(x_{i-1})}{(x_i - x_{i-1})} - \frac{f_i''(x_{i-1})(x_i - x_{i-1})(x_i - x)}{6}$$
$$+ \frac{f(x_i)}{(x_i - x_{i-1})} - \frac{f_i''(x_i)(x_i - x_{i-1})}{6}(x_i - x_{i-1})$$

In this equation, the unknowns second derivatives are evaluated using:

$$(x_i - x_{i-1}) f(x_{i+1}) + 2(x_{i+1} - x_{i-1}) f(x_i) + (x_{i+1} - x_i) f(x_{i+1}) =$$
$$-\frac{6}{(x_{i+1} - x_i)}(f(x_{i+1}) - f(x_i)) + \frac{6}{(x_i - x_{i-1})}(f(x_{i-1}) - f(x_i))$$

Bilinear Interpolation

This is a two-dimensional interpolation. The variables in the two dimensions are: x, y. The function (bilinear) is $f(x_i, y_i)$. A simple product approach for developing a bilinear function is: We have values at four points: $f(x_1, y_1)$, $f(x_2, y_1)$, $f(x_1, y_2)$, $f(x_2, y_2)$. We have to interpolate between $f(x_1, y_1)$ and $f(x_2, y_2)$. First, we hold y fixed and change x linearly. Using Lagrange form.

$$f(x_i, y_1) = \frac{(x_i - x_2)}{(x_1 - x_2)} f(x_1, y_1) + \frac{(x_i - x_1)}{(x_2 - x_1)} f(x_2, y_1)$$

At (x_i, y_2) is: $f(x_1, y_2) = \dfrac{(x_i - x_2)}{(x_1 - x_2)} f(x_1, y_2) + \dfrac{(x_i - x_1)}{(x_2 - x_1)} f(x_2, y_2)$

In y direction:

$$f(x_i, y_i) = \frac{(y_i - y_2)}{(y_1 - y_2)} f(x_i, y_1) + \frac{(y_i - y_1)}{(y_2 - y_1)} f(x_i, y_2)$$

Substituting $f(x_i, y_1)$ and $f(x_i, y_2)$ in $f(x_i, y_i)$ equation gives:

$$f(x_i, y_i) = \frac{(x_i - x_2)}{(x_1 - x_2)} \frac{(y_i - y_2)}{(y_1 - y_2)} f(x_1, y_1) + \frac{(x_i - x_1)}{(x_2 - x_1)} \frac{(y_1 - y_2)}{(y_1 - y_2)} f(x_2, y_1)$$
$$+ \frac{(x_i - x_2)(y_i - y_1)}{(x_1 - x_2)(y_2 - y_1)} f(x_1, y_2) + \frac{(x_i - x_1)(y_i - y_2)}{(x_2 - x_1)(y_2 - y_1)} f(x_2, y_2)$$

Example: Given the temperatures of a two-dimensional plate (heated plate) as:

$$T(2, 1) = 70; T(9, 1) = 67.5$$

$$T(2, 6) = 65; T(9, 6) = 80$$

Estimate temperature at $x_i = 5$ and $y_i = 4$.

Substituting these given values in $f(x_i, y_i)$ equation:

$x_i = 5;\quad y_i = 4\quad f(x_1, y_1) = 70\quad f(x_1, y_2) = 65$

$x_1 = 2;\quad y_1 = 1\quad f(x_2, y_1) = 67.5\quad f(x_2, y_2) = 80$

$x_2 = 9\quad y_2 = 6$

$$f(5, 4) = \frac{(5-9)\ (4-6)}{(92-9)\ (1-6)} \times 70 + \frac{(5-2)(4-6)}{(9-2)(1-6)} \times 67.5 + \frac{(5-9)(4-1)}{(2-9)(6-1)} \times 65 + \frac{(5-2)(4-1)}{(9-2)(6-1)} \times 80$$

$$= 70.4 \text{ Ans.}$$

EXERCISES

1. Derive an expression for the evaluation of constants m and c of the straight line $y = ax + c$ by the method of least squares which is fitted to n data points. $(x_1, y_1), (x_2, y_2), \ldots (x_n, y_n)$.
2. Solve the following equations by Gauss–Seidel method:

$$8x - 3y + 2z = 20$$

$$6x + 3y + 12z = 35$$

$$4x + 11y - z = 33$$

3. Use the power method to find the largest eigenvalue of the matrix.

$$A = \begin{bmatrix} 1 & 1 & -1 \\ 1 & 2 & 1 \\ 2 & -1 & 1 \end{bmatrix}$$

4. Find the eigenvalues and corresponding eigenvectors of

$$\begin{bmatrix} 5 & -2 & 0 \\ -2 & 6 & 2 \\ 0 & 2 & 7 \end{bmatrix}$$

5. Fit a parabola: $(y = a + bx + c\,x^2)$ for the following data.

x	1	2	3	4	5
y	1250	1400	1650	1950	2300

6. (a) Derive the Newton's forward interpolation formula.

   (b) Given that $f(20) = 24; f(24) = 32; f(28) = 35; f(32) = 40$

   Find $f(25)$ using Newton's forward interpolation formula.

7. Find $y(0.08)$ using Adam's predictor method and corrector method.

   Given that: $\frac{dy}{dx} = x^2 + y; y(0) = 1$, by solving initial values $y(0.02)$, $y(0.04)$, $y(0.06)$ by Runge–Kutta method by taking $h = 0.02$.

8. Solve $y'' = x + y$. For $y(0.25)$, $y(0.50)$, $y(0.75)$ by taking $h = 0.25$ by shooting method. Given that $y(0) = 0$, $y(1) = 1$ and initial guesses for $y_1'(0)$ and $y_2'(0)$ are 6 and 8.

9. Solve by Galerkin method: $\frac{d^2y}{dx^2} + y + x = 0; y(0) = 0; y(1) = 0$

10. Use Gerschgorin circle theorem to estimate the eigenvalues of the matrix.

$$\begin{bmatrix} 10 & 1 & 0 & 1 \\ 0.2 & 8 & 0.2 & 0.2 \\ 1 & 1 & 2 & 1 \\ 1 & -1 & -1 & -11 \end{bmatrix}$$

11. Determine the largest eigenvalue and corresponding eigenvector of the matrix.

$$\begin{bmatrix} 2 & 1 & 0 \\ -1 & 2 & -1 \\ 0 & 1 & 2 \end{bmatrix}$$

12. Given the differential Eq. $\frac{8d^2y}{dx^2} - \frac{x^2dy}{dx} + 2y^2 = 0$ and the initial condition that at $x = 0$, $y = 1$ and $\frac{dy}{dx} = 0$. Find the values of y when $x = 2$ and $x = 3$.

13. Using any numerical method, above the differential equation, $\frac{dy}{dx} = x + 0.1\,y^2; x(0) = 1.8$, and $y(0) = 0.x$ varies from 1.8 to 3 with increment of 0.3 (dx).

14. Using any numerical method given above, solve the differential equation, $\frac{dy}{dx} = x + 0.1y^2$; $x_0 = 0$, $y_0 = 0.50$ in the interval (0, 1) taking $dx = 0.20$.

15. Derive Simpson's rule for numerical interation. The following table shows the instantaneous rate of flow of crude oil through a pipeline. The flow rate (ϕ) is given in thousands of Kg/h. The time t is given in minutes.

ϕ	2.7	2.1	1.8	1.8	2.0	2.3	2.6	3.0	3.6	4.2	4.7
t	0	12	24	36	48	60	72	84	96	108	120

Using Simpson's rule, calculate the total quantity of oil flowing in 120-minute period covered by the data.

16. Given the differential equation: $\dfrac{d^2y}{dx^2} = y^2 + xy + x^2$ and the initial conditions that at $x = 1$, $y = 4$ and $\dfrac{dy}{dx} = 5$. Find the value of y when $x = 1.5$ and $x = 2.0$.

17. The differential equation: $\dfrac{10dy}{dx} = x^2 + y^2$ is to pass through the points $x = 0$; $y = 1$. Find the values of y corresponding to x from 0 to 0.50 with 0.10 increment.

18. Solve $y'' = \dfrac{1}{2}(x + y + y' + 2)$; $y(0) = 0$; $y'(0) = 0$. Choosing $dx = 0.20$ for $x = 0$ to 1.0.

❑❑❑

5

Laplace Transforms and Solutions, Other Mathematical Methods

Laplace Transforms, Its Applications

$f(t)$ Function	$f(s)$ Laplace Transform
1	$\frac{1}{S}$
t	$\frac{1}{S^2}$
$\frac{t^{n-1}}{n-1!}$	$\frac{1}{S^n}$
$\frac{1}{\sqrt{\pi t}}$	$\frac{1}{(S)^{1/2}}$
$2\sqrt{\frac{t}{\pi}}$	$\frac{1}{S^{3/2}}$
e^{at}	$\frac{1}{S-a}$
te^{at}	$\frac{1}{(s-a)^2}$
$\frac{1}{(n-1)!}t^{n-1}e^{at}$	$\frac{1}{(S-a)^n}$
$\frac{1}{(S^2+a^2)}$	$\left(\frac{1}{a}\right)\sin at$

(Contd...)

$\frac{S}{S^2+a^2}$	$\cos at$
$\frac{1}{S^2-a^2}$	$\left(\frac{1}{a}\right)\sinh at$
$\frac{S}{S^2-a^2}$	$\cosh at$
$\frac{1}{(S-a)^2+\omega^2}$	$\left(\frac{1}{\omega}\right)e^{at}\sin\omega t$
$\frac{S-a}{(S-a)^2+\omega^2}$	$\left(\frac{1}{\omega}\right)e^{at}\cos\omega t$
$\frac{1}{S(S^2+\omega^2)}$	$\left(\frac{1}{\omega^2}\right)(1-\cos\omega t)$
$\frac{1}{2}\frac{(e^{at}-e^{bt})}{(\pi t^3)^{1/2}}$	$\sqrt{(S-a)}-\sqrt{(S-b)}$
$J_0(at)$	$\frac{1}{(S^2+a^2)^{1/2}}$
$\frac{1}{\sqrt{\pi t}}e^{at}(1+2at)$	$\frac{S}{(S-a)^{3/2}}$
$\frac{t}{2\omega}\sin\omega t$	$\frac{S}{(S^2+\omega^2)^2}$
$\frac{1}{\omega^2}(1-\cos\omega t)$	$\frac{1}{S(S^2+\omega^2)}$
$\frac{1}{\omega^2}(\omega t-\cos\omega t)$	$\frac{1}{S^2(S^2+\omega^2)}$
$\frac{1}{2\omega}(\sin\omega t+\omega t\cos\omega t)$	$\frac{S^2}{(S^2+\omega^2)^2}$
$\frac{t^{K-1}e^{at}}{\Gamma(K)}$	$\frac{1}{(S-a)^K}$
$\frac{1}{(a-b)(ae^{at}-be^{bt})}$	$\frac{S}{(S-a)(S-b)}$
$\left(\frac{1}{\omega}\right)\sin\omega t$	$\frac{1}{(S^2+\omega^2)}$

5.1 LAPLACE TRANSFORM METHODS

$$L = \int_0^\infty e^{-St}\,dt$$

where L is Laplace transform operator.

$$Lf(t) = \int_0^\infty f(t)\,e^{-St}\,dt = f(S)$$

$$f(t) = L^{-1} f(S)$$

1. $f(t) = 1$

$$Lf(t) = \int_0^\infty e^{-St} f(t)\,dt = \int_0^\infty e^{-St}\,1\,dt = \left.\frac{e^{-St}}{S}\right|_0^\infty = -\frac{0}{S} - \left(-\frac{1}{S}\right) = \frac{1}{S}$$

2. $L(x^n)$

$$f(t) = x^n; \quad Lf(t) = \int_0^\infty e^{-St} x^n\,dt = \int_0^\infty e^{-St} t^n\,dt \qquad \text{(Here } x = t\text{)}$$

Putting $St = p,\ dp = Sdt,\ t = p/S$

$$= \int_0^\infty \frac{p^n}{S^n} e^{-p} S^{-1}\,dp = \int_0^\infty p^n \frac{e^{-p}\,dp}{S^{n+1}} = \frac{\Gamma(n+1)}{S^{n+1}}$$

For n to be positive integer $\Gamma(n+1) = n!$

$$f(S) = \frac{n!}{S^{n+1}}$$

3. $L(e^{ax})$ $f(t) = e^{at}$

$$f(S) = \int_0^\infty e^{at} e^{-St}\,dt = \int_0^\infty e^{-(S-a)t}\,dt$$

$$= \left.-\frac{1}{(S-a)} e^{-(S-a)t}\right|_0^\infty = \frac{1}{(S-a)}$$

4. $L(\sin at)$; $\quad f(t) = \sin at$

$$f(S) = \int_0^\infty \frac{e^{-St}(e^{iat} - e^{-iat})\,dt}{2i}; \quad \sin at = \frac{(e^{iat} - e^{-iat})}{2i}$$

$$\cos \omega t = \frac{(e^{iat} + e^{-iat})}{2}; \quad \sinh at = \frac{(e^{at} - e^{-at})}{2}$$

$$\cosh at = \frac{(e^{at} + e^{-at})}{2}; \; f(s) = \frac{1}{2i}\left(\int_0^\infty e^{-(S-ia)t}dt - \int_0^\infty e^{-(S+ia)t}dt\right)$$

$$= \frac{1}{2i}\left(-\frac{e^{-(S-ia)t}}{(S-ia)}\Bigg|_0^\infty + \frac{e^{-(S+ia)t}}{(S+ia)}\Bigg|_0^\infty\right) = \frac{1}{2i}\left(\frac{1}{(S-ia)} - \frac{1}{(S+ia)}\right)$$

$$= \frac{1}{2i}\left(\frac{2ia}{S^2+a^2}\right) = \frac{a}{(S^2+a^2)}$$

5. $L(\cos at)$; $\quad f(t) = \cos at$

where a is a constant term.

$$f(S) = \int_0^\infty \frac{e^{-St}(e^{iat} + e^{-iat})\,dt}{2} = \frac{1}{2}\left(\int_0^\infty e^{-St}(e^{iat} + e^{-iat})\,dt\right)$$

$$= \frac{1}{2}\left(\int_0^\infty e^{-(S-ia)t} + \int_0^\infty e^{-(S-ia)t}\right)dt = \frac{1}{2}\left(-\frac{1}{S-ia} + \frac{1}{S+ia}\right)$$

$$= \frac{2S}{2(S^2+a^2)} = \frac{S}{S^2+a^2} \quad \textbf{Ans.}$$

6. $L(t \sin at)$

$$f(t) = t \sin at; \; \sin at = \frac{e^{iat} - e^{-iat}}{2i}; \; L(t) = \frac{1}{S^2}$$

$$L\left(\frac{t(e^{iat} - e^{-iat})}{2i}\right) = L\left(\frac{te^{iat}}{2i}\right) - L\left(\frac{te^{-iat}}{2i}\right)$$

$$L\left(\frac{te^{iat}}{2i}\right) = \frac{1}{(S-ia)^2 2i}; \quad L\left(\frac{te^{-iat}}{2i}\right) = \frac{-1}{(S+ia)^2\, 2i}$$

Adding both: $\dfrac{1}{2i(S-ia)^2} - \dfrac{1}{2i(S+ia)^2} = \dfrac{4i\,aS}{2i(S^2+a^2)} = \dfrac{2aS}{(S^2+a^2)}$

7. $L(e^{-3t}(2\cos 5t - 3\sin 5t)) = 2L(e^{-3t}\cos 5t) - 3L(e^{-3t}\sin 5t)$

$$= \frac{2(S+3)}{(S+3)^2+5^2} - \frac{3S}{(S+3)^2+5^2} = \frac{2S-9}{S^2+6S+34}$$

8. $Lf(t) = t/T$ when $0 < t < T$

$= 2$ when $t > T$

$$Lf(t) = \int_0^T e^{-St}\frac{t}{T}\,dt + \int_T^\infty e^{-St}2\,dt = \frac{1}{T}\int_0^T te^{-St}\,dt + 2\int_T^\infty e^{-St}\,dt$$

Using $u\,dV = uv - V\,du$ relation.

$$= \frac{1}{T}\left(\frac{te^{-St}}{S}\Bigg|_0^T - \int_0^T \frac{e^{-St}}{S}\,dt\right) + 2\int_0^\infty e^{-St}\,dt$$

$$= \frac{1}{T}\left(\frac{Te^{-St}}{-S} - 0 - \frac{e^{-St}}{S^2}\Bigg|_0^T\right) + \left(\frac{0-e^{-ST}}{-S}\right)^2$$

$$= -\frac{e^{-St}}{S} - \frac{(e^{-ST}-1)}{TS^2} + \frac{2e^{-ST}}{S}$$

$$f(S) = \frac{1-e^{-ST}}{TS^2} - \frac{e^{-ST}}{S} \quad \textbf{Ans.}$$

9. $f(x)$ $\quad f(t) = \cos\left(t - \frac{2\pi}{3}\right)$ for $t > \pi/3$

$$= 0 \quad \text{for } t < 2\pi/3$$

where T is period.

$$Lf(t) = \int_0^T \frac{e^{-St} f(t)\,dt}{(1-e^{-ST})} \quad \text{for a periodic function of period } T.$$

$$Lf(t) = \int_0^T e^{-St} f(t)\,dt + \int_T^{2T} f(t)\,e^{-St}\,dt + \int_{2T}^{3T} f(t)\,e^{-St}\,dt + \ldots$$

In the second integration putting $t = u + T$, in the third $t = u + 2T$

Then, $$L(f(t)) = \int_0^T e^{-St} f(t)\,dt + \int_0^T e^{-S(u+T)} f(u+T)\,du + \int_0^T e^{-S(u+2T)} f(u+2T)\,du + \ldots$$

$$= \int_0^T e^{-St} f(t)\,dt + e^{-ST}\int_0^T e^{-Su} f(u)\,du + e^{-2ST}\int_0^T e^{-Su} f(u)\,du$$

Since $f(u) = f(u+T) = f(u+2T)$ etc.

$$= \int_0^T e^{-St} f(t)\,dt + e^{-ST}\int f(t)\,e^{-St}\,dt + e^{-ST}\int e^{-St} f(t)\,dt$$

$$= (1 + e^{-ST} + e^{-2ST} + \ldots)\int_0^T e^{-St} f(t)\,dt$$

$$= \int_0^\infty \frac{e^{-St} f(t)\,dt}{(1-e^{-ST})}$$

10. Laplace transform of the periodic function,

$$f(t) = \sin \omega t, \quad 0 < t < \pi/\omega$$
$$= 0, \quad \pi/\omega < t < 2\pi/\omega$$

Since $f(t)$ is a periodic function with period of $2\pi/\omega$,

$$Lf(t) = \frac{1}{1-e^{-2\pi S/\omega}} \int_0^{2\pi/\omega} e^{-St} f(t)\, dt$$

$$= \frac{1}{1-e^{-2\pi S/\omega}} \left(\int_0^{\pi/\omega} e^{-St} \sin \omega t\, dt + \int_{\pi/\omega}^{2\pi/\omega} e^{-St} \times 0 \times dt \right)$$

$$= \frac{\omega e^{-\pi S/\omega} + \omega}{(1-e^{-2\pi S/\omega})(S^2+\omega^2)} = \frac{\omega}{(1-e^{-\pi S/\omega})(S^2+\omega^2)}$$

11.
$$LJ_0(x) = L\left(1 - \frac{x^2}{2^2} + \frac{x^4}{2^2 \cdot 4^2} - \ldots\right)$$

$$LJ_0(x) = \frac{1}{S} - \frac{1 \times 2!}{2^2 S^3} + \frac{1 \times 4}{2^2 \cdot 4^2 \cdot S^5}$$

$$= \frac{1}{S}\left(1 - \frac{1}{2}\frac{1}{S^2} + \frac{1 \cdot 3}{2 \cdot 4}\frac{1}{S^4} - \frac{1 \cdot 3 \cdot 5}{2 \cdot 4 \cdot 6}\frac{1}{S^6} + \ldots\right)$$

$$= \frac{1}{S}\left(1 + \frac{1}{S^2}\right)^{-1/2} = \frac{1}{(S^2+1)^{1/2}} \quad \textbf{Ans.}$$

$$J_0'(x) = -J_1(x)$$

$$L(Erf(x)) = \frac{2}{\pi}\int_0^{\sqrt{x}} e^{-t^2}\, dt = \frac{2}{\sqrt{\pi}}\left(x^{1/2} - \frac{x^{3/2}}{3} + \frac{x^{5/2}}{5 \cdot 2}\right)$$

$$= \frac{2}{\sqrt{\pi}}\int_0^{\sqrt{x}} \left(1 - t^2 + \frac{t^4}{2!} - \frac{t^6}{3!} + \ldots\right) dt$$

$$= \frac{2}{\sqrt{\pi}}\left(x^{1/2} - \frac{x^{3/2}}{3} + \frac{x^{5/2}}{5 \cdot 2} - \frac{x^{7/2}}{7 \cdot 3} + \ldots\right)$$

Taking L.T.: $\frac{1}{S^{3/2}} - \frac{1}{2S^{5/2}} + \frac{1}{2} \times \frac{3}{4}\frac{1}{S^{7/2}} \ldots$

$$= \frac{1}{S^{3/2}}\left(1 - \frac{1}{2S} + \frac{1}{2} \times \frac{3}{4}\frac{1}{S^2} + \ldots\right)$$

$$= \frac{1}{S^{3/2}}\left(1 - \frac{1}{S}\right)^{-1/2} = \frac{1}{S(S^2+1)^{1/2}}$$

12. Laplace transforms of unit step function (Figure 5.1)

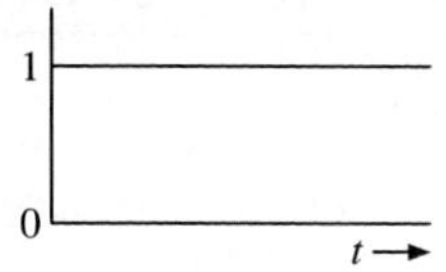

Fig. 5.1

$$L\,(u(t)) = \int_0^\infty e^{St}\, f(t)\, dt$$

$$= \int_0^\infty e^{St}\, 1\, dt = -\frac{1}{S} e^{-St}\Big|_0^\infty = \frac{1}{S}$$

5.1.1 Laplace Transforms of Unit Impulse Function

If a is the magnitude of impulse and b is the time duration, then impulse amount is: $a \times b$

Laplace transform of this impulse is ab

Laplace transform of sine function (Fig. 5.2)

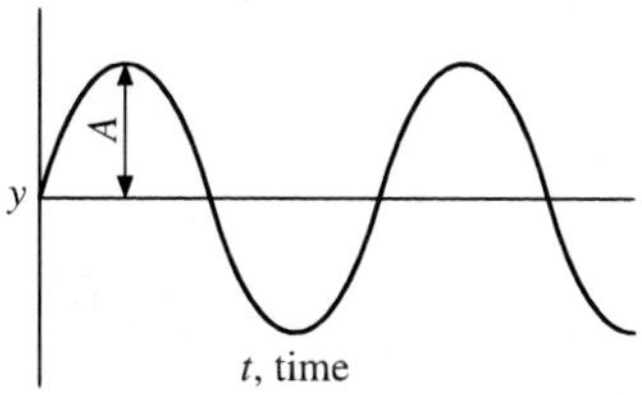

Fig. 5.2

$$L\,[y(t)] = L\,(A \sin \omega t) = y(S) = \frac{LA\,(e^{i\omega t} - e^{-i\omega t})}{2i}$$

$$y(S) = \frac{1}{2}\left(\int_0^\infty e^{-St} e^{i\omega t}\, dt - \int_0^\infty e^{-St} e^{-i\omega t}\, dt\right)$$

$$= \frac{1}{2i}\left(\int_0^\infty e^{-(S-i\omega)t}\, dt - \int_0^\infty e^{-(S+i\omega)t}\, dt\right)$$

$$= \frac{1}{2i}\left(\frac{-e^{-(S-i\omega)\,t}}{(S-i\omega)}\Big|_0^\infty + \frac{e^{-(S+i\omega)t}}{(S+i\omega)}\Big|_0^\infty\right)$$

$$= \frac{1}{2i}\left(\frac{1}{(S-i\omega)} - \frac{1}{(S+i\omega)}\right)$$

$$= \frac{1}{2i}\,\frac{2ia}{S^2+\omega^2} = \frac{a}{(S^2+\omega^2)}$$

5.2 LAPLACE TRANSFORM OF UNIT STEP FUNCTION WITH LAG TIME

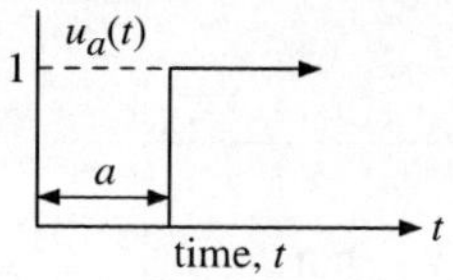

Fig. 5.3

The Laplace transform of $u_a(t)$ is

$$L\,(u_a(t)) = \int_0^\infty e^{-St} \times u_a(t)\,dt = \int_0^a e^{-St}\,0\,dt + \int_a^\infty e^{-St}\,1\,dt = -\frac{1}{S}e^{-St}\Big|_a^\infty = \frac{e^{-aS}}{S}$$

13. $$\mathrm{L}\,(f''(t)) = S^2 f(S) - Sf(0) - f'(0)$$

$$\int_0^\infty e^{-St} f''(t)\,dt\ ;\ \text{Integrating by parts, } d(uV) = u\,dV + V\,du$$

$$\int V\,du = uV - \int u\,dV = uV - \int u\,dV\,;$$

Here, $V = e^{-St}$, $\quad dV = -Se^{-St}\,dt;\ du = f;\ u = f'$

$$e^{-St} f'(t)\Big|_0^\infty - \int_0^\infty Sf'\,e^{-St}\,dt = -f'(0) - S\int_0^\infty e^{-St} f'(t)\,dt$$

Using $\quad V = e^{-St},\ dV = -Se^{-St}\,dt;\ du = f';\ u = f = -f'(0) + S\left(fe^{-St} + \int f\,Se^{-St}\,dt\right)$

$$= -f'(0) - Sf(t)\,e^{-St}\Big|_0^\infty + \int_0^\infty S^2 f(t)\,e^{-St}\,dt = S^2 f(S) - f'(0) - Sf(0).$$

14. Inverse Laplace transforms:

Example 5.1 Inverse Laplace transforms of: (*a*) $\log\left(\frac{1+S}{S}\right)$; (*b*) $\frac{1}{S^2(S^2+a^2)}$; (*c*) $\frac{S}{(S+a)^2}$; (*d*) $\frac{S}{S^2(S+2)}$

Solution: (a) $$\log\left(\frac{1+S}{S}\right) = \log\,(1+S) - \log S = \int_S^\infty \frac{dS}{S+1} - \int_S^\infty \left(\frac{1}{S}\right) dS;$$

$$L\left(\frac{f(t)}{t}\right) = \int_S^\infty f(S)\,dS$$

So, $$f(S) = -\frac{1}{S}.$$

Hence, $f(t) = -1;$

$$L^{-1}\left(-\int_{S}^{\infty}\left(\frac{1}{S}\right)dS\right) = L^{-1}\left(\left(-\frac{1}{S}\right)1\right) = -1$$

$$L^{-1}\left(\int_{S}^{\infty}\frac{dS}{S+1}\right) = L^{-1}\left(\frac{1}{S+1}\right) = e^{-t};$$

$$\int_{S}^{\infty}\frac{dS}{S+1} - \int_{S}^{\infty}\frac{1}{S}\,dS = \frac{e^{-t}-1}{t}$$

Example 5.2 Inverse Laplace transform of: $\dfrac{1}{S^2(S^2+a^2)}$

Solution: $\left(\frac{1}{S^2} - \frac{1}{S^2+a^2}\right)\frac{1}{a^2} = \left(\frac{1}{a^2}\right)\left(L^{-1}\left(\frac{1}{S^2}\right)\right) - L^{-1}\left(\frac{1}{S^2+a^2}\right)$

$$= \left(\frac{1}{a^2}\right)\left(t - \frac{1}{a}\sin at\right) = \frac{t}{a^2} - \frac{1}{a^3}\sin at$$

Example 5.3 Laplace inverse: $\dfrac{S}{(S+a)^2}$.

Solution: Inverse of $\dfrac{1}{(S+a)^2}$ is te^{-at}

$$S \text{ is differential: } \frac{d}{dt}(te^{-at}) = e^{-at} - tae^{-at}$$

Example 5.4 Inverse Laplace transform of: $\dfrac{1}{S^2(S+2)}$.

Solution: By partial fractions:

$$\frac{A}{S} + \frac{B}{(S+2)} + \frac{C}{S^2} = A\,(S+2)\,S^2 + BS^2 + CS\,(S+2) = 1$$

For $S = 0$, no solution.

For $S = -2$, $4\ B = 1;\ B = \frac{1}{4}$

For $S = 1$ gives $3A + 3C = \frac{3}{4}$

For $S = -1$, $A - C = \frac{1}{4}$

Solving both gives $A = \frac{1}{2},\ C = -\frac{1}{4}$ and $B = \frac{1}{4}$

$$\frac{1}{(S+2)S^2} = \frac{1}{2S} + \frac{1}{S+2} - \frac{1}{4S^2}$$

Taking inverse Laplace transforms of: $\frac{1}{2S} + \frac{1}{S+2} - \frac{1}{4S^2}$ gives: $\frac{1}{2} - e^{-2t} - \frac{1}{4}t$ **Ans.**

Example 5.5 Find inverse Laplace transform of: $\frac{2S+18}{S^2+9} + \frac{24}{S^3} - \frac{30\,(S)^{1/2}}{S^4}$

Solution: Using the background of: $L\,(\sin at) = \frac{a}{S^2+a^2}$

$$L\,(\cos at) = \frac{S}{S^2+a^2};\quad L\left(\frac{1}{S^n}\right) = \frac{t^{n-1}}{n!}$$

$$\frac{1}{S^{n+1/2}} = \frac{2^n t^{n-1/2}}{1\cdot 3\cdot (2n-1)\sqrt{\pi}}$$

$$L^{-1}\left(\frac{2S}{S^2+9}\right) = 2\cos 3t;\quad L^{-1}\left(\frac{18}{S^2+9}\right) = 6\sin 3t;$$

$$L^{-1}\left(\frac{24}{S^3}\right) = 4t^3;\quad L^{-1}\frac{30}{(S^3)^{1/2}} = \frac{8t^{5/2}\times 30}{15\sqrt{\pi}}$$

$$y\,(t) = 2\cos 3t + 6\sin 3t + 4t^3 - \frac{16t^{5/2}}{\sqrt{\pi}} \quad \textbf{Ans.}$$

5.3 SOLVING EQUATIONS USING LAPLACE TRANSFORMS

Example 5.6 Solve: $\frac{d^2y}{dx^2} + 4y = 3;\ y(0) = y'\,(0) = 1.$

Solution:

$$L\left(\frac{d^2y}{dx^2}\right) = S^2y(S) - Sy(0) - y'(0) = S^2y(S) - S - 1$$

$$L\,(y) = y(S);\quad L(3) = \frac{3}{S}$$

$$S^2y(S) - S - 1 + 4y(S) = \frac{3}{S};\quad y(S) = \frac{3+S+S^2}{S(S^2+4)}$$

$$y(S) = \frac{3}{S(S^2+4)} + \frac{1}{(S^2+4)} + \frac{S}{S^2+4}$$

$$L\,(\sin ax) = \frac{a}{S^2+a^2};\quad L\,(\cos ax) = \frac{S}{S^2+a^2}$$

$$\frac{1}{S^2+4} = \frac{1}{2}\left(\frac{2}{S^2+2^2}\right) = \frac{1}{2}\sin 2x\,; \quad \frac{S}{S^2+4} = \cos 2x$$

$$\frac{3}{S(S^2+4)} = 3\int_0^x \frac{1}{2}\sin 2x\,dx = \frac{3}{4}\times(-\cos 2x)\Big|_0^x = \frac{3}{4}(1-\cos 2x)$$

$$y(t) = \cos 2x + \frac{1}{2}\sin 2x + \frac{3}{4}(1-\cos 2x)$$

Example 5.7 Using Laplace transforms, solve the following differential equation:

$$y'' + 2y' + 5y = e^t \sin t;\ y(0) = 0,\ y'(0) = 1$$

Solution:

$$L(y'') = S^2y(S) - Sy(0) - y'(0) = S^2y(S) - 1$$

$$L(2y') = 2Sy(S) - 2y(0) = 2Sy(S);\ L(y) = y(S)$$

$$L(e^t \sin t) = \frac{1}{(S+1)^2+1}$$

The given differential equation in Laplacian form is

$$S^2y(S) - 1 + 2Sy(S) + 5y(S) = \frac{1}{(S+1)^2+1}$$

$$y(S) = \frac{S^2+2S+3}{(S^2+2S+2)(S^2+2S+5)}$$

$$= \frac{1}{3}\left(\frac{1}{(S+1)^2+1^2}\right) + \frac{2}{3}\left(\frac{1}{(S+1)^2+2^2}\right)$$

Taking Inverse Laplace transforms

$$y(t) = \frac{1}{3}e^{-t}\sin t + \frac{1}{3}e^{-t}\sin 2t = \frac{1}{3}e^{-t}(\sin t + \sin 2t) \quad \textbf{Ans.}$$

Example 5.8 Solving unsteady state heat conduction equation in a semi-infinite medium.

Solution: The unsteady state heat conduction equation is

$$\frac{dT}{dt} = C^2\frac{d^2T}{dx^2} \qquad \ldots(1)$$

where T is temperature;

t is time;

x is distance;

C^2 is a constant.

At $t = 0$, $T = T_0$ for all x values.

Taking Laplace transforms on both sides

$$L\left(C^2\frac{d^2T}{dx^2}\right) = ST(S) - T = C^2\frac{d^2T(S)}{dx^2} \qquad \ldots(2)$$

$$L\left(\frac{dT}{dt}\right) = ST(S) - T(0) \qquad \ldots(3)$$

The given equation becomes:

$$C^2\frac{d^2T(S)}{dx^2} = ST(S) - T(0) \qquad \ldots(4)$$

Rearranging: $\dfrac{d^2T(S)}{dx^2} - \dfrac{S}{C^2}T(S) = -\dfrac{T_0}{C^2}$; Taking $q^2 = \dfrac{S}{C^2}$

$$(D^2 - q^2)\,T(S) = -\frac{T_0}{C^2}\text{ ; Complimentary solution is } D^2 - q^2 = 0$$

$$T\,(\text{C.S.}) = Ae^{-qx} + Be^{qx}$$

Particular integral is

$$\frac{-T_0}{-C^2q^2\left(1-\dfrac{D^2}{q^2}\right)} = \frac{T_0}{C^2}\frac{C^2}{S}\left(1-\frac{D^2}{q^2}\right)^{-1} = \frac{T_0}{S}\left(1+\frac{D^2}{q^2}+\frac{D^4}{q^4}+\ldots\right) = \frac{T_0}{S} \qquad \ldots(5)$$

Total solution: $$T(S) = Ae^{-qx} + Be^{qx} + \frac{T_0}{S} \qquad \ldots(6)$$

The constants A and B are calculated by boundary conditions.

At $x = 0$; $$T = T_1 \qquad \ldots(7)$$

$$T(S) = \frac{T_1}{S} \qquad \ldots(8)$$

Substituting in Eq. (6) $\dfrac{T_1}{S} = A + \dfrac{T_0}{S}$; $A = \dfrac{T_1 - T_0}{S}$

As x approaches infinity, T is finite, so $B = 0$

$$T = \frac{T_1 - T_0}{S}e^{-\sqrt{S/C^2}\,x} + \frac{T_0}{S} \qquad \ldots(9)$$

By inverse Laplace transform, Eq. (9) becomes:

$$T = (T_1 - T_0)\operatorname{erfc}\left(\left(\frac{x}{2}C\right)t\right) + T_0 \qquad \ldots(10)$$

Example 5.9 Laplace transforms in chemical reaction engineering:

(a) In chemical reactions: $A \xrightarrow{K_1} R \xrightarrow{K_2} S$

$$\frac{dC_A}{dt} = -K_1 C_A \qquad \text{...(1)}$$

$$\frac{dC_R}{dt} = K_1 C_A - K_2 C_R \qquad \text{...(2)}$$

$$\frac{dC_S}{dt} = K_2 C_R \qquad \text{...(3)}$$

Initial conditions are $C_A = C_{AO};\ C_{RO} = C_{BO} = 0$

Taking Laplace transforms of Eq. (1), we get

$$SC_A - C_{AO} = -K_1 C_A(S);\ C_{AO}(S) = \frac{C_{AO}}{S + K_1} \qquad \text{...(4)}$$

Taking Laplace transforms of Eq. (2):

$$SC_R(S) = K_1 C_A(S) - K_2 C_R(S);\ C_R(S) = \frac{K_1 C_A(S)}{(S + K_2)} \qquad \text{...(5)}$$

Substituting C_A from Eq. (4) gives

$$C_R = \frac{K_1 C_{AO}}{(S + K_1)(S + K_2)}$$

By partial fractions:

$$C_R = \frac{K_1 C_{AO}}{K_2 - K_1}\left(\frac{1}{S + K_1} - \frac{1}{S + K_2}\right) \qquad \text{...(6)}$$

Taking inverse Laplace transforms of Eq. (4) gives

$$C_A(t) = C_{AO}\, e^{-K_1 t} \qquad \text{...(7)}$$

Taking inverse Laplace transforms of Eq. (6) gives

$$C_R(t) = \frac{K_1 C_{AO}}{K_2 - K_1}(e^{-K_1 t} - e^{-K_2 t}) \qquad \text{...(8)}$$

C_S is obtained by the relation

$$C_A + C_B + C_S = C_{AO};\ C_S = C_{AO} - C_A - C_B \qquad \text{...(9)}$$

When $K = K_1 = K_2$, we get

$$C_R = \frac{KC_{AO}}{(S + K)^2} \qquad \text{...(10)}$$

Taking inverse Laplace transform of Eq. (10)

$$C_R = KC_{AO}t\, e^{-Kt}$$

(b) $A \xrightarrow{K_1} R \underset{K_{21}}{\overset{K_2}{\rightleftarrows}} S$ **chemical reaction**

The rate equations are

$$\frac{dC_A}{dt} = -K_1 C_A \qquad \text{...(11)}$$

$$\frac{dC_R}{dt} = K_1 C_A - K_2 C_R + K_{21} C_S \qquad \text{...(12)}$$

$$\frac{dC_S}{dt} = K_2 C_R - K_{21} C_S \qquad \text{...(13)}$$

With initial conditions: $C_A = C_{AO}$; $C_{BO} = C_{CO} = 0$; K_{21} is the rate constant for the reverse reaction,

$$S \longrightarrow R.$$

Taking Laplace transform of Eq. (11) is

$$SC_A - C_{AO} = -K_1 C_A;\; C_A = \frac{C_{AO}}{S + K_1}$$

Taking Laplace transforms of Eq. (12)

$$SC_R = K_1 C_A - K_2 C_R + K_{21} C_S \qquad \text{...(14)}$$

Taking Laplace transforms of Eq. (13)

$$SC_S = K_2 C_R - K_{21} C_S;\; C_S = \frac{K_2 C_R}{S + K_{21}} \qquad \text{...(15)}$$

Substituting Eqs. (14), and (15) in Eq. (13), we get

$$C_R = K_1 C_{AO} \frac{(S + K_{21})}{[S(S + K_1)(S + K_2 + K_{21})]}$$

In partial fractions,

$$C_R = \frac{K_2 C_{AO}}{(K_1 + K_{21})S} + \frac{(K_1 - K_2)C_{AO}}{(K_2 + K_{21} - K_1)(S + K_1)} + \frac{K_1 K_2 C_{AO}}{(K_2 + K_{21})(K_1 - K_2 - K_{21})} \times \frac{1}{(S + K_2 + K_{21})} \qquad \text{...(16)}$$

Taking inverse Laplace transforms of Eq. (16)

$$C_R = \frac{K_2 C_{AO}}{(K_2 + K_{21})} + \frac{(K_1 - K_2)C_{AO}}{(K_2 + K_{21} - K_1)} e^{-K_1 t}$$

$$+\frac{K_1K_2C_{AO}\,e^{-(K_2+K_{21})t}}{(K_2+K_{21})(K_1-K_2-K_{21})} \quad \ldots(17)$$

(c) For the reaction: $A \underset{K_{11}}{\overset{K_1}{\rightleftarrows}} R \xrightarrow{K_2} S$

$$\frac{dC_A}{dt} = -K_1C_A + K_{11}C_R \quad \ldots(18)$$

$$\frac{dC_R}{dt} = K_1C_A - K_{11}C_R - K_2C_R \quad \ldots(19)$$

$$\frac{dC_S}{dt} = K_2C_R \text{ from Eq. (3)}$$

Taking Laplace transforms for the above differential equations

$$SC_A - C_{AO} = -K_1C_A + K_{11}C_R, \text{ giving}$$

$$C_A = \frac{C_{AO}}{(S+K_1)} + \frac{K_{11}C_R}{(S+K_1)} \quad \ldots(20)$$

$$SC_R = K_1C_A - K_{11}C_R - K_2C_R \quad \ldots(21)$$

(Since $C_{RO} = 0$)

$$SC_S = K_2C_R \text{ (since } C_{SO} = 0) \quad \ldots(22)$$

After substituting C_A from Eq. (20) in Eq. (21) and rearranging gives

$$C_R = \frac{K_2C_{AO}}{[S^2 + (K_1 + K_{11} + K_2)\,S + K_1K_2]}$$

or
$$C_R = \frac{K_2C_{AO}}{[(S-a)\,(S-b)]} \quad \ldots(23)$$

a, b are the roots of the equation of: $S^2 + (K_1 + K_{11} + K_2)S + K_1K_2 = 0$

$$a = \left[\frac{-(K_1+K_{11}+K_2)+((K_1+K_2+K_{11})^2-4K_1K_2)^{1/2}}{2}\right]$$

$$b = \frac{-(K_1+K_{11}+K_2)-[(K_1+K_{11}+K_2)^2-4K_1K_2]^{1/2}}{2}$$

Taking inverse Laplace transforms of Eq. (23)

$$C_R = K_2C_{AO}\,(a-b)\,(e^{at} - e^{bt}) \quad \ldots(24)$$

C_A is obtained by substituting C_R from Eq. (23) into Eq. (20) and taking inverse Laplace transforms,

$$C_A = \frac{C_{AO}}{(S+K_1)} + \frac{K_1C_R}{(S+K_1)} = \frac{C_{AO}}{(S+K_1)} + \frac{K_1K_{11}C_{AO}}{(S-a)\,(S-b)\,(S+K_1)}$$

$$C_A = \frac{C_{AO}}{S+K_1} + \frac{K_1K_{11}C_{AO}}{(a-b)(a+K_1)(S-a)} + \frac{1}{(b-a)(b+K_1)(S-b)} + \frac{1}{(K_1-a)(K_2-b)(S+K_1)}$$

Taking inverse Laplace transforms,

$$C_A = C_{AO}e^{-K_1t} + \frac{K_1K_{11}C_{AO}e^{at}}{(a-b)(a+K_1)} + \frac{e^{bt}}{(b-a)(b+K_1)} + \frac{e^{-K_1t}}{(K_1-a)(K_2-b)} \quad ...(25)$$

C_S is obtained from Eq. (9) by using C_R from Eq. (24) C_A from Eq. (25).

(d) $A \underset{K_{11}}{\overset{K_1}{\rightleftarrows}} R \underset{K_{21}}{\overset{K_2}{\rightleftarrows}} S$

$$\frac{dC_A}{dt} = -K_1C_A + K_{11}C_R \text{ from Eq. (18)}$$

$$\frac{dC_R}{dt} = K_1C_A - K_{11}C_R - K_2C_R + K_{21}C_S \quad ...(26)$$

$$\frac{dC_S}{dt} = K_2C_R - K_{21}C_S \quad ...(27)$$

Taking Laplace transforms of the above equations.

$$SC_A(S) - C_{AO} = -K_1C_A(S) + K_{11}C_R(S)$$

$$C_A(S) = \frac{C_{AO}}{(S+K_1)} + \frac{K_{11}C_R}{(S+K_1)} \quad ...(28)$$

$$SC_S - C_{SO} = K_2C_R(S) - K_{21}C_S(S)$$

$$C_S(S) = \frac{K_2C_R(S)}{(S+K_{21})} \quad ...(29)$$

$$SC_R(S) - C_{RO} = K_1C_A - K_{11}C_R - K_2C_R + K_{21}C_S$$

$$C_R(S) = \frac{K_1C_A}{(1+K_{11}+K_2)} + \frac{K_{21}C_S}{(1+K_{11}+K_2)}$$

Substituting $C_A(S)$ and $C_R(S)$ from Eqs. (28) and (29),

$$C_R(S) = \frac{K_1C_{AO}(S+K_{21})}{S[S^2+(K_1+K_{11}+K_2+K_{21})S+(K_1K_2+K_{11}K_{21}+K_1K_{21})]}$$

or

$$C_R(S) = \frac{K_1C_{AO}(S+K_{21})}{[(S-a)S(S-b)]} \quad ...(30)$$

a and b are the roots of the quadratic equation

$$S^2 + (K_1 + K_2 + K_{11} + K_{21})S + K_1K_2 + K_{11}K_{21} + K_1K_{21} = 0$$

$$\text{Here, } a = \frac{-(K_1 + K_2 + K_{11} + K_{21}) + [(K_1 + K_{11} + K_2 + K_{21})^2 - 4 \times 1 \times (K_1K_2 + K_{11}K_{21} + K_1K_{21}]^{1/2}}{2}$$

$$b = \frac{-(K_1 + K_2 + K_{11} + K_{21}) - [(K_1 + K_2 + K_{11} + K_{21})^2 - S \times 1 \times (K_1K_2 + K_{11}K_{21} + K_1K_{21})]^{1/2}}{2}$$

After expanding in partial fractions,

$$C_R(S) = K_1C_{AO}\left[\frac{K_2}{abS} + \frac{(a+K_{21})}{a(a-b)(S-a)} + \frac{(b+K_{21})}{b(b-a)(S-b)}\right] \quad \ldots(31)$$

Taking inverse Laplace transforms of Eq. (31)

$$C_R = K_1C_{AO}\left[\frac{K_{21}}{abS} + \frac{(a+K_{21})\,e^{at}}{a(a-b)} + \frac{(b+K_2)\,e^{bt}}{b(b-a)}\right] \quad \ldots(32)$$

Substituting from Eq. (30) in Eq. (28), we get

$$C_A = \frac{C_{AO}}{S-K_1} + \frac{K_{11}K_1C_{AO}(S+K_2)}{S(S+K_1)(S-a)(S-b)} \quad \ldots(33)$$

The second term is expanded in partial fractions as

$$K_1K_{11}C_{AO}\left(\begin{array}{l}\dfrac{K_{21}}{K_1abS} + \dfrac{(K_1 - K_{21})}{K_1(K_1-a)(K_1-b)(S+K_1)} \\ + \dfrac{(a+K_{21})}{a(a+K_1)(a-b)(S-a)} + \dfrac{(b+K_2)}{b(b+K_1)(b-a)(S-b)}\end{array}\right) \quad \ldots(34)$$

Taking inverse Laplace transforms of Eq. (31)

$$C_R = K_1C_{AO}\left(\frac{K_2}{ab} + \frac{(a+K_{21})\,e^{at}}{a(a-b)} + \frac{(b+K_{21})\,e^{bt}}{b(b-a)}\right) \text{ from Eq. (32)}$$

Substituting from Eq. (30) (C_S) in Eq. (28), we get

$$C_A = \frac{C_{AO}}{(S+K_1)} + \frac{K_{11}K_1C_{AO}(S+K_2)}{S(S+K_1)(S-a)(S-b)} \quad \ldots(33)$$

The expansion of second term is given in Eq. (34)

Taking inverse Laplace transforms, we get

$$C_A = C_{AO}\,e^{-K_1t} + K_1K_{11}C_{AO}\left(\begin{array}{l}\dfrac{K_{21}}{K_1ab} + \dfrac{(K_1 - K_2)\,e^{K_1t}}{K_1(K_1-a)(K_1-b)} \\ + \dfrac{(a+K_{21})\,e^{at}}{a\,(a+K_1)(a-b)} + \dfrac{(b+K_{21})\,e^{bt}}{b\,(b+K_1)(b-a)}\end{array}\right) \quad \ldots(35)$$

It is seen that when $K_{11} = 0$ and $K_{21} = 0$; C_A becomes $C_{AO}e^{-K_1t}$

$$C_S = C_{AO} - C_A - C_R \text{ from Eq. (9)}$$

C_S is given by Eq. (9) and C_R by Eq. (32). C_A by Eq. (35)

(e) Parallel Reactions: Refer Fig. 5.4

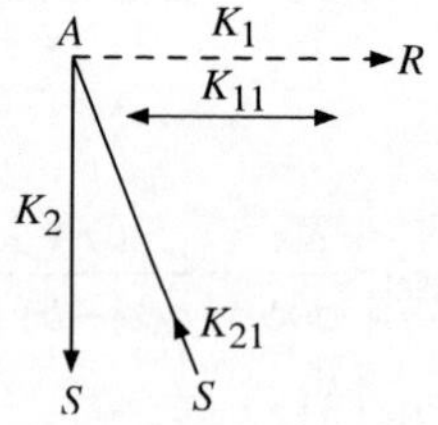

Fig. 5.4

$$\frac{dC_A}{dt} = -K_1C_A + K_{11}C_R - K_2C_A + K_{21}C_S \quad \ldots(36)$$

$$\frac{dC_R}{dt} = K_1C_A - K_{11}C_R \quad \ldots(37)$$

$$\frac{dC_S}{dt} = K_2C_A - K_{21}C_S \quad \ldots(38)$$

Initial conditions are: $C_A = C_{AO}$; $C_{RO} = C_{SO} = 0$ at time $t = 0$

Taking Laplace transforms of the above equations

$$SC_A - C_{AO} = -(K_1 + K_2)\, C_A + K_{11}C_R + K_{21}C_S \quad \ldots(39)$$

$$S_{CR} - C_{RO} = K_1C_A - K_{11}C_R$$

or
$$C_R = \frac{K_1C_A}{S + K_{11}} \quad \ldots(40)$$

Similarly,
$$C_S = \frac{K_2C_A}{(S + K_{21})} \quad \ldots(41)$$

Substituting Eq. (40) and Eq. (41) in Eq. (39)

$$SC_A - C_{AO} = -(K_1 + K_2)\, C_A + \frac{K_{11}K_1C_A}{S + K_{11}} \frac{+K_{21}K_2C_A}{S + K_{21}} \quad \ldots(42)$$

Solving for C_A: $C_A = \dfrac{C_{AO}(S + K_{11})(S + K_{21})}{S\,[S^2 + S(K_1 + K_2 + K_{11} + K_{21}) + K_1K_{21} + K_2K_{11} + K_{11}K_{21}]}$ ...(43)

Taking a, b as the roots of the quadratic equation,

$$(S^2 + S\,(K_1 + K_2 + K_{11} + K_{21}) + K_1K_{21} + K_2K_{11} + K_{11}K_{21}) = 0$$

$$a = \frac{-(K_1 + K_2 + K_{11} + K_{21}) + [(K_1 + K_2 + K_{11} + K_{21})^2 - 4 \times 1 \times (K_1K_2 + K_2K_{11} + K_{11}K_{21})]^{1/2}}{2}$$

$$b = \frac{-(K_1 + K_2 + K_{11} + K_{21}) - [(K_1 + K_2 + K_{11} + K_{21})^2 - 4 \times 1 \times (K_1K_{21} + K_2K_{11} + K_{11}K_{21})]^{1/2}}{2}$$

We get, $$C_A = \frac{C_{AO}(S + K_{11})(S + K_{21})}{(S - a)\, S\, (S - b)} \qquad \ldots(44)$$

After putting in partial fractions

$$C_A = \frac{C_{AO}K_{11}K_{21}}{abS} + \frac{C_{AO}(a + K_1)(a + K_2)}{a(a - b)(S - a)} + \frac{C_{AO} \times (b + K_{11})(b + K_{21})}{b(b - a)(S - b)} \qquad \ldots(45)$$

Taking inverse Laplace transforms of Eq. (45) gives

$$C_A = \frac{C_{AO}K_{11}K_{21}}{ab} + \frac{C_{AO}(a + K_{11})(a + K_{21})e^{at}}{a(a - b)} + \frac{C_{AO}(b + K_{11})(b + K_{21})\, e^{bt}}{b(b - a)} \qquad \ldots(46)$$

Substituting Eq. (44) in Eq. (40) gives

$$C_R = \frac{K_1C_{AO}\,(S + K_{21})}{(S(S - a)(S - b))} \qquad \ldots(47)$$

Expressing Eq. (47) in partial fractions,

$$C_R = \frac{C_{AO}K_1K_{21}}{abS} + \frac{C_{AO}K_1(b + K_{21})}{b(b - a)(S - b)} + \frac{C_{AO}K_1(a + K_{21})}{a(a - b)(S - a)} \qquad \ldots(48)$$

After taking inverse Laplace transforms of Eq. (48):

$$C_R = \frac{C_{AO}K_1K_{21}}{ab} + \frac{C_{AO}K_1(a + K_{21})e^{at}}{a(a - b)} + \frac{C_{AO}K_1(b + K_{21})\, e^{bt}}{b(b - a)} \qquad \ldots(49)$$

$$C_S = \frac{K_2C_{AO}(S + K_{11})}{[S(S - a)(S - b)]} \qquad \ldots(50)$$

After partial fractions, we get

$$C_S = \frac{C_{AO}K_2K_{11}}{abS} + \frac{C_{AO}K_2(a + K_{11})}{a(a - b)(S - a)} + \frac{C_{AO}K_2(b + K_{11})}{b(b - a)(S - b)} \qquad \ldots(51)$$

Taking inverse Laplace transforms of Eq. (51)

$$C_S = \frac{C_{AO}K_2K_{11}}{ab} + \frac{C_{AO}K_2(a + K_{11})\, e^{at}}{a(a - b)} + \frac{C_{AO}K_2(b + K_{11})e^{bt}}{b(b - a)} \qquad \ldots(52)$$

Reference: *Indian Chemical Engineer Journal*, A. Sood, Section-A, Vol. 46, No. 1, Jan–Mar. 2004.

Example 5.10 Laplace transform method in Tubular Flow Reactor: Transient behaviour of a wall catalysed tubular flow reactor.

Mathematical analysis: The irreversible reaction, A → Products is considered at the wall of the tubular reactor. The assumptions are:

1. The physical properties such as diffusivity remain constant throughout the reactor.
2. Isothermal conditions prevail through the reactor.
3. The velocity is uniform across the cross section of the reactor.

Solution: The reactant enters at uniform concentration of C_0. Initially, the system is at steady state condition or the fluid concentration is zero. At time $t = 0$, the reactant starts flowing in and simultaneously, the reaction occurs at the catalytic wall. (Fig. 5.5):

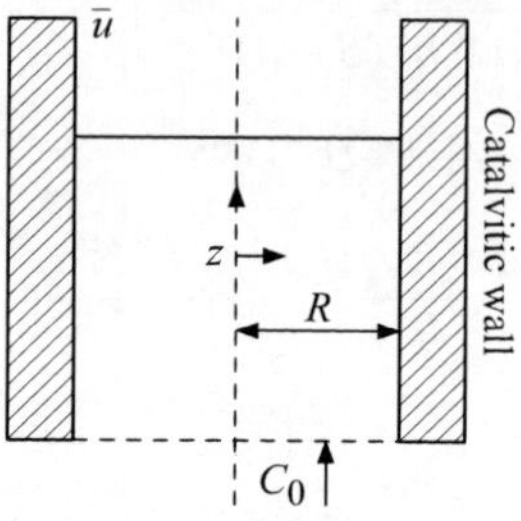

Fig. 5.5

The partial differential equation describing the unsteady state mass balance, for the reactor can be written as:

The boundary and initial conditions are

$$C(r, 0, t) = C \quad \ldots(1)$$

$$C_0\,(r, z, 0) = 0 \quad \ldots(2)$$

$$\frac{dC}{dt} + \bar{u}\frac{dC}{dz} = D\left(\frac{d^2C}{dr^2} + \frac{1}{r}\frac{dC}{dr} + \frac{d^2C}{dz^2}\right) \quad \ldots(1a)$$

$$-D\frac{dC}{dr}(R, z, t) = KC\,(R, z, t);\; C\,(0, z, t) \text{ is finite and } C\,(r, \infty, t) \text{ is finite} \quad \ldots(2a)$$

The following dimensionless groups are introduced.

$$\omega = \frac{C}{C_0};\; \xi = -\frac{r}{R};\; \zeta = \frac{z}{R};$$

$$\tau = \frac{Dt}{R^2};\; Pe = 2\bar{u}\frac{R}{D} \quad \ldots(3)$$

Equations ($1a$) and ($2a$) become, in terms of the dimensionless variables,

$$\frac{d\omega}{dT} + \frac{Pe}{2}\frac{d\omega}{d\zeta} = \frac{d^2\omega}{d\xi^2} + \frac{1}{\xi}\frac{d\omega}{d\xi} + \frac{d^2\omega}{d\zeta^2} \quad \ldots(4)$$

$$w(\xi, 0, \tau) = 1; \; w(\xi, \zeta, 0) = 0;$$

$$\frac{dw}{d\xi} = (1, \zeta, \tau) = \beta\omega; \; w(\xi, \infty, T) \text{ is finite} \qquad \ldots(5)$$

$$B = K\frac{R}{D} \text{ is Damkohler number.}$$

Case 1: Axial diffusion included

Equation (4) together with B.Cs and I.Cs (Eq. (5)) are solved by the combination of the Laplace transform and separation of variable techniques.

Defining the Laplace transform with respect to r as

$$\omega_S = \int_0^\infty e^{-ST}\, \omega\, d\tau \qquad \ldots(6)$$

and taking the Laplace transform of Eq. (4) with respect to τ as

$$S\omega_S + \frac{Pe}{2}\frac{d\omega_S}{d\zeta} = \frac{d^2\omega_S}{d\xi^2} + \frac{d^2\omega_S}{d\zeta^2} + \left(\frac{1}{\xi}\right)\frac{d^2\omega_S}{d\xi^2} \qquad \ldots(7)$$

The boundary conditions now become

$$\omega_S = (\xi, 0) = \frac{1}{S};$$

$$\frac{-d\omega_S}{d\xi}(1, \xi) = \beta w_S(1, \xi); \; w_S(0, \xi) \text{ is finite}; \; w_S(\xi, \infty) \text{ is finite} \qquad \ldots(8)$$

Solving Eq. (7) by separating the variables and using the finiteness of the concentration as $\xi \to \infty$, the solution is

$$\omega_S = (C_1 J_0(\lambda\xi) + C_2 Y_0(\lambda\xi) \exp\left(-\frac{1}{4}(Pe^2 + 16(S + \lambda^2)^{1/2} - Pe)\xi\right) \qquad \ldots(9)$$

From the condition that ω_S is finite at, $\xi = 0$, $C_2 = 0$ since Y_0 tends to become infinite as $\xi \to 0$.

From the conditions at the wall, values are given by the transcendental equation,

$$BJ_0(\lambda) - J_1(\lambda) = 0 \qquad \ldots(10)$$

There are infinite values of satisfying Eq. (10), hence the solution of Eq. (7) becomes,

$$\omega_S = B_1 J_0(\lambda_1\xi)\, \text{Exp}\left(-\frac{1}{4}(Pe^2 + 19(S + \lambda_1^2)^{1/2} - Pe)\xi\right) \qquad \ldots(11)$$

From the condition at the entrance, that $\xi = 0$, $\omega_S = \frac{1}{S}$ and the orthogonal property of the Bessel function, the constants B_1 are given by

$$B_1 = \frac{\beta^2}{S\lambda_1 J_1(\lambda_1)(\lambda^2 + \beta^2)} \qquad \text{...(12)}$$

The transform of the mean or cup mixing concentration is given by

$$\omega_S = 2\int_0^1 \omega_S \, \xi \, d\xi \qquad \text{...(13)}$$

Combining Eqs. (11), (12) and (13)

$$\omega_S = 2\frac{\beta^2}{3}\sum_{i=1}^{\infty} \frac{1}{\lambda_1^2(\lambda_1^2 + \beta^2)} \text{Exp}\left(-\frac{1}{4}(Pe^2 + 16(S + \lambda_1^2)^{1/2} - Pe)\xi\right) \qquad \text{...(14)}$$

The inverse Laplace transform of Eq. (14) gives the mean concentration $\bar{w}$ as a function of ξ and τ.

$$\omega = \sum_{i=1}^{\infty} \frac{2\beta^2}{1} \frac{1}{\lambda_1^2(\lambda_1^2 + \beta^2)} \int_0^{\tau} \xi \exp\left(\frac{-1}{4}(Pe^2 + 16(S + \lambda_1^2)^{1/2} - Pe)\xi\right) dr \qquad \text{...(15)}$$

On performing the inverse transformation, the radial mean concentration is given by

$$\omega = \sum_{i=1}^{\infty} \frac{4B^2 e^{Pe}/4}{(\pi\lambda_1^2(\lambda_1^2 + \beta^2))^{1/2}} \int_0^{\tau} \text{Exp}\left(\frac{\xi^2}{4\pi^2} + \left(\left(\frac{Pe^2}{16} + \lambda_1^2\right)r^2\right)\right) T'^{-1.5} dT' \qquad \text{...(16)}$$

Case 2: No axial diffusion

The term $\frac{d^2C}{dz^2}$ is neglected in Eq. (1), the dimensional form of equation is

$$\frac{d\omega}{d\tau} + \frac{Pe}{2}\frac{d\omega}{d\xi} = \frac{d^2\omega}{d\xi^2} + \frac{1}{\omega}\frac{d\omega}{d\xi} \qquad \text{...(17)}$$

The initial and boundary conditions are the same as in Eq. (5). Equation (17) is solved by the double Laplace transform method.

Defining Laplace transform with respect to x by

$$\omega_{sp} = \int_0^{\infty} e^{-PS} \omega_S \, d\xi \qquad \text{...(18)}$$

and taking the double Laplace transform of Eq. (17), with respect to τ and ξ, using Eq. (16) and Eq. (18)

$$\left(\frac{d^2\omega_{sp}}{d\xi^2} + \frac{1}{\xi}\frac{d\omega_{sp}}{d\xi} - \left(S + P\frac{Pe}{2}\right)\omega_{sp} = \frac{Pe}{2S}\frac{1}{\left(S + P\frac{Pe}{2}\xi\right)}\right) = 0 \qquad \text{...(19)}$$

The solution of Eq. (19) is

$$\omega_{sp} = \frac{Pe}{2S}\frac{1}{\left(S+P\frac{Pe}{2}\right)} + C_3 I_0\left(\left(S+P\frac{Pe}{2}\xi\right)^{1/2} + C_4 K_0\left(S+P\frac{Pe}{2}\xi\right)^{1/2}\right) \quad \text{...(20)}$$

Since ω_{sp} is finite as $\xi \to 0$, $C_4 = 0$. Otherwise K_0 tends to become infinite. Using the boundary condition at the wall,

$$\bar{\omega}_{sp} = \frac{Pe}{2S\left(S+\frac{PPe}{2}\right)}\left(1-\frac{\beta I_0\left(S+P\frac{Pe}{2}\xi\right)^{1/2}}{\left(S+P\frac{Pe}{2}\right)^{1/2}\left(\begin{array}{c}S+P\frac{Pe}{2}I_1\left(S+P\frac{Pe}{2}\right)^{1/2}\\ +\beta I_0\times\left(S+P\frac{Pe}{2}\right)^{1/2}\end{array}\right)}\right) \quad \text{...(21)}$$

Equation (21) is inverted using the method of residues. Equation (21) is inverted with respect to P, Eq. (21) is rearranged as

$$\frac{1}{S\left(\frac{2S}{Pe}+P\right)}\left(1-\frac{2\beta\, I_1\left(\frac{Pe}{2}\left(\frac{2S}{Pe}+P\right)^{1/2}\right)}{\begin{array}{c}\left(\frac{Pe}{2}\left(\frac{2S}{Pe}+P\right)^{1/2}\right)\left(\frac{Pe}{2}\left(\frac{2S}{Pe}+P\right)^{1/2}\right)\\ I_1\left(\left(\frac{Pe}{2}\left(\frac{2S}{Pe}+P\right)^{1/2}\right)+BI_0\right)\times\left(\left(\frac{Pe}{2}\left(\frac{2S}{Pe}+P\right)^{1/2}\right)\right)\end{array}}\right) \quad \text{...(a)}$$

Using the exponential rule of Laplace transform, the inverse is

$$\frac{e^{-2S\xi/Pe}}{S}L^{-}\left(\frac{1}{P}\right)\left(1-\frac{2\beta I_1\left(\left(P\frac{Pe}{2}\right)^{1/2}\right)}{\left(P\frac{Pe}{2}\right)^{1/2}I_1\left(P\frac{Pe}{2}\right)^{1/2}+BI_0\left(P\frac{Pe}{2}\right)^{1/2}}\right) \quad \text{...(b)}$$

Let $$\bar{f}(P) = \frac{1}{P}\left(\frac{I_1\left(P\frac{Pe}{2}\right)^{1/2}\Big/\left(P\frac{Pe}{2}\right)^{1/2}}{\left(P\frac{Pe}{2}\right)^{1/2}I_1\left(P\frac{Pe}{2}\right)^{1/2}+\beta I_0\left(P\frac{Pe}{2}\right)^{1/2}}\right) \quad \text{...(c)}$$

The inverse transform can be written as

$$f(\xi) = \frac{1}{2\beta} \sum \frac{2\beta \, \text{Exp}\,(-2\lambda_1^2 \xi / Pe)}{\lambda_1^2 \, (\lambda_1^2 + \beta^2)} \qquad \text{...(d)}$$

where $\lambda = j\left(P \dfrac{Pe}{2}\right)^{1/2}$ and $j = (-1)^{1/2}$ ...(e)

Thus, the inverse Laplace transform of Eq. (21) becomes with respect to P as

$$\omega_S = \frac{\exp\left(\dfrac{-2S\xi}{Pe}\right)}{S} \sum \frac{4\beta^2 \exp\left(\dfrac{-2\lambda_1^2}{Pe}\right)}{\lambda_1^2 \left(\lambda_1^2 + \beta^2\right)} \qquad \text{...(f)}$$

The radial mean concentration is obtained by inverting equation with respect to S

$$\bar{\omega} = \frac{\sum 4\beta^2 \exp\left(\dfrac{-2\lambda_1^2 \xi}{Pe}\right)}{\lambda_1^2 \, (\lambda_1^2 + \beta^2)} L^{-1}\left(\frac{e^{-2\xi/Pe}}{S}\right) \qquad \text{...(g)}$$

Now, $$L^{-1}\left(\frac{e^{-2\xi S/Pe}}{S}\right) = 0 \quad \text{for } 0 < \tau < 2\xi/Pe \qquad \text{...(h)}$$

$$= 1 \quad \text{for } \tau > 2\xi/Pe$$

The mean concentration is

$$\bar{\omega} = \frac{\sum 4\beta^2 \exp\,(-2\lambda_1^2 \xi / Pe)}{\lambda_1^2 \, (\lambda_1^2 + \beta^2)} \quad \text{for} \quad \tau > 2\xi/Pe = 0 \qquad \text{...(i)}$$

for $0 < \tau < 2\xi/Pe$, $\bar{\omega}$ is zero.

(*Reference:* Indian Chemical Engineer Journal, J. Raghuraman, Oct–Dec. 1973, Vol. 4.)

5.4 LAPLACE TRANSFORMS IN INSTRUMENTATION AND PROCESS CONTROL

Example 5.11 Overall gain $(\text{A.R.})_0$ = Product of individual gains or magnitudes.

Overall angle is: Summation of the individual transfer function's angles.

e.g., $$G(S) = \frac{6Se^{-2S}}{(2S+1)\,(S-1)\,(3S+1)}$$

For S: Magnitude: $S = i\omega$; $[(\text{R.P.})^2 + (\text{I.P.})^2]^{1/2} = (0^2 + \omega^2)^{1/2} = \omega$

Angle = $\tan^{-1}$ (Imaginary part/Real part) = $\tan^{-1}(\omega/0)$

$= \tan^{-1}(\infty) = 90^\circ$

For e^{-2S}: $e^{+i\omega} = \cos\omega + i\sin\omega;\; e^{-i\omega} = \cos\omega - i\sin\omega$:

$$e^{-2i\omega} = \cos 2\omega - i\sin 2\omega$$

$$\text{Magnitude} = ((\text{R.P.})^2 + (\text{I.P.})^2)^{1/2} = ((\cos 2\omega^2 + \sin 2\omega^2)^{1/2} = 1$$

Angle: $\tan^{-1}$ (Imaginary part/Real part) = $\tan^{-1}(-\sin 2\omega/\cos 2\omega) = \tan^{-1}(-\tan 2\omega)$

$$= -2\omega \text{ Rad.} = -180 \times \frac{2\omega}{2\pi} \text{ degrees.}$$

For 6: Magnitude = 6; Angle = 0 degree.

For $G(i\omega)$: $$\text{Magnitude} = \frac{6 \times \omega \times 1}{(4\omega^2+1)^{1/2}(\omega^2+1)^{1/2}(9\omega^2+1)^{1/2}}$$

For $\dfrac{1}{S-1}$: $$\frac{1}{i\omega - 1} = \frac{i\omega + 1}{i^2\omega^2 - 1} = \frac{1 + i\omega}{-1(1+\omega^2)}$$

$$\text{Magnitude} = ((\text{R.P.})^2 + (\text{I.P.})^2)^{1/2} = \left(\frac{1}{(1+\omega^2)^2} + \frac{\omega^2}{(1+\omega^2)^2}\right)^{1/2}$$

$$= \frac{1}{(1+\omega^2)^{1/2}}$$

Angle: $\tan^{-1}$ (Imaginary part/Real part)

$$= \tan^{-1}\left(\frac{-\omega/(1+\omega^2)}{-1/(1+\omega^2)}\right) = \tan^{-1}(\omega)$$

$$\text{Angle overall} = 0 + 90^\circ - \frac{180 \times 2\omega}{(2\pi)} - \tan^{-1}(2\omega) + \tan^{-1}\omega - \tan^{-1}3\omega.$$

For $\dfrac{1}{TS+1}$: Magnitude is

$$\frac{1}{(Ti\omega+1)} = \frac{Ti\omega - 1}{T^2\omega^2 i^2 - 1}$$

$$= \frac{(1 - Ti\omega)}{(1+\omega^2T^2)} = \frac{1}{1+\omega^2T^2} - \frac{Ti\omega}{1+\omega^2T^2}$$

$$\text{Magnitude} = ((\text{R.P.})^2 + (\text{I.P.})^2)^{1/2} = \frac{1}{(1+\omega^2T^2)^{1/2}}$$

$$\text{Angle} = \tan^{-1}\left(\frac{I.P.}{R.P.}\right) = \tan^{-1}\left[\left(\frac{-\omega T}{1+\omega^2T^2}\right)\bigg/\left(\frac{1}{1+\omega^2T^2}\right)\right]$$

$$= \tan^{-1}(-\omega T)$$

Example 5.12 Obtaining Transfer fuctions:

Transfer functions for a stirred tank chemical reactor. Reaction is of first order, and isothermal. F = Feed rate; V = Volume of reactor; C = Concentration; t = Time; K = First order rate constant. (Fig. 5.6):

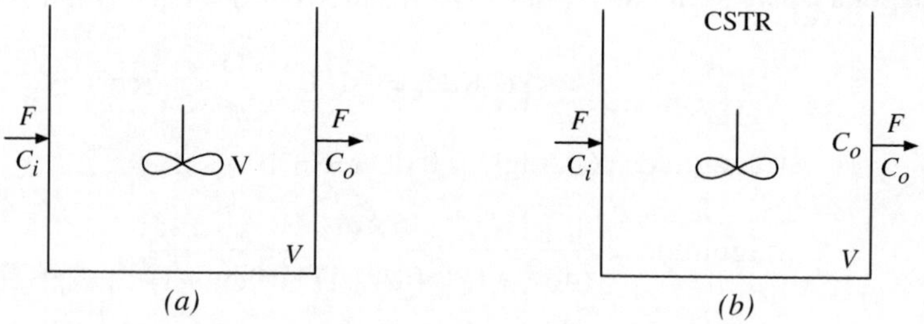

Fig. 5.6

Using unsteady state component balance for the CSTR:

$$F \times C_i - F \times C_o - KVC_o = V\frac{dC_o}{dt} \quad \ldots(1)$$

At steady state, Eq. (1) is

$$FC_{is} - FC_{os} - KVC_{os} = V\,\frac{dC_{os}}{dt} \quad \ldots(2)$$

Equations (1) and (2) give

$$F(C_i - C_{is}) - F(C_o - C_{os}) - KV\,(C_o - C_{os}) = \frac{d}{dt}(C_o - C_{os})\,V \quad \ldots(3)$$

Using deviation variables as

$$C_i - C_{is} = C_i'; \; C_o - C_{os} = C_o'$$

Equation (3) becomes

$$FC_i' - FC_o' - KVC_o' = V\frac{dC_o'}{dt} \quad \ldots(4)$$

Taking Laplace transforms on both sides

$$FC_i'\,(S) - FC_o'(S) - KVC_o'(S) = SVC_o'(S)$$

Collecting common terms gives

$$C_o'(S)\,(SV + F + KV) = FC_i'(S)$$

$$\frac{C_o'(S)}{C_i'(S)} = \frac{F}{(SV + F + KV)} = \frac{\dfrac{F}{(F+KV)}}{\dfrac{SV}{F+KV}+1} = \frac{R}{TS+1}$$

5.5 OTHER TRANSFORMS

The integral transform of a function $f(x)$ denoted by $I(f(x))$ is defined by

$$\bar{f}(S) = \int_{x_1}^{x_2} f(x)\, K(S, x)\, dx$$

$K(S, x)$ is the kernel of the transform and is a function of S and $x \cdot f(x)$ is the inverse transform of $\bar{f}(S)$.

1. Laplace transform: When $K(S, x) = e^{-Sx}$, it leads to

$$\bar{f}(S) = \int_{0}^{\infty} f(x)\, e^{-Sx}\, dx$$

2. Fourier transform: When $K(S, x) = e^{iSx}$ then

$$F(S) = \int_{-\infty}^{\infty} f(x)\, e^{iSx}\, dx$$

3. Melin transform: When $K(S, x) = x^{S-1}$

$$M(S) = \int_{0}^{\infty} f(x)\, x^{S-1}\, dx$$

4. Fourier sine transform: When $K(S, x)$ is a sine function.
5. Fourier cosine transform: When Kernal, $K(S, x)$ is a cosine function.
6. Hankel transform: When $K(S, x)$ is a Bessel function.
7. Z-transform: $U(z) = \sum_{n=0}^{\infty} u_n z^{-n}$

5.6 Z-TRANSFORM

Z transforms are used in discrete data analysis or sampled data analysis. A discrete solution is expressed by difference equation. Z-transform is written as

$$z(u_n) = U(z) = \sum_{n=0}^{\infty} u_n z^{-n}$$

Inverse z-transform is

$$z^{-1}(U(z)) = u_n$$

Standard z-transform is

$$z(a^n) = \frac{z}{z-a} \qquad \text{...(1)}$$

Proof: $$z(a^n) = \sum_{n=0}^{\infty} a^n z^{-n} = \sum \left(\frac{a}{z}\right)^n$$

Expanding by series: $1 + \left(\frac{a}{z}\right) + \left(\frac{a}{z}\right)^2 + \ldots = \dfrac{1}{1-\left(\frac{a}{z}\right)} = \dfrac{z}{z-a}$ **Ans.**

$$z(n) = \frac{z}{(z-1)^2} \quad \text{...(2)}$$

$$z(n^2) = \frac{z^2+z}{(z-1)^2} \quad \text{...(3)}$$

$$z(n^3) = \frac{z^3+4z^2+z}{(z-1)^4} \quad \text{...(4)}$$

$$z(na^n) = \frac{az}{(z-a)^2} \quad \text{...(5)}$$

$$z\,(\cos n\theta) = \frac{z(z-\cos\theta)}{(z^2-2z\cos\theta+1)} \quad \text{...(6)}$$

$$z\,(\sin n\theta) = \frac{z\sin\theta}{(z^2-2z\cos\theta+1)} \quad \text{...(7)}$$

$$z(a^n \cos n\theta) = \frac{z(z-a\cos\theta)}{(z^2-2az\cos\theta+a)^2} \quad \text{...(8)}$$

$$z(a^n \sin n\theta) = \frac{az\sin\theta}{(z^2-2az\cos\theta+a^2)} \quad \text{...(9)}$$

$$z\left(\frac{1}{n!}\right) = e^{1/z} \quad \text{...(10)}$$

$$z\,(\sinh n\theta) = \frac{\sin n\theta}{z^2-2z\cosh\theta+1} \quad \text{...(11)}$$

$$z\,(\cosh n\theta) = \frac{z(z-\cosh\theta)}{(z^2-2z\cosh\theta+1)} \quad \text{...(12)}$$

Inverse of: $\dfrac{z^2+z}{(z+2)\,(z+4)}$

$$U(z) = \frac{z^2+z}{(z+2)\,(z+4)};$$

$$\frac{U(z)}{z} = \frac{z+1}{(z+2)(z-4)} = \frac{A}{(z+2)} + \frac{B}{(z-4)}$$

By partial fractions

$$\frac{U(z)}{z} = \frac{1}{6(z+2)} + \frac{5}{6(z-4)}$$

$$U(z) = \frac{z}{6(z+2)} + \frac{5z}{6(z-4)}\text{; taking inverse } z \text{ transforms.}$$

$$U_n = \frac{1}{6}(-2)^n + \frac{5}{6}(4)^n \quad \textbf{Ans.}$$

5.7 APPLICATIONS OF Z-TRANSFORMS

(1) Using z-transform solve the difference equation

$$y_{n+2} - 3y_{n+1} + 2y_n = u_n \text{ with } y_0 = 0;\ y_1 = 1 \text{ and } u_n = 1 \text{ for } n = 0, 1, 2, 3, \ldots$$

Taking z-transforms on both sides

$$z^2 (y(z) - y_0 - y_1 z^{-1}) - 3z\,[y(z) - y_0] + 2y(z) = \frac{z}{(z-1)}$$

Using $y_0 = 0$; $y_1 = 1$ values.

$$(z^2 - 3z + 2)\, y(z) = \frac{z}{z-1} + z = \frac{z^2}{z-1};\quad \frac{y(z)}{z} = \frac{z}{(z-1)(z-2)(z-1)^2}$$

$$\frac{y(z)}{z} = \frac{A_1}{z-1} + \frac{B_1}{z-2} + \frac{C_1}{(z-1)^2};$$

By partial fractions

$$A_1 = \frac{-7}{4};\ B_1 = 2;\ C_1 = \frac{-3}{4};\quad \frac{y(z)}{z} = \frac{-7}{4(z-1)} + \frac{2}{(z-2)} - \frac{3}{4(z-1)^2}$$

$$y(z) = \frac{-7z}{4(z-1)} + \frac{2z}{(z-2)} - \frac{-3z}{4(z-1)^2}$$

Taking inverse z transforms of z:

$$y_n = -\frac{7}{4} + 2(2)^n - \frac{3\,n(1)}{4 \times (-1)}$$

$$y_n = -\frac{7}{4} + 2(2)^n + \frac{3n}{4} \quad \textbf{Ans.}$$

(2) Using z-transform, solve the difference equation

$$U_{n+2} - 3u_{n+1} = 2u_n = 3n + 5$$

Taking z-transforms on both sides,

$$z^2\,(u(z) - u_0 - u_1 z^{-1}) - 3z\,(u(z) - u_0) + 2u(z) = \frac{3z}{(z-1)^2} + \frac{5z}{z-1}$$

or
$$u(z)\,(z^3 - 3z + 2) = \frac{5z^2 - 2z}{(z-1)^2} + u_0(z^2 - 3z) + u_1 z$$

$$u(z) = \frac{5z^2 - 2z}{(z-1)(z-2)(z-1)^2} + \frac{u_0(z^2 - 3z)}{(z-1)(z-2)} + \frac{u_1 z}{(z-1)(z-2)}$$

Taking inverse z transforms

$$u_n = \frac{z^{-1}(5z^2 - 2z)}{(z-1)^2\,(z-1)(z-2)} + \frac{u_0 z^{-1}(z^2 - 3z)}{(z-1)(z-2)} + \frac{u_1 z^{-1}(z)}{(z-1)(z-2)}$$

Using
$$z(1) = \frac{z}{z-1};\; z(n) = \frac{z}{(z-1)^2}$$

$$z(n^2) = \frac{z^2 + z}{(z-1)^3};\; z(n^3) = \frac{z^3 + 4z^2 + z}{(z-1)^4}$$

$$\frac{5z^2 - 2z}{(z-2)(z-1)^3} = \frac{A(z^2 + z)}{(z-1)^3} + \frac{B(z)}{(z-1)^2} + \frac{Cz}{z-2}$$

$$A = -\frac{3}{2};\; B = \frac{41}{6};\; C = 8$$

By partial fractions

$$\frac{5z^2 - 2z}{(z-2)(z-1)^3} = \frac{3(z^2 + z)}{2(z-1)^3} + \frac{41z}{6(z-1)^2} + \frac{8z}{(z-2)}$$

$$\frac{z^2 - 3z}{(z-1)(z-2)} = \frac{Az}{z-1} + \frac{Bz}{z-2};\; A = +2;\; B = -1;$$

$$\frac{z^2 - 3z}{(z-1)(z-2)} = \frac{2z}{z-1} - \frac{z}{z-2}$$

$$\frac{z}{(z-1)(z-2)} = \frac{Az}{z-1} + \frac{Bz}{z-2};$$

$$A = -1;\; B = 1$$

Giving
$$\frac{z}{(z-1)(z-2)} = \frac{z}{z-2} - \frac{z}{z-1}$$

u_n is inverse transform of

$$\frac{-3}{2}\frac{(z^2+z)}{(z-1)^3}+\frac{41z}{6(z-1)^2}+\frac{3z}{z-2}+\frac{2z}{z-1}-\frac{z}{z-2}+\frac{z}{z-2}-\frac{z}{z-1}$$

$$= -\frac{3}{2}\frac{(z^2+z)}{(z-1)^3}+\frac{41}{6}\frac{z}{(z-1)^2}+\frac{3z}{(z-2)}-\frac{z}{z-1}$$

$$u_n = -\frac{3}{2}z(n^2)+\frac{41}{6}z(n)+3(2)^n-1 \quad \textbf{Ans.}$$

5.8 DIFFERENCE EQUATIONS

These are used for sampled data or to discrete data systems. Laplace transforms are used for continuous data systems.

A difference equation is a relation between the differences of an unknown function at one or more general values of the argument.

e.g., $$\Delta y_{n+1} + y_n = 2 \qquad ...(1)$$

$$\Delta y_{n+1} = y_{n+2} - y_{n+1}$$

So Eq. (1) becomes

$$y_{n+2} - y_{n+1} + y_n = 2.$$

This is another form of the difference equation.

Example 5.13 Given, $y_n = C\,2^n + d(-2)^n$; obtain the difference equation, eliminating the constants, C and d.

Solution:

$$y_n = C\,2^n + d(-2)^n;$$

$$y_{n+1} = 2C\,2^n - 2d(-2)^n$$

$$y_{n+2} = 4C\,2^n + 4d(-2)^n$$

Eliminating C and d, giving

$$y_{n+2} - 4y_n = 0$$

which is the described desired equation.

$$\begin{vmatrix} y_n & 1 & 1 \\ y_{n+1} & 2 & -2 \\ y_{n+2} & 4 & 4 \end{vmatrix} = 0$$

Linear, difference equations with constant coefficients is of the form

$$y_{n+r} + a_1 y_{n+r-1} + a_2 y_{n+r-2} + \dots \dots + a_n y_n = f(n)$$

where $a_1, a_2, \dots, a_n$ are constants.

If $u_1(n) \cdot u_2(n), \dots u_r(n)$ be the r independent solutions of the equation

$$y_{n+r} + a_1 y_{n+r-1} + \dots + a_n y_n = 0$$

then its complementary solution is

$$C_1 u_1(n) + C_2 u_2(n) + \dots + C_r u_r(n).$$

where $C_1, C_2, \dots C_r$ are arbitrary constants. If particular solution is P.I., then the complete solution is: C.S. + P.I.

5.8.1 Rules for Complementary Function

Case 1: e.g., $y_{n+2} + by_{n+1} + cy_n = 0$; Second order.

This is written as $(E^2 + bE + c)y_n = 0$

i.e., $E^2 + bE + c = 0;$

The roots are $(E - \lambda_1)(E - \lambda_2)\, y_n = 0$

$$(E - \lambda_1)\, y_n = 0; \;\; (E - \lambda_2) y_n = 0$$

The solutions are

(a) If roots are real and distinct

$$y_n = C_1 \lambda_1^n; \;\; y_2 = C_2 \lambda_2^n$$

General solution is $y_n = C_1 \lambda_1^n + C_2 \lambda_2^n$

(b) If roots are real and equal $\lambda_1 = \lambda_2$

$$(E - \lambda_1)^2 y_n = 0; \;\; y_n = (\lambda_1)^n Z_n$$

where $Z_n = C_1 + C_2^n$

(c) If roots are imaginary $\lambda_1 = \alpha + i\beta; \;\; \lambda_2 = \alpha - i\beta$

$$y_n = r^n (C_1 \cos n\theta + C_2 \sin n\theta)$$

$$r = (\alpha^2 + \beta^2)^{1/2}; \;\; \theta = \tan^{-1}\left(\frac{\beta}{\alpha}\right)$$

In general:

Roots	*Solution*
1. Real and Distinct Roots $(\lambda_1, \lambda_2, \dots)$	$y_n = C_1\lambda_1^n + C_2\lambda_2^n + C_3\lambda_3^n + \dots$
2. 2 Real and Equal Roots $(\lambda_1, \lambda_2, \dots)$	$y_n = (C_1 + C_2 n)\, \lambda_1^n + C_3\lambda_3^n + \dots$
3. 3 Real and Equal Roots $(\lambda_1, \lambda_2, \dots)$	$y_n = (C_1 + C_2 n + C_3 n^2)\lambda_1^n + \dots$
4. $\alpha + i\beta, \alpha - i\beta$ imaginary pair	$y_n = r^n (C_1 \cos n\theta + C_2 \sin n\theta)$

Example 5.14 Of complimentary

Solution:

$$(\Delta^2 - 3\Delta + 2) = 0$$

$$(E^2 - 3E + 2)\, y_n = 0$$

$$E^2 - 3E + 2 = 0$$

The roots are:

$$(E - 2)\,(E - 1) = 0$$

$$\lambda_1 = 2;\ \lambda_2 = 1$$

$$y_n = C_1 2^n + C_2 1^n \text{ **Ans.**}$$

5.9 FINDING THE PARTICULAR INTEGRAL

The particular integral is given by $\dfrac{f(n)}{\phi(E)}$

where $\phi(E) = (E^n + a_1 E^{n-1} + a_2 E^{n-2} + \ldots\ \ldots\ a_n)$

Case 1: When $f(n) = a^n$; $P(I) = \dfrac{a^n}{\phi(E)}$; a is a constant.

Put $E = a$; P.I. $= \dfrac{a^n}{\phi(a)}$; $\phi(a)$ = should not be zero. For $\phi(a) = 0$, then

(1) $(E - a)y_n = a^n$; P.I. $= \dfrac{1}{E - a} a^n = na^{n-1}$

(2) $(E - a)^2 y_n = a^n$; P.I. $= \dfrac{1}{(E - a)^2} a^n = n(n - 1)\,\dfrac{a^{n-2}}{2}$

(3) $(E - a)^3 y_n = a^n$; P.I. $= \dfrac{1}{(E - a)^3} a^n = n(n - 1)\,(n - 2)\,\dfrac{a^{n-3}}{3}$

Example 5.15 $y_{n+2} - 3y_{n+1} + 2y_n = 4^n$

Solution:

$$(E^2 - 3E + 2)y_n = 4^n;$$

$$E^2 - 3E + 2 = 0;\ (E - 2)\,(E - 1) = 0$$

$$\text{C.S.} = C_1 2^n + C_2 1^n = C_2 + C_1 2^n$$

$$\text{P.I.} = \frac{4^n}{E^2 - 3E + 2};\quad \text{Put } E = 4$$

Giving

$$\frac{4^n}{16 - 12 + 2} = \frac{4^n}{6}$$

Complete solution: $C_2 + C_1 2^n + \dfrac{4^n}{6}$ **Ans.**

Example 5.16 Solve $y_{n+2} - 2y_{n+1} + y_n = n^2 3^n$

Solution:

$$(E^2 - 2E + 1)y_n = 0$$

$$(E^2 - 2E + 1) = 0$$

$$(E - 1)(E - 1)y_n = 0$$

Complimentary solution is

$$(C_1 + C_2 n)\, 1^n = C_1 + C_2 n$$

Particular solution is

$$\frac{1}{(E-1)^2} 3^n n^2 = \frac{n^2 3^n}{(3E-1)^2} = 3^n - \frac{1n^2}{(1+3\Delta)^2}$$

$$= 3^n (1 + 3\Delta)^{-2} [n(n-1) + n]$$

$$= 3^n (1 - 6\Delta + 18\Delta^2)(n^2 + n)$$

$$= 3^n (n^2 + n - 6 \times (2n + 1) + 18 \times 2)$$

$$= 3^n (n^2 - 11n + 30)$$

Complete solution is

$$y_n = C_1 + C_2 n + 3^n (n^2 - 11n + 30) \quad \textbf{Ans.}$$

5.10 FORMULATION OF DIFFERENCE EQUATIONS FOR CHEMICAL ENGINEERING PROBLEMS

An equation of form:

$$a_0 y_0 + a_1 y_1 + a_2 y_2 + \ldots + a_n y_n = \phi(x) \qquad \ldots(1)$$

is a finite difference equation if $y = f(x)$ and x can take values of $x_0, x_0 + h, \ldots, x_0 + nh$. Where $a_0, a_1, a_2, \ldots, a_n$ are constants or functions of the independent variable. $\phi(x)$ can be a constant or a function of x. Equation (1) can be written as

$$a_0 f(x_0) + a_1 f(x_0 + h) + \ldots + a_n f(x_0 + nh) = \phi(x)$$

or

$$(a_n E^n + a_{n-1} E^{n-1} + \ldots + a_0) f(x_0) = \phi(x) \qquad \ldots(2)$$

Since

$$f(x_0 + rh) = E^r f(x_0); \quad E = 1 + \Delta$$

$$(P_n \Delta^n + P_{n-1} \Delta^{n-1} + \ldots + P_0) f(x_0) = \phi(x) \qquad \ldots(3)$$

Equations (1), (2) and (3) are different forms of the same finite difference equation.

Example 5.17 For the stagewise absorber shown in Fig. 5.6, formulate the difference equations to find compositions.

L = Liquid fluid flow rate, solute free oil, kg moles/hr.

X = Liquid composition in kg moles of solute/kg mole of solute free oil

Y = Vapour composition in moles of solute per mole of solute free gas.

Solution: Using Fig. 5.7, the solute balance over plate m is

$$LX_{m-1} + GY_{m+1} = LX_m + Y_mG_m \quad ...(1)$$

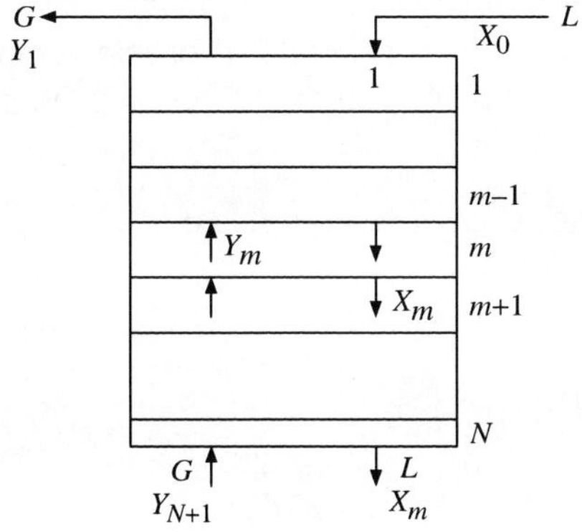

Fig. 5.7

Equation (1) can be rearranged as:

$$L(X_m - X_{m-1}) = G(Y_{m+1} - Y_m) \quad ...(2)$$

The equilibrium relation is

$$Y_m = K_mX_m; \; Y_{m+1} - Y_m = \left(\frac{L}{G}\right)(X_m - X_{m-1}) \quad ...(3)$$

Replacing X_m and X_{m-1} by $\frac{Y_m}{K_m}$ and $\frac{Y_{m-1}}{K_{m-1}}$ in Eq. (3)

$$Y_{m+1} - Y_m = \frac{LY_m}{GK_m} - \frac{LY_{m-1}}{GK_{m-1}}$$

Using $\frac{L}{K_mG} = A_m$

$$Y_{m+1} - Y_m = A_mY_m - A_{m-1}Y_{m-1}$$

$$Y_{m+1} = Y_m(1 + A) - AY_{m-1} \quad ...(4)$$

Rearranging: $$Y_{m+1} - (1 + A)Y_m + AY_{m-1} = 0 \quad ...(5)$$

is the difference equation.

Equation (5) is

$$[E^2 - (1 + A)E + A]Y_m = 0$$

This has only complimentary solution.

$$E^2 - (1 + A)E + A = 0. \text{ The roots are } \frac{(1+A) \pm (1-A)}{2}; \; E = 1 \text{ or } A.$$

Solution is $$Y_m = C_1 1^m + C_2 A^m = C_1 + C_2 A^m \textbf{ Ans.}$$

Example 5.18 In the chemical reaction: $A \xrightarrow{K_1} B \xrightarrow{K_2} C$ where K_1 and K_2 are reaction rate constants. The reaction takes place in a series of CSTRs at constant temperature. Let the concentration of A leaving any CSTR be C_{An}, for B it be C_{Bn} and for C be C_{Cn}. Obtain the concentration of A and B in each CSTR in the n battery.

Solution: The material balance over stage n reactor gives:

Component A: $$\frac{dC_A}{dt} = K_1 C_A; \; C_{A,\,n-1} - C_{A,\,n} = K_1 C_{A,\,n} t \quad \text{...(1)}$$

where t is holding time, in the CSTR.

Component B: $$\frac{dC_B}{dt} = K_1 C_A - K_2 C_B;$$

$$C_{B,\,n-1} - C_{B,\,n} = K_2 C_{B,\,n} t - K_1 C_{A,\,n} t \quad \text{...(2)}$$

Let $$K_1 t = x \quad \text{and} \quad K_2 t = y$$

Then, $$(1 + x)C_{A,\,n} - C_{A,\,n-1} = 0 \quad \text{...(3)}$$

$$[E(1 + x) - 1]\, C_{A,\,n-1} = 0 \quad \text{...(4)}$$

$$(E(1 + x) - 1) = 0; \quad E = \frac{1}{1+x}$$

$$C_{B,\,n-1} = K_2\, C_{B,n}\, t - C_1\, P_1 n \quad \text{...(5)}$$

Substituting $C_{A,\,n}$ from Eq. (5) into Eq. (2); let $\frac{1}{1+x} = p_1$

$$C_{B,\,n-1} = K_2 C_{B,\,m}\, t - C_1 p_1{}^n; \; ((1 + y)E - 1)\, C_{B,\,n-1} = xB_1 p_1{}^n$$

Complimentary solution is

$$(1 + y)E - 1 = 0; \; E = \frac{1}{(1+y)}$$

$$C_{B,\,n-1} = C_2 p_2{}^{n-1}; \quad C_{B,\,n} = C_2 p_2{}^n \quad \text{...(6)}$$

where C_2 is constant and $p_2 = \frac{1}{1+y}$

Particular solution for $C_{B, n-1}$ is

$$\frac{xC_1p_1^n}{(1+y)E-1} = \frac{xC_1p_1^n}{(1+y)p_1-1} = \frac{xC_1p_1^n}{(y-x)}$$

Complete solution is $$C_{B, n} = C_2p_2^n + \frac{xC_1p_1^n}{(y-x)}$$

$$C_{A, n} = C_1\left(\frac{1}{1+x}\right)^n$$

where C_1 and C_2 are evaluated by the boundary conditions.

Example 5.19 In an extraction battery shown in Fig. 5.8 nicotine is extracted from water using kerosene as solvent in a countercurrent battery.

R = Flow rate of water phase; E = Flow rate of kerosene phase (extract phase)

N = Number of stages; X_n = Nicotine concentration in raffinate phase leaving nth stage

Y_n = Nicotine concentration in extract phase leaving stage n: $Y_n = mX_n$

H = Hold up of kerosene per stage

h = Hold up of water per stage. Fig. 5.8:

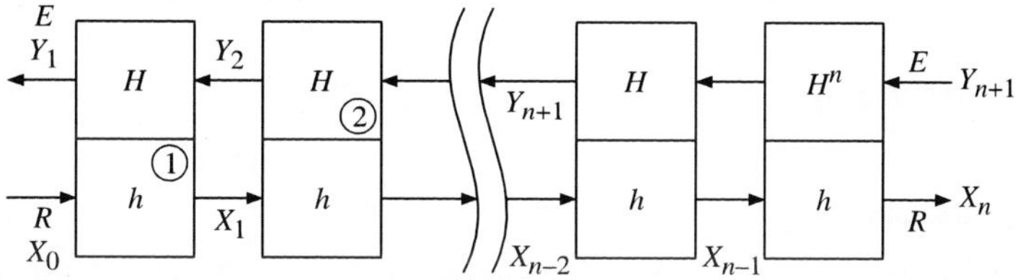

Fig. 5.8

Solution: Nicotine balance on stage n is

$$R \times X_{n-1} + E \times Y_{n+1} - R \times X_n - E \times Y_n = h\frac{dX_n}{dt} + H\frac{dY_n}{dt} \quad \ldots(1)$$

Equilibrium relation is $$Y_n = mX_n \quad \ldots(2)$$

Eliminating Y_n, Y_{n+1} between Eqs. (1) and (2)

$$AX_{n+1} - (A + 1)\, X_n + X_{n-1} = B\frac{dX_n}{dt} \quad \ldots(3)$$

Here, $$A = \frac{mE}{R};\quad B = \frac{h+mH}{R} \quad \ldots(4)$$

At steady state, $\frac{dX_n}{dt} = 0$; Eq. (3) becomes

$$AX_{n+1} - (A+1)X_n + X_{n-1} = 0 \quad \ldots(5)$$

$$[E^2A - (A+1)E + 1]X_{n-1} = 0$$

$$[E^2A - (A+1)E + 1] = 0$$

$$E^2A - AE - E + 1 = 0$$

$$(AE - 1)(E - 1) = 0$$

$$AE = 1;\ E = \frac{1}{A};\ E = 1 \text{ are the roots.}$$

The complementary solution is

$$X_{n-1} = C_1^{n-1} + C_2\left(\frac{1}{A}\right)^{n-1}$$

or

$$X_n = C_1 + C_2\left(\frac{1}{A}\right)^n$$

where C_1 and C_2 are constant terms, obtained from the process conditions.

5.11 STATISTICAL METHODS

5.11.1 Least Squares Method

Straight line fit: $y = m \times x + c$ for n points.

For n points, squares of the error is

$[y_1 - (mx_1 + c]^2 + [(y_2 - (mx_2 + c)]^2 + \ldots + [(y_n - (mx_n + c)]^2$ is minimum. Variables are m and c.

$$\frac{dS}{dc} = 0 = -2[y_1 - (mx_1 + c)] - 2[y_2 - (mx_2 + c)] - \ldots - 2(y_n - (mx_n + c) = 0 \quad \ldots(1)$$

$$\frac{dS}{dm} = 0 = -2(y_1 - (mx_1 + c)]\, x_1 - 2[y_2 - (mx_2 + c)]x_2 \ldots 2[y_n - (mx_n + c)]x_n = 0 \quad \ldots(2)$$

Solving Eqs. (1) and (2) simultaneously gives

$$mc + (\Sigma\, x)m = \Sigma\, y \quad \ldots(3)$$

$$(\Sigma\, x)c + (\Sigma\, x^2)m = \Sigma\,(xy) \quad \ldots(4)$$

Solving Eqs. (3) and (4) gives m and c values.

Example 5.20 Using least squares method, fit a straight line to the data below:

w	50	70	100	120
p	12	15	21	25

Solution: $y = mx + c; \quad p = bw + a$

$$nc + (\Sigma x)\, m = \Sigma y; \quad (\Sigma x)c + (\Sigma x^2)m = \Sigma (xy)$$

n	$w = x$	y	x^2	xy
1	50	12	2500	600
2	70	15	4900	1050
3	100	21	10000	2100
4	120	25	14400	3000
	$\Sigma x = 340$	$\Sigma y = 73$	$\Sigma x^2 = 31{,}800$	$\Sigma xy = 6750$

$$4c + 340\, m = 73; \quad 340\, c + 31800\, m = 6750$$

Solving both equations, $m = 0.1724; \quad c = 3.596$

$$P = 0.1724\, w + 3.596 \text{ **Ans.**}$$

Parabola: $y = ax^2 + bx + c$

$$S \text{ (sum of the squares)} = [y_1 - (ax_1^2 + bx_1 + c)]^2 + [y_2 - (ax_2^2 + bx_2 + c)]^2 + \ldots + [y_n - (ax_n^2 + bx_n + c)]^2$$

Obtaining, $\dfrac{dS}{da} = 0; \quad \dfrac{dS}{db} = 0; \quad \dfrac{dS}{dc} = 0$

giving relations of,

$$na + (\Sigma x)b + (\Sigma x^2)\, c = \Sigma y; \quad (\Sigma x)a + (\Sigma x^2)b + (\Sigma x^3)c = \Sigma xy$$

$$(\Sigma x^2)a + (\Sigma x^3)b + (\Sigma x^4)c = \Sigma (x^2 y)$$

Solving the above three equations gives: a, b, c values.

Example 5.21 Find the best values of a and b in the equation given below.

$y^2 + a^2x^2 - 2abx - 2by - 2axy + b^2 = 0$; x and y relation is

x	2.24	4.0	23.0	56.1	90.0	144.0	210.0	272.0	324.0
y	376	360	256	182	132	81	42.2	25	9

Solution: The given equation is rearranged in the linear form as:

$$y^2 + a^2x^2 - 2abx - 2by - 2axy + b^2 = 0$$

or $$(y + ax)^2 - 2abx - 2by - 4axy + b^2 = 0$$

or $$(y + ax)^2 - 2b\,(y + ax) + b^2 - 4axy = 0$$

or $$(y + ax - b)^2 = 4axy$$

or $$(y + ax - b) = 2(ax)^{1/2}\,(y)^{1/2}$$

or $$[\,y + ax - 2(ax)^{1/2}\,(y)^{1/2}] = b$$

or $$[(y)^{1/2} - (ax)^{1/2}]^2 = b$$

or $$(y)^{1/2} = (b)^{1/2} + (a)^{1/2}\,(x)^{1/2}$$

or $$Y = A + BX$$

where $Y = (y)^{1/2}$ and $X = (x)^{1/2}$; $A = (b)^{1/2}$; $B = (a)^{1/2}$

Now the equation is of straight line form.

Now applying the least squares method:

$$mA + B\,\Sigma\,X_i = \Sigma\,Y_i$$

$$A\,\Sigma\,X_i + B\,\Sigma\,X_i^2 = \Sigma\,X_i\,Y_i;$$

From the given data:

$X = (x)^{1/2}$	$Y = (y)^{1/2}$	$X^2 = x$	XY
1.497	19.39	2.24	29.02
2.000	18.97	4.00	37.95
4.800	16.00	23.00	76.73
7.490	13.49	56.10	101.05
9.487	11.49	90.00	109.00
12.00	9.00	144	108.00
14.49	6.496	210	94.14
16.49	5.00	272	82.46
18.00	3.00	324	54.00
Σ 86.254	Σ 102.84	Σ 1125.34	Σ 692.35

$$9A + 86.254B = 102.84;\;\; 86.254A + 1125.34B = 692.35$$

Solving simultaneously $\quad B = -0.9817 = (a)^{1/2};\; a = -0.9638$

$$A = 20.835 = (b)^{1/2};\; b = 434.12;\; B = -0.9817$$

The final equation is $\quad (y)^{1/2} = 20.835 - 0.9817\,(x)^{1/2}$

5.11.2 Maximum Percentage Error

Example 5.22 What is the maximum percentage error in f in the below given equation, $P = \dfrac{4flV^2}{2gD}$

$$P = \pm 2\% \text{ error};\quad l = \pm 2\% \text{ error};$$

$$V = \pm 1\% \text{ error};\quad D = \pm 1\% \text{ error}.$$

Solution:

$$f = \frac{gDP}{2lV^2}$$

$$df = \frac{gD\,dP}{2lV^2} + \frac{dD\,Pg}{2lV^2} - \frac{VlgPD\,dV}{l^2V^4} - \frac{0.5V^2g\,PD\,dl}{l^2V^4}$$

Dividing by the f equation,

$$\frac{df}{f} = \frac{dP}{P} + \frac{dD}{D} - \frac{2dV}{V} - \frac{-dl}{l}$$

$$= \pm 2\% \pm 1\% \pm 2 \times 1\% \pm 2\%$$

$$\text{Maximum error} = 2 + 1 + 2 + 2 = 7\%$$

5.11.3 Definitions in Statistics

1. Sample mean $(\bar{x}) = \sum \dfrac{x_i}{n} = (x_1 + x_2 + \ldots + x_n)/n$

   where n is the total number of experiments/observations, x_i $(i = 1, 2, \ldots, n)$ are the variables.

2. Standard deviation: $S^2 = \dfrac{1}{n-1}\sum (x_i - \bar{x})^2$; where $\bar{x}$ is arithmetic mean.

3. Geometric mean: (G.M.) $= (x_1 \times x_2 \times x_3 \ldots)^{1/n}$

4. Harmonic mean (H.M.): $= \dfrac{1}{\dfrac{1}{n}\left(\dfrac{1}{x_1} + \dfrac{1}{x_2} + \ldots \dfrac{1}{x_n}\right)}$

Example 5.23 A train travels at 50 km/hr in the first hour and 60 km/hr in the second hour and 80 km/hr in the third hour. Calculate the average speed in km/hour, using the three means.

Solution:

(a) Arithmetic mean; speed: $\dfrac{50 + 60 + 80}{3} = 63.3$ km/hour

(b) Geometric mean; speed: $(50 + 60 + 80)^{1/3} = 62.15$ km/hour

(c) Harmonic mean (H.M.): $\dfrac{1}{\dfrac{1}{3}\left(\dfrac{1}{50} + \dfrac{1}{60} + \dfrac{1}{80}\right)} = 61.02$ km/hour

5. Mean deviation (M.D.): $= \frac{1}{n}\sum f_i \,|x_i - A|$ where A is mean, median, or mode.

$$\text{M.D.} = \frac{1}{n}\sum f_i \,|(x_i - \bar{x})|$$

where n is the total number of observations, x_i are observations and $\bar{x}$ is mean value.

6. Standard deviation (S.D.): $\sigma = \sum \frac{[(x_i - \bar{x})^2]^{1/2}}{n}$

Square of the standard deviation is called variance.

7. Coefficient of variation: The ratio of standard deviation to the mean deviation is called the coefficient of variation.

8. Relations: Mean deviation = 4/5th of standard deviation.

Example 5.24 The following are the wickets taken by two bowlers in cricket matches. Who is a better bowler and who is more consistent?

A	5	0	1	6	2	0
B	4	2	1	2	2	2

Solution: For A: Arithmetic mean ($\bar{x}$) $= \frac{5+0+1+6+2+0}{6} = 2.33$

For B: Arithmetic mean $= \frac{4+2+1+2+2+2}{6} = 2.17$

For A: Standard deviation $= (x_i - \bar{x})^2$

$$= (5 - 2.33)^2 + (0 - 2.33)^2 + (1 - 2.33)^2 + (6 - 2.33)^2 + (2 - 2.33)^2 + (0 - 2.33)^2$$

$$= 7.13 + 5.43 + 1.77 + 13.47 + 0.11 + 5.43 = 33.34$$

$$\text{S.D.} = \left(\frac{33.34}{6}\right)^{1/2} = 2.36$$

For B: S.D. is $(x_i - \bar{x})^2 = (4 - 2.17)^2 + (2 - 2.17)^2 + (1 - 2.17)^2 + (2 - 2.17)^2 + (2 - 2.17)^2 + (2 - 2.17)^2$

$$= 3.35 + 0.029 + 1.37 + 0.029 + 0.029 = 4.95;\ \text{S.D.} = 0.91$$

$$\text{Coefficient of variation of } A = \frac{\text{S.D. of } A}{\text{M.D. of } A} = \frac{2.36}{2.33} = 1.013$$

$$\text{Coefficient of variation of } B = \frac{0.91}{2.17} = 0.42$$

For A: M.D. = 2.33; C.V. = 1.013

For B: M.D. = 2.17; C.V. = 0.42

A is a better player than B; since M.D. of A is greater than M.D. of B.

B is more consistent than A, since C.V. of B is less than C.V. of A.

9. Standard error of estimate

Chi-square distribution: $\left(\frac{x-\mu_2}{\sigma}\right)^2$ is chi-square variate; $\frac{x-\mu}{\sigma}$ is standard variate; x is random variable; μ = mean value; σ = standard deviation.

$$\text{Chi-square} = \frac{(x_1-\mu_1)^2}{\sigma_1} + \frac{(x_2-\mu_2)^2}{\sigma_2} + \ldots = \sum_{i=1}^{v} \frac{(x_i-\mu_i)^2}{\sigma_i}$$

V = Independent random variables; Chi-square distribution is

Probability function is
$$P(t) = \frac{1}{2V^{v/2}\,\Gamma(V/2)} e^{-V^2/2} (\chi^2)^{(V/2)-1}$$

$\Gamma(V) = (V-1)!\ \ 0 < \chi < \infty$, if V is a positive integer.

$\lceil V = (V-1)\lceil(V-1)$ for V greater than 1.

5.11.4 Applications of χ^2 Distribution (chi^2)

(1) Chi-square test for goodness of fit (2) Chi-square test for independence of attributes (3) to test if the population has a specific value of the variance, σ^2 (4) to test the equality of population proportions.

Constants of χ^2 distribution test with V density function:

Mean = V Mode = V–2 Variance = –2V

$\mu = 0$ $\mu_1 = 2V$ $\mu_j = 48V + 12V^2$

$\beta_1 = \frac{8}{V} = \frac{\mu_1^2}{\mu_2^2}$ $\mu_3 = 8V$; $\mu_4 = 48V + 12V^2$

$\beta_2 = \frac{\mu_4}{\mu_2^2} = 3 + \frac{12}{V}$ $\gamma_1 = \left(\frac{8}{V}\right)^{1/2}$; $\gamma_2 = \beta_2 - 3 = \frac{12}{V}$

(5) Pearson's coefficient of skewness:

$$\frac{(\text{Mean} - \text{Mode})}{\text{S.D.}} = \frac{V-(V-2)}{(2V)^{1/2}}$$

(6) For large V, the standard variate is

$$\chi^2 \text{ (chi-square)} = \frac{-V}{(2V)^{1/2}}$$

Goodness of fit:

$$\chi^2 = \sum_{i=1}^{n} \frac{(O_1 - E_1)^2}{E_1} + \frac{(O_2 - E_2)^2}{E_2} + \ldots$$

follows v^2 distribution data with $V = n - 1, d, f$, where O_1, O_2 are the observed frequencies and E_1, E_2, ... are theoretical frequencies or expected frequencies. $O_1, O_2 \ldots$ are under some theory or hypothesis.

$$\sum_{i=1}^{n} O_i = \sum_{i=1}^{n} E_i = N;$$

$$\sum O = \sum E = N;$$

$$\sum (O - E) = \sum O - \sum E = +N - N = 0$$

$$\chi^2 = \sum \left(\frac{O^2}{E} \right) - N$$

Example 5.25 Given below (a) are the frequencies of distribution and (b) the frequencies of normal distribution having same: mean, standard deviation and total frequency as in (a). Apply χ^2 test of goodness of fit.

(a) O:	13	66	220	493	792	924	792	495	220	66	13	12	1
(b) E:	17	66	210	484	799	944	799	484	210	66	17	15	2
$O - E$:	–4	0	10	9	–7	–20	–7	11	10	0	–4	–3	–1
$(O - E)^2$	16	0	100	81	49	400	49	121	100	0	16	9	1
$\frac{(O-E)^2}{E}$	0.941	0	0.476	0.167	0.061	0.438	0.061	0.25	0.476	0	0.94	0.6	0.50
S. No.	1	2	3	4	5	6	7	8	9	10	11	12	13

Solution: $\sum(O - E) = -6;\quad \sum (O - E)^2 = 942;\quad \frac{(O-E)^2}{E} = 4.91$

$$\text{Chi-square} = \sum \left[\frac{(O-E)^2}{E} \right] = 4.91$$

Here $n = 13$; degrees of freedom = 13 – 1 – 2 – 2 = 8

Tabulated values of chi-square at 8 d.f. at 5% level of significance is: 15.507. Since calculated value of chi square = 4.91 is much less than the tabulated value, viz., 15.507. It is highly non-significant. Hence, we may accept the null hypothesis at 5% level of significance and conclude that the normal distribution is a good fit to the given distribution.

5.12 STUDENTS' *t* DISTRIBUTION

For V greater than 30, t distribution tends to standard normal distribution. $V = n - 1$; n is sample size or number.

$$t = \frac{\bar{x} - \mu}{(S^2 / n)^{1/2}}; \quad \bar{x} = \frac{\Sigma x}{n}; \quad S^2 = \frac{1}{n-1} \sum (x - \bar{x})^2$$

The applications of t distribution are:

1. t-test for the significance of single mean, population variance is unknown.
2. t-test for the significance of the difference between two sample means, population variance are equal but unknown.
3. t-test for the significance of an observed sample correlation coefficient.

Example 5.26 A soap manufacturing company was distributing a particular brand of soap through a large number of retail shops. Before advertisement, the mean sales per week per shop was 140 hundreds. After advertisement a sample of 26 shops was taken and the mean sales was found to be 147 hundreds. Standard deviation is 10. Is the advertisement effective or not?

Solution: $n = 26$; $\bar{x} = 147$ hundreds; $s = 10$ hundreds.

Null Hypothesis: H_0: $\mu = 140$ hundreds. The deviation between $\bar{x}$ and μ is due to fluctuations of sampling. That means, the advertisement is not effective.

Alternate hypothesis, H_1: $\mu > 140$

Under null hypothesis, H_0, the test statistic is:

$$t = \frac{\bar{x} - \mu}{S / (n)^{1/2}} = \frac{\bar{x} - \mu}{s / (n-1)^{1/n}}; \quad t_{n-1} = t_{25}$$

$$t = \frac{147 - 140}{10 / (25)^{1/2}} = \frac{7 \times 5}{10} = 3.5$$

Tabulated value of t for 25 d.f. at 5% level of significance for single (right) tail test is: 1.708, that is, t_{25} (0.05) = 1.708 (from t-tables). Since the calculated t is greater than the table value, so the advertisement is effective in increasing the sales. If s is 25, then calculated t value (1.4) is less than the table value of 1.708. Then advertisement is not effective.

5.12.1 *t*-test for Difference of Means

$$t = \frac{\bar{x} - \bar{y}}{S \times \left(\frac{1}{n_1} + \frac{1}{n_2}\right)^{1/2}}; \quad \bar{x} = \frac{\Sigma x}{n_1}; \quad \bar{y} = \frac{\Sigma y}{n_2};$$

$$S^2 = \frac{1}{n_1 + n_2 - 2}\left[\sum (x - \bar{x})^2 + (y - \bar{y})^2\right]$$

Example 5.27 The nicotine content in milligrams of two samples of tobacco were found to be as follows:

Sample *A*	24	27	26	21	25	–
Sample *B*	27	30	28	31	36	22

Can it be said that the two samples come from normal populations having the same mean?

Solution: If μ_1 is not equal to μ_2 (means),

Computation of sample means and standard deviations

Sample *A*			Sample *B*		
x	$d = x - A = x - 25$	d^2	y	$D = y - B = y - 30$	D^2
24	–1	1	27	–3	9
27	+2	4	30	0	0
26	+1	1	28	–2	4
21	–4	16	31	1	1
25	0	0	36	6	36
			22	–8	64
Total	–2	22	Total	–6	114

$$\bar{x} = A + \sum \frac{d}{n_1} = 25 - \frac{2}{5} = 24.6$$

$$\bar{y} = B + \frac{\sum D}{n_2} = 30 + \frac{(-6)}{6} = 29$$

$$\sum (x - \bar{x})^2 = \sum d^2 - \frac{\left(\sum d\right)^2}{n_1}$$

$$\sum (y - \bar{y})^2 = \sum D^2 - \frac{\left(\sum D\right)^2}{n_2}$$

$$22.4 - \frac{4}{5} = 21.2; \; 114 - \frac{36}{6} = 108$$

$$S^2 = \frac{1}{n_1 + n_2 - 2} \sum (x - \bar{x})^2 + \sum (y - \bar{y})^2$$

$$= \frac{21.2 + 108}{9} = 14.36$$

$$t = \frac{\bar{x} - \bar{y}}{\left[S^2\left(\frac{1}{n_1} + \frac{1}{n_2}\right)\right]^{1/2}} = \frac{24.6 - 29}{\left[14.36 \times \left(\frac{1}{5} + \frac{1}{6}\right)\right]^{1/2}} = -1.92$$

The tabulated value of t for 9 density function $(n_1 + n_2 - 2)$ at 5% level of significance from tables of t is 2.262. Since calculated value (1.92) is less than 2.262, we conclude that the samples come from normal population with the same mean.

5.13 PAIRED *t*-TEST FOR DIFFERENCE OF MEANS

1. The sample sizes are equal, i.e., $n_1 = n_2 = \ldots = n$ (say)
2. The sample observations $(x_1, x_2, \ldots x_n)$ and $(y_1, y_2, \ldots y_n)$ are not completely independent but they were dependent in pairs, i.e., $(x_1, y_1), (x_2, y_2), \ldots$. Correspond to the first, second, … nth unit respectively.

$$t = \frac{\bar{d}}{S/(n)^{1/2}} = \frac{d}{(S^2/n)^{1/2}}$$

$$d = x - y;\ \bar{d} = \frac{1}{n}\sum d;\ S^2 = \frac{1}{n-1}\left[\sum d^2 - \frac{(\sum d^2)}{n}\right]$$

Example 5.28 The sales data of a washing machine in six shops before and after a concessional price are as under:

Shops	*A*	*B*	*C*	*D*	*E*	*F*
Before	53	28	31	48	50	42
After	58	29	30	55	56	45

Is the concessional scheme a success or not at 5% level of significance?

Solution:

Shop	*A*	*B*	*C*	*D*	*E*	*F*	Total
x	53	28	31	48	50	42	
y	58	29	30	55	56	45	
$d = x - y$	–5	–1	1	–7	–6	–3	$\Sigma d = -21$
d^2	25	1	1	49	36	9	$\Sigma d^2 = 121$

$$\bar{d} = \frac{\Sigma d}{n} = -\frac{21}{6} = -3.5;$$

$$S^2 = \frac{1}{n-1}\left[\sum d^2 - \frac{\left(\sum d\right)^2}{n}\right]$$

$$= \frac{1}{5}\left(121 - \frac{441}{6}\right) = 9.5;$$

$$t = \frac{-3.5}{\left(\frac{9.5}{6}\right)^{1/2}} = -2.78$$

The tabulated value of t for 5 degrees of freedom at 5% level of significance is: 2.02 for one tailed test. Here calculated t, 2.78 is greater than the table value of 2.02, so the concessional scheme is a success.

5.13.1 *t*-test for Sample Correlation Coefficient

$$t = \frac{r}{(1-r^2)^{1/2}} \times (n-2)^{1/2}$$

r = Observed sample correlation coefficient;

n = Sample size; $n - 2$ is degrees of freedom.

Example 5.29 A random sample of 27 pairs of observations from a normal population gives a correlation coefficient of 0.42. Is it that the variables in the population are uncorrelated?

Solution: $n = 27;\ r = 0.42;\ t = \frac{r\,(n-2)^{1/2}}{(1-r^2)^{1/2}} = \frac{0.42 \times (27.2)^{1/2}}{(1-0.42^2)^{1/2}} = 2.31$

The tabulated value of t for 25 degrees of freedom and 5% level of σ significance for a two tailed test is 2.06. Since calculated t (2.31) is greater than tabulated t (2.06), so the variables are correlated in the population.

5.14 F* STATISTIC

$$F = \frac{X / n_1}{Y / n_2}; \quad F \text{ is } F \text{ statistic}$$

Fischer's *Z*-distribution:

$$Z = \frac{1}{2}\log_e F$$

or

$$Z = \text{Exp}\ (2Z) = e^{2Z}$$

Relation between t, F, and Chi-square distribution:

$$F = t^2 = \left(\frac{\bar{x} - \mu}{S / (n)^{1/2}}\right)^2$$

Relation between F and chi-square is:

$$nF = \chi^2$$

Example 5.30 Can the following two samples be regarded as coming from the same sample normal population?

Sample	Size	Sample mean	Sum of the squares of deviations from the mean
1	10	12	120
2	12	15	314

Solution: Equality of means is tested by applying t test. Equality of variance is tested by F-test.

t-test assumes, $\sigma_1^2 = \sigma_2^2$;

We are given:

$$n_1 = 10;\ n_2 = 12;$$

$$\bar{x} = 12;\ \bar{y} = 15;$$

$$\sum (x-\bar{x})^2 = 120;\ \sum (y-\bar{y})^2 = 314$$

F-test:

$$S_x^2 = \frac{1}{n_1 - 1}\sum (x-\bar{x})^2$$

$$= \frac{120}{9} = 13.33;\ S_y^2 = \frac{1}{n_2-1}\sum (y-\bar{y})^2$$

$$= \frac{314}{11} = 28.55$$

$$F = \frac{S_y^2}{S_x^2} = \frac{23.55}{13.33} = 2.14$$

Tabulated $F_{0.05}(n_2 - 1, n_1 - 1) = F_{0.05}(11, 9) = 3.10$ (from F tables)

Since tabulated $F(3.10)$ is greater than calculated $F(2.14)$, so null hypothesis of equality of population variances may be accepted.

So, $\sigma_1^2 = \sigma_2^2$

t-test:

$$t = \frac{\bar{x} - \bar{y}}{\left[S^2\left(\frac{1}{n_1} + \frac{1}{n_2}\right)^{1/2}\right]};\quad t_{n_1+n_2-2}\ (t_{10+12-2} = t_{20})$$

$$S^2 = \frac{1}{n_1+n_2-2}\left[\sum ((x-\bar{x})^2 + \sum (y-\bar{y})^2)\right]$$

$$= \frac{1}{20}(120 + 314) = 21.7$$

$$t = \frac{12 - 15}{\left[21.7\left(\frac{1}{10} + \frac{1}{12}\right)\right]^{1/2}} = -1.54$$

$t_{0.5}(20)$ for two tailed test from Tables is: 2.086 which is t_{obs}.

Since t_{obs} (2.086) is greater than $|t|$ (1.54), so $\mu_1 = \mu_2$ (means).

Since F test and t test show, $\mu_1 = \mu_2$ and $\sigma_1 = \sigma_2$, so that the samples are drawn from the same normal population.

5.15 THEORY OF ATTRIBUTES, CONSISTENCY OF DATA

Attribute is quality or characteristics. Some examples are: beauty, honesty, gender, health, smoking, drinking, etc. This deals with qualitative characteristics which cannot be measured quantitatively.

The presence of attributes in the units of the population is denoted by the capital letters of the English alphabet say A, B, C, D, etc., and their absence is denoted by the corresponding small letters of the Greek alphabet say $\alpha, \beta, \gamma, \delta, \ldots$ etc.

For example, if c denotes males then γ denotes females.

Example 5.31 Given, $N = 2000$; $(A) = 1500$; $(B) = 100$ and $(AB) = 350$, test the consistency of the data.

Solution: Given, $N = 2000$; $(A) = 1500$; $(B) = 100$; and $(AB) = 350$

(AB) is common to (A) and (B) or combined (A), (B) qualities. (A) is one quality; (B) is second quality (another). To test the consistency of the data, we will find the ultimate class frequencies, i.e., (αB), $(A\beta)$ and $(\alpha\beta)$

$(\alpha B) = (B) - (AB) = 100 - 350 = -250$. Since (αB) is less than zero, the given data are not consistent.

Example 5.32 In a report on consumer preference, it was given that out of 500 persons surveyed, 400 preferred variety A; 380 preferred variety B; and 270 liked both A and B. Is the data consistent or no?

Solution: Given, $N = 500$; $(A) = 400$; $(B) = 380$; $(AB) = 270$

$$(\alpha\beta) = N - (A) - (B) + (AB)$$

$$= 500 - 400 - 380 + 270 = -10$$

Since each frequency is to be positive, therefore, the given data are inconsistent.

Example 5.33 Given, $N = 2000$; $(A) = 1500$; $(B) = 100$ and $(AB) = 350$; test the consistency of data.

Solution: Given, $N = 2000$; $(A) = 1500$; $(B) = 100$ and (AB) 350;

To test the consistency of data, we find the class frequencies: (αB); $(A\beta)$ and $(\alpha\beta)$.

Then, $(\alpha B) = (B) - (AB) = 100 - 350 = -250$; since (αB) is less than zero (–250), the given data is inconsistent.

5.16 COMPLEX VARIABLES

$$u = x + iy; \quad x = \text{real part}; \quad y = \text{imaginary part};$$

$$u = \text{complex variable.}$$

$$\text{Magnitude of a complex variable} = ((\text{Real part}^2 + \text{Imaginary part}^2))^{1/2}$$

$$\text{Angle of a complex number} = -\tan^{-1} (\text{Imaginary part/Real part})$$

These relations are used in frequency response analysis of process control.

EXAMPLES IN CHEMICAL ENGINEERING

Example 5.34 Find the amplitude ratio and angle of the transfer function , $G(S) = \dfrac{5}{(S+1)}$

Solution: Putting, $S = i\omega$; $G(i\omega) = \dfrac{5}{1+i\omega}$

Multiplying and dividing by $\dfrac{1}{(1-i\omega)}$ gives

$$\frac{5(1-i\omega)}{(1^2 - i^2\omega^2)} = \frac{5(1-i\omega)}{(1+\omega^2)} = \frac{5-5i\omega}{1+\omega^2}$$

Amplitude ratio (magnitude of complex number):

$$[(\text{Real part})^2 + (\text{Imaginary part})^2]^{1/2} = \left[\left(\frac{5}{1+\omega^2}\right)^2 + \left(\frac{-5\omega}{1+\omega^2}\right)^2\right]^{1/2}$$

$$= \frac{5}{[(1+\omega^2)]^{1/2}}$$

$$\text{Angle (phase angle)} = \tan^{-1}\left(\frac{\text{Imaginary part}}{\text{Real part}}\right)$$

$$= \tan^{-1}\left[\frac{\dfrac{-\omega}{1+\omega^2}}{\dfrac{1}{1+\omega^2}}\right]$$

$$= \tan^{-1}(-\omega) = -\tan^{-1}(\omega) \quad \textbf{Ans.}$$

Example 5.35 Find magnitude (A.R.) and angle of the transfer function,

$$G(S) = \frac{6S \times e^{-2S}}{(2S+1)(S-1)(3S+1)} = G(S)$$

Solution: Overall gain (A.R.O.) is the product of the individual gains (magnitudes).

Overall angle is summation of the individual transfer functions angles.

For S: Magnitude: $S = i\omega$; $(\text{R.P}^2 + \text{Img. } P^2)^{1/2}$

$$= (0^2 + \omega^2)^{1/2} = \omega$$

Angle: $\tan^{-1}\left(\frac{\text{Imaginary part}}{\text{Real part}}\right) = \tan^{-1}\left(\frac{\omega}{0}\right) = \infty;$

$$\tan(90^\circ) = \infty. \text{ So angle} = 90^\circ$$

For e^{-2S}: $e^{-2i\omega}$; $e^{-i\omega} = \cos i\omega - i \sin i\omega$

$$e^{-2i\omega} = \cos 2i\omega - i\sin 2i\omega$$

$$\text{Magnitude} = [(\text{R.P}^2 + (\text{I.P}^2]^{1/2} = (\cos 2\omega^2 + \sin 2\omega^2)^{1/2} = 1$$

$$\text{Angle} = \tan^{-1}\left(\frac{\text{I.P.}}{\text{R.P.}}\right) = \tan^{-1}\left(\frac{-\sin 2\omega}{\cos 2\omega}\right)$$

$$= \tan^{-1}(-\tan 2\omega) = -2\omega \text{ Rad.}$$

$$= \left(\frac{-180}{\pi}\right) \times 2\omega \text{ degrees.}$$

For 6: Magnitude is 6. Angle is zero degrees.

For $G(i\omega)$:
$$\text{Magnitude} = \frac{6 \times \omega \times 1}{(4\omega^2+1)^{1/2}(\omega^2+1)^{1/2}(9\omega^2+1)^{1/2}}$$

For $\frac{1}{S-1}$, putting $S = i\omega$,
$$\frac{1}{i\omega - 1} = \frac{i\omega + 1}{(i^2\omega^2 - 1)} = \frac{-1}{1+\omega^2} - \frac{i\omega}{1+\omega^2}$$

$$\text{Magnitude} = (\text{R.P.}^2 + \text{I.P.}^2)^{1/2} = \left[\frac{1}{(1+\omega^2)^2} + \frac{\omega^2}{(1+\omega^2)^2}\right]^{1/2} = \frac{1}{(1+\omega^2)^{\frac{1}{2}}}$$

$$\text{Angle} = \tan^{-1}\left(\frac{\text{I.P.}}{\text{R.P.}}\right)$$

$$= \tan^{-1}\left(\frac{\frac{-\omega}{1+\omega^2}}{\frac{-1}{1+\omega^2}}\right) = \tan^{-1}(\omega)$$

5.17 VECTOR ANALYSIS

1. For equation of **continuity of a fluid:** $\frac{d\rho}{dt} = \rho \text{ div } V$

$$\text{div } V = \frac{dV_x}{dx} + \frac{dV_y}{dy} + \frac{dV_z}{dz} = \nabla \cdot V$$

For incompressible fluid div $V = 0$

2. Curl: If $V = \Omega \times R$; $\omega = \omega_1 i + \omega_2 j + \omega_3 k$; $R = xi + yj + zk$

W = Angular velocity

R = Radius vector

$$V = \begin{vmatrix} i & j & k \\ \omega_1 & \omega_2 & \omega_3 \\ x & y & z \end{vmatrix}$$

$$\text{Curl } V = \begin{vmatrix} i & j & k \\ \frac{d}{dx} & \frac{d}{dy} & \frac{d}{dz} \\ (\omega_2 z - \omega_3 y) & (\omega_3 x - \omega_1 z) & (\omega_1 y - \omega_2 x) \end{vmatrix}$$

$$= i(\omega_1 + \omega_1) + j(\omega_2 + \omega_2) + k(\omega_3 + \omega_3) = 2\Omega;$$

$$\Omega = \text{Curl } V/2$$

If curl of a velocity vector is zero, then irrotational flow. The curl of any vector gives a measure of the angular velocity at any point of the vector field.

3. Conversion of coordinates:

$$\text{div } V = \frac{dV_1}{dx} + \frac{dV_2}{dy} + \frac{dV_3}{dz}$$

$$\text{div } V = \nabla \cdot V = \left(\frac{di}{dx} + \frac{dj}{dy} + \frac{dk}{ds} \right) \cdot (V_1 i + V_2 j + V_3 k)$$

$$= \frac{dV_1}{dx} + \frac{dV_2}{dy} + \frac{dV_3}{dz}; \quad V \text{ is a vector.}$$

$$\text{Grad } f = \left(df \frac{i}{dx} + df \frac{j}{dy} + df \frac{k}{dz} \right); \quad f \text{ is a scalar function.}$$

$$\text{Curl of a vector } (V) = V(x, y, z)$$

$$= V_1 i + V_2 j + V_3 k$$

$$\text{Curl } V = \nabla\times V = \begin{bmatrix} i & j & k \\ \frac{d}{dx} & \frac{d}{dy} & \frac{d}{dz} \\ V_1 & V_2 & V_3 \end{bmatrix}$$

$$= \left(\frac{dV_3}{dy} - \frac{dV_2}{dz}\right)i + \left(\frac{dV_1}{dz} - \frac{dV_3}{dx}\right)j + \left[k\left(\frac{dV_2}{dx} - \frac{dV_1}{dy}\right)\right]$$

5.18 ORTHOGONAL CURVILINEAR COORDINATES

Coordinates Conversion Using Vector Analysis

$$x = x(u, v, w);\ y = y(u, v, w);\ z = z(u, v, w) \quad \ldots(1)$$

If Eq. (1) is solved, u, v, w in terms of x, y, z then

$$u = u(x, y, z);\ v = v(x, y, z);\ w = w(x, y, z)$$

Arc element length is $\quad ds^2 = h_1^2 du^2 + h_2^2 dv^2 + h_3^2 dw^2 \quad \ldots(2)$

Area element is: $\quad ds_w = |du \times dv|;\ ds_v = |du \times dw|;\ ds_u = |dv \times d_w|$

$$ds_w = h_1 h_2\, du\, dv;\ ds_v = h_1 h_3\, dudw;$$

$$ds_u = h_2 h_3\, dvdw$$

Volume element: $\quad dv = (h_1 du\, T_u)\ (h_2\, dv\, T_v)\ (h_3 dw\, T_w)$

$$= h_1 h_2 h_3\, du\, dv\, dw \quad \ldots(3)$$

Cylindrical Coordinates: r, θ, z are the variables.

$$x = r\cos\theta;\ y = r\sin\theta;\ z = z$$

Arc element 3: $\quad (ds)^2 = (dx)^2 + (dy)^2 + (dz)^2$

$$= (dr)^2 + (r\, d\theta)^2 + dz^2$$

So, $\quad h_1 = 1;\ h_2 = r;\ h_3 = 1$

Volume element, $dv = r\, drd\theta dz$;

$$dR = \left(\frac{dR}{du}\right)du + \left(\frac{dR}{dv}\right)dv + \left(\frac{dR}{dw}\right)dw$$

$$\frac{dR}{dw} = h_3 T_w;\ \frac{dR}{du} = h_1 T_u;\ \frac{dR}{dv} = h_2 T_v$$

h_1, h_2, h_3 are scalar factors, T_u, T_v, T_ω are unit tangent vectors.

$$dR = h_1 duT_u + h_2 dvT_v + h_3 dwT_w \quad \ldots(4)$$

$$\Delta f = \frac{T_u df}{h_1 du} + \frac{T_v df}{h_2 dv} + \frac{T_w df}{h_3 dw}$$ (gradient f) where f is a scalar function.

$$\nabla\cdot F = \frac{1}{h_1 h_2 h_3}\left[\frac{\partial}{\partial u}(h_2 h_3 f_1) + \frac{\partial}{\partial v}(h_3 h_1 f_2) + \frac{\partial}{\partial w}(h_1 h_2 f_3)\right]$$

$$\Delta \times F = \begin{bmatrix} \dfrac{T_u}{h_2 h_3} & \dfrac{T_v}{h_3 h_1} & \dfrac{T_w}{h_1 h_2} \\ \dfrac{\partial}{\partial u} & \dfrac{\partial}{\partial v} & \dfrac{\partial}{\partial w} \\ h_1 f_1 & h_2 f_2 & h_3 f_3 \end{bmatrix}$$

where $F = f_1 T_u + f_2 T_v + f_3 T_w$

$$\nabla^2 f = \nabla \cdot (\nabla f) = \nabla \cdot \left(\frac{T_u}{h_1} \frac{\partial f}{\partial u} + \frac{T_v}{h_2} \frac{\partial f}{\partial u} + \frac{T_w}{h_3} \frac{\partial f}{\partial w} \right)$$

$$\nabla^2 f = \frac{1}{h_1 h_2 h_3} \sum \frac{\partial}{\partial u} \left(\frac{1}{h_1} \frac{\partial f}{\partial u} h_2 h_3 \right)$$

In cylindrical coordinates $x = r\cos\theta;\ y = r\sin\theta;\ z = z$

$$(ds)^2 = (dx)^2 + (dy)^2 + (dz)^2 = (dr)^2 + (r\,d\theta)^2 + (dz)^2;$$

$$h_1 = 1;\ h_2 = r;\ h_3 = 1$$

$$\nabla^2 f = \frac{1}{r}\left[\frac{\partial}{\partial r}\left(r \frac{\partial f}{\partial r} \right) + \frac{\partial}{\partial \theta}\left(\frac{1}{r} \frac{\partial f}{\partial \theta} \right) + \frac{\partial}{\partial z}\left(r \frac{\partial f}{\partial z} \right) \right]$$

$$= \frac{\partial^2 f}{\partial r^2} + \frac{1}{r}\frac{\partial f}{\partial r} + \frac{1}{r^2}\frac{\partial^2 f}{\partial \theta^2} + \frac{\partial^2 f}{\partial z^2}$$

5.19 IN SPHERICAL COORDINATES

$$x = r\sin\theta\cos\phi;\ y = r\sin\theta\sin\phi;\ z = r\cos\phi;$$

$$(ds)^2 = (dx)^2 + (dy)^2 + (dz)^2$$

$$= (dr)^2 + r^2 (d\theta)^2 + (r\sin\theta)^2 (d\phi)^2$$

$$h_1 = 1;\ h_2 = r;\ h_3 = r\sin\theta;$$

$$\nabla f = \frac{T_u}{h_1}\frac{\partial f}{\partial u} + \frac{T_v}{h_2}\frac{\partial f}{\partial u} + \frac{T_\omega}{h_3}\frac{\partial f}{\partial \omega}$$

$$\nabla f = \frac{\partial f}{\partial r} T_r + \frac{1}{r}\frac{\partial f}{\partial \theta} T_\theta + \frac{1}{r\sin\theta}\frac{\partial f}{\partial \phi} T_\phi$$

$$F = f_1 T_u + f_2 T_v + f_3 T_\omega$$

$$\nabla \cdot f = \frac{1}{h_1 h_2 h_3}\left[\frac{\partial}{\partial u}(h_2 h_3 f_1) + \frac{\partial}{\partial v}(h_3 h_1 f_2) + \frac{\partial}{\partial \omega}(h_1 h_2 f_3) \right]$$

$$\nabla \cdot f = \frac{1}{r^2 \sin\theta}\left[\frac{\partial}{\partial r}(r^2 \sin\theta\, f_1) + \frac{\partial}{\partial \theta}(r\sin\theta\, f_2) + \frac{\partial}{\partial \phi}(r f_3) \right]$$

Expressions for Curl F;

where $F = f_1 T_u + f_2 T_v + f_3 T_\omega$

$$\nabla \times F = \begin{bmatrix} \dfrac{T_u}{h_2 h_3} & \dfrac{T_v}{h_3 h_1} & \dfrac{T_\omega}{h_1 h_2} \\ \dfrac{\partial}{\partial u} & \dfrac{\partial}{\partial v} & \dfrac{\partial}{\partial \omega} \\ h_1 f_1 & h_2 f_2 & h_3 f_3 \end{bmatrix}$$

$$\nabla \times F = \begin{vmatrix} \dfrac{T_r}{r^2 \sin\theta} & \dfrac{T_\theta}{r \sin\theta} & \dfrac{T_\phi}{r} \\ \dfrac{\partial}{\partial r} & \dfrac{\partial}{\partial \theta} & \dfrac{\partial}{\partial \phi} \\ f_1 & r f_2 & r \sin\theta f_3 \end{vmatrix}$$

$$\nabla^2 F = \frac{1}{h_1 h_2 h_3}\left[\frac{\partial}{\partial u}\left(\frac{h_2 h_3}{h_1}\frac{\partial f}{\partial r}\right) + \frac{\partial}{\partial v}\left(\frac{h_3 h_1}{h_2}\frac{\partial f}{\partial v}\right) + \frac{\partial}{\partial \omega}\left(\frac{h_1 h_2}{h_3}\frac{\partial f}{\partial \omega}\right)\right]$$

$$\nabla^2 F = \frac{1}{r^2 \sin\theta}\left[\frac{\partial}{\partial r}\left(r^2 \sin\theta \frac{\partial f}{\partial r}\right) + \frac{\partial}{\partial \theta}\left(r\frac{\sin\theta}{r}\frac{\partial f}{\partial \theta}\right) + \frac{\partial}{\partial \phi}\left(\frac{\partial f}{r \sin\theta\, \partial \phi}\right)\right]$$

$$= \frac{\partial^2 f}{\partial r^2} + \frac{1}{r^2}\frac{\partial^2 f}{\partial \theta^2} + \frac{1}{r^2 \sin^2\theta}\frac{\partial^2 f}{\partial \phi^2} + \frac{2}{r}\frac{\partial f}{\partial r} + \frac{\cot\theta}{r^2}\frac{\partial f}{\partial \phi}$$

5.20 REGRESSION

Linear regression, least square regression, polynomial regression.

Linear Regression: The simplest example of least square regression is fitting a straight line to a paired set of observation (x_1, y_1), (x_2, y_2), (x_3, y_3).... (x_n, y_n). The mathematical equation is: $y = I + mx + e$ where I is intercept, m is slope and e is error.

Requirement for best fit: One method for best fit of line through the data is to minimize the sum of the residual errors of all given points.

$$\sum_{i=1}^{n} e_i = \sum_{i=1}^{n} (y_i - I_i - m_i x_i)^2$$

Other methods like: $\sum_{i=1}^{n} (y_i - I_i - m_i x_i)$ and using absolute values are $\sum_{i=1}^{n} e_i = \sum_{i=1}^{n} |y_i - I_i - m_i x_i|$ are worse than minimising the residual errors method.

The selected one is: $S = \sum_{i=1}^{n} e_i^2 = \sum_{i=1}^{n} (y_{i,\text{ measured}} - y_{i,\text{model}})^2$

$$S = \sum_{i=1}^{n} (y_i - I_i - m\,x_i)^2$$

I and *m* values are calculated by minimising *S* as *dS/dm* and *dS/dI* as zeroes.

Least Square Regression of Fitting a Straight Line

$$dS/dm = -2\sum (y_i - I - m\,x_i)$$

$$dS/dI = -2\sum ((y_i - I - m\,x_i)x_i)$$

Equating to zeroes give: $0 = \sum y_i - \sum I - \sum m\,x_i$

$$0 = \sum y_i\,x_i - \sum I x_i - \sum m x_i^2$$

$$\sum I = nI;\ nI + \left(\sum x_i\right)m = \sum y_i$$

$$\left(\sum x_i\right)I + \left(\sum x_i^2\right)m = \sum x_i\,y_i$$

Solving simultaneously gives, $m = n\dfrac{\sum x_i\,y_i - \sum x_i \times \sum y_i}{n\sum x_i^2 - \left(\sum x_i\right)^2}$;

$$I = \frac{\sum y_i - \left(\sum x_i\right)m}{n}$$

For problem please see chapter 5 (statistical methods).

S (sum of the squares of the errors) $= \sum e_i^2 = \sum_{i=1}^{n} (y_i - I - m\,x_i)^2$

Standard error of the estimate $(S_{y/x}) = \sqrt{\dfrac{S}{(n-2)}}$

$S_t = \sum (y_i - \bar{y})^2$ where $\bar{y}$ is average (mean) value.

$r^2 = \dfrac{S_t - S}{S_t}$ where r^2 is coefficient of determination and *r* is correlation coefficient. $\left(= \sqrt{r^2}\right)$.

r^2 indicates that the fraction of original uncertainty is explained by linear model.

Linearizing non-linear equations: Examples are: $y = a_1\,e^{b_1 x}$; $y_2 = a_2\,x^{b_2}$; $y = a_3\dfrac{x}{(b_3 + x)}$ where a_1, b_1, a_2, b_2, a_3 and b_3 are constants. Equation $y = a_1\,e^{b_1 x}$ is linearized by taking logarithms on both sides ($\ln y = \ln a_1 + b_1\,x \ln e$). Here $\ln e = 1$ giving: $\ln y = \ln a_1 + b_1 x$ which is of linear form (straight line form). The other equation ($y = a_2\,x^{b_2}$) is linearized by taking logarithms on both sides, as: $\log y = \log a_2 + b_2 \log x$.

The equation $y = a_3\dfrac{x}{(b + x)}$ is linearized by inverting it, to give:

$$1/y = \frac{(b_3 + x)}{a_3 x} = (b_3/a_3)\,(1/x) + (1/a_3)$$

A plot of $(1/y)$ versus $(1/x)$ is linear, with slope of (b_3/a_3) and intercept of $1/a_3$.

Polynomial Regression

Some data cannot be represented by a straight line relation. For these a curve can represent to fit the data. One method to perform this is to fit polynomials called polynomial regression. The least square method can be extended to polynomial method.

For example: For second order polynomial: $y = a_0 + a_1 x + a_2 x^2 + e$

$$S = \sum_{i=1}^{n} (y_i - I - m_1\, x_i - m_2\, x_i^2)$$

$$dS/dI = -2\sum (y_i - I - m_1\, x_i^2)^2$$

$$dS/dm_1 = -2\sum x_i\,(y_i - I - m_1\, x_i - m_2\, x_i^2)$$

$$dS/dm_2 = -2\sum x_i^2 (y_i - I - m_1\, x_i - m_2\, x_i^2)$$

The above equations are equated to zeroes for minimising giving:

$$(n)I + \left(\sum x_i\right) m_1 + \left(\sum x_i^2\right) m_2 = \sum y_i$$

$$\left(\sum x_i\right) I + \left(\sum x_i\right)^2 m_1 + \left(\sum x_i^3\right) m_2 = x_i\, y_i$$

$$\left(\sum x_i^2\right) I + \left(\sum x_i^3\right) m_1 + \left(\sum x_i^4\right) m_2 = \sum x_i^2\, y_i$$

$1, m_1, m_2$ are calculated by solving the above equations simultaneously. For mth. order polynomial, the equation is: $y = a_0 + a_1\, x + a_2\, x^2 + \ldots + a_m\, x_m + e^{\mathrm{m}}$

Standard error is $(S_{y/x}) = \sqrt{\dfrac{S}{n(m+1)}}$

r^2 (correlation coefficient) $= \dfrac{S_t - S}{S_t}$

where $S_t = (y_i - \overline{y})^2$

5.21 LINEAR ALGEBRAIC EQUATIONS

Gauss Elimination Method

The set of equations are (linear):

$$a_{11}\, x_1 + a_{12}\, x_2 + a_{13}\, x_3 + \ldots\ldots + a_{1n}\, x_n = b_1 \qquad \ldots 5.1\text{ a}$$

$$a_{21}x_1 + a_{22}x_2 + a_{23}x_3 + ...+ a_{2n}x_n = b_2 \quad ...\ 5.1\ b$$

$$a_{31}x_1 + a_{32}x_2 + ...+ a_{3n}\cdot x_n = b_3 \quad ...\ 5.1\ c$$

$$\vdots$$

$$a_{n1}x_1 + a_{n2}x_2 + a_{n3}x_3 + ...+ a_{nn}x_n = b_n \quad ...\ 5.1\ n$$

The solution contains eliminating unknowns and then solution through back substitution.

Forward eliminating of unknowns: The first step is to eliminate the first variable x_1, from second to the last equation. To do this, we have to multiply Eq. 5.1 a by a_{21}/a_{11} to give:

$$a_{21}x_1 + \frac{a_{21}}{a_{11}} a_{12}x_2 .. + \frac{a_{21}}{a_{11}} a_{1n}x_n = \frac{a_{21}}{a_{11}} b_1 \quad ...\ 5.2$$

Now this equation is subtracted (5.2) from Eq. (5.1 b) to give:

$$\left(a_{22} - \frac{a_{21}}{a_{11}} a_{12}\right)x_2 + ... + \left(a_{2n} - \frac{a_{21}}{a_{11}} a_{1n}x\right)x_n = \frac{b_2 - a_{21}\ b_1}{a_{11}}$$

or $$a_{22}'\ x_2 + ... + a_{2n}'\ x_n = b_2'$$

This procedure is repeated for the left equations. For example: Equation 5.1 a can be multiplied by a_{31}/a_{11} and the resulting equation is subtracted from Eq. 5.1 c. This procedure results for the remaining equations as:

$$a_{11}x_1 + a_{12}x_2 + a_{13}x_3 + ...\ a_{1n}x_n = b_1 \quad ...\ 5.3\ a$$

$$a'_{22}x_2 + a'_{23} + ...\ a'_{2n}x_n = b_2' \quad ...\ 5.3\ b$$

$$a'_{32}x_2 + a'_{33} + ... + a'_{3n}x_n = b_3 \quad ...\ 5.3\ c$$

$$\vdots$$

$$a'_{n2}x_2 + a'_{n3} + ... + a'_{nn}x_n = b_n' \quad ...\ 5.3\ n$$

Now repeating the above to eliminate the second variable from Eq. 5.3c through 5.3n equations. To do multiply by a'_{32}/a_{22}' and subtract from 5.3 c. Doing same operation for the, remaining equations, gives;

$$a_{11}x_1 + a_{12}x_2 + a_{13}x_3 + ... + a_{1n}x_n = b_1$$

$$a'_{22}x_2 + a'_{23}x_3 + ... + a'_{2n}\ x_n = b_2''$$

$$a'_{33}x_3 + \qquad + a''_{3n}\ x_n = b_3''$$

$$\vdots$$

$$a_{n3}''\ x_3 + ... + a_{nn}''\ x_n = b_n''$$

This procedure is continued for the remaining equations. The final set of equations are:

$$a_{11}x_1 + a_{12}x_2 + a_{13}x_3 + ... + a_{1n} = x_n = b_1 \quad ...\ 5.4\ a$$

$$a'_{22}x_2 + a'_{23}x_3 + ... + a'_{2n}x_n = b'_2 \quad ...\ 5.4\ b$$

$$a''_{33}\, x_3 + ... + a''x_n = b_3''$$

....

....

$$a_{nn}^{\,n-1}\, x_n = b_n^{\,n-1} \quad \text{... 5.4 n}$$

$$x_n = b_n^{\,n-1}/a_{nn}^{\,n-1} \quad \text{... 5.5}$$

This x_n value can be used into $(n-1)$ equation to obtain for x_{n-1}. This procedure can be represented by the following formula.

$$x_i = \frac{b_i^{i-1} - \sum_{j=1+1}^{n} a_{ij}^{i-1}\, x_j}{a_{ii}^{i-1}} \quad \text{for } i = n-1, n-2, ... 1$$

Example: Solve using Gauss elimination method.

$$x_1 + x_2 - x_3 = -3 \quad \text{... 5.5a}$$

$$6x_1 + 2x_2 + 2x_3 = 2 \quad \text{... 5.5 b}$$

$$-3x_1 + 4x_2 + x_3 = 1 \quad \text{... 5.5 c}$$

Step-1: Multiplying Eq. 5.5 a by 6 and subtracting from Eq. 5.5 b gives:

$$-4x_2 + 8x_3 = 20 \quad \text{... 5.5 d}$$

Step-2: Multiplying Eq. 5.5 a by –3 and subtracting from Eq. 5.5 c gives:

$$7x_2 - 2x_3 = -8 \quad \text{... 5.5 e}$$

Solving Eqs. 5.5 d and 5.5 e gives: $x_2 = -0.5$ and $x_3 = 2.25$.

Substituting x_2 and x_3 values in Eq. 5.5 a gives: $x_1 = -0.25$.

The values of x_1, x_2, x_3 are confirmed in Eqs. 5.5 b and 5.5 c.

$x_1 = -0.25$; $x_2 = -0.5$; $x_3 = 2.25$ **Ans.**

If division by zero comes anywhere, then the method does not work.

Jordan Method

The elimination method gives an identity matrix.

Example:

$$x_1 + x_2 - x_3 = -3 \quad \text{... 5.5 a}$$

$$6x_1 + 2x_2 + 2x_3 = 2 \quad \text{... 5.5 b}$$

$$-3x_1 + 4x_2 + x_3 = 1 \quad \text{... 5.5 c}$$

***Solution*:** Step-1: First writing the coefficients and right-hand side terms as:

$$\begin{bmatrix} 1 & 1 & -1 & -3 \\ 6 & 2 & 2 & 2 \\ -3 & 4 & 1 & 1 \end{bmatrix}$$

Step-2: Normalize the first row, if it is not. Normalized by dividing with x_1 coefficient throughout.

Step-3: x_1 term can be eliminated from second row by multiplying 1st row by 6 and subtracting from second row, giving:

$$\begin{bmatrix} 1 & 1 & -1 & -3 \\ 0 & -4 & 8 & 20 \\ 0 & -7 & 2 & 8 \end{bmatrix}$$

Step-4: Normalizing second row gives:

$$\begin{bmatrix} 1 & 1 & -1 & -3 \\ 0 & 1 & -2 & -5 \\ 0 & -7 & 2 & 8 \end{bmatrix}$$

Step-5: Removal of x_2 from 3rd row, by multiplying row 2 by –7 and subtracting from 3rd row gives:

$$\begin{bmatrix} 1 & 1 & -1 & -3 \\ 0 & 1 & -2 & -5 \\ 0 & 0 & 1 & 2.25 \end{bmatrix}$$

Step-6: The x_3 term can be removed from 2nd row by multiplying 3rd row by –2 and subtracting from 2nd row giving as:

$$\begin{bmatrix} 1 & 1 & -1 & -3 \\ 0 & 1 & 0 & -0.5 \\ 0 & 0 & 1 & 2.25 \end{bmatrix}$$

Step-7: x_2 term is removed from 1st row by subtracting 2nd row, x_3 term is removed from first tow giving:

$x_1 = -0.25$; $x_2 = -0.50$; $x_3 = 2.25$ Ans.

Gauss Seidel Method

This is an iterative process. The set of equations to be solved are:

$$[A]\,(X) = (B)$$

Taking a 3rd order matrix (3 × 3) set of equations, if the diagonal elements are all non-zero, the first equation can be solved from x_1. The second for x_2 and the third for x_3 to give:

$$x_1 = b_1 - \left(\frac{a_{12}\ x_2 - a_{13}x_3}{a_{11}}\right);\ x_2 = b_2 - \left(\frac{a_{21}\ x_1 - a_{23}x_3}{a_{22}}\right)$$

$$x_3 = b_3 - \left(\frac{a_{31}\ x_1 - a_{32}x_2}{a_{33}}\right)$$

The solution is started by assuming zero values for unknown x_1, x_2 and x_3.

Example: $10x_1 + 2x_2 - x_3 = 27$; $-3x_1 - 6x_2 + 2x_3 = -61.5$

$x_1 + x_2 + 5x_3 = -21.5$

***Solution*:** $x_1 = (27 - 2x_2 + x_3)/10 = 2.7 - 0.2x_2 + 0.1x_3$

$x_2 = (-61.5 + 3x_1 - 2x_3)/-6 = (61.5 - 3x_1 + 2x_3)/6$

$x_3 = (-21.5 - x_1 - x_2)/5 = (-21.5 - x_1 - x_2)/5$

Iteration 1: $x_1 = 2.7 - 0 - 0$ (Assuming x_2 and x_3 as zeroes in the beginning) = 2.7

$x_2 = (61.5 - 3 \times 2.7 + 2 \times 0)/6 = 8.9$ (Assuming x_3 as zero in the beginning).

$x_3 = (-21.5 - 2.7 - 8.9)/5 = -6.62$

Iteration 2: $x_1 = 2.7 - 0.2 \times 8.9 - .662 = 0.26$

$x_2 = (61.5 - 0.78 + 13.24)/6 = 7.9$

$x_3 = (-21.5 - 0.26 - 7.9)/5 = -5.94$

It is reducing.

The iterations are continued until desired accuracy is reached.

LU decomposition and Matrix Inversion

LU decomposition: This method avoids dividing by zero during calculations. This is another method to solve a set of linear equations like Gauss elimination method. The system of linear equations is:

$$[A]\ (X) = (B) \qquad \text{... 5.6}$$

The above equations are rearranged as: $[A]\ (X) - (B) = 0$ ... 5.7

Equation 5.2 can be expressed as an upper triangular system as:

$$\begin{bmatrix} u_{11} & u_{12} & u_{13} \\ 0 & u_{22} & u_{23} \\ 0 & 0 & u_{33} \end{bmatrix} \begin{pmatrix} x_1 \\ x_2 \\ x_3 \end{pmatrix} = \begin{pmatrix} d_1 \\ d_2 \\ d_3 \end{pmatrix} \qquad \text{... 5.8}$$

This is the manipulations that occurs in Gauss elimination process. The elimination is used to reduce the system to upper triangular form. Equation 5.2 is expressed in matrix notation and rearranged to give:

$$[U]\ (X) - (D) = 0 \qquad \text{... 5.9}$$

Assuming a lower diagonal matrix with 1's on the diagonal,

$$[L] = \begin{bmatrix} 1 & 0 & 0 \\ 1_{21} & 1 & 0 \\ 1_{31} & 1_{32} & 1 \end{bmatrix} \quad \text{... 5.10}$$

which has the property that: $[L]\{[U](X) - (D)\} = [A](X) - (B)$ ... 5.11

If the equation agrees, then:

$$[L] = [A] \quad \text{... 5.12}$$

$$[L](D) = (B) \quad \text{... 5.13}$$

Solutions are based on Eqs. 5.9, 5.12 and 5.13.

Gauss Elimination using LU Decomposition

The forward elimination method reduces the original coefficient matrix $[A]$ to upper triangular form of:

$$[U] = \begin{bmatrix} a_{11} & a_{12} & a_{13} \\ 0 & a'_{22} & a'_{23} \\ 0 & 0 & a''_{33} \end{bmatrix} \quad \text{... 5.14}$$

The matrix $[L]$ is produced during the steps. For three equation system,

$$\begin{bmatrix} a_{11} & a_{12} & a_{13} \\ a_{12} & a_{22} & a_{23} \\ a_{31} & a_{32} & a_{33} \end{bmatrix} \begin{pmatrix} x_1 \\ x_2 \\ x_3 \end{pmatrix} = \begin{pmatrix} b_1 \\ b_2 \\ b_3 \end{pmatrix}$$

The first step in Gauss eliminating is to multiply row 1 by a factor $f_{21} = a_{21}/a_{11}$ and substract the result from second row to eliminate a_{21}.

Similarly row 1 is multiplied by $f_{31} = a_{31}/a_{11}$ and the result subtracted from 3rd row to eliminate a_{31}. The next step, is to multiply the modified second row by: $f_{32} = a'_{32}/a'_{22}$ and subtract the result from 3rd row to eliminate a'_{32}. After elimination the $[A]$ can be written as:

$$\begin{matrix} a_{11} & a_{12} & a_{13} \\ f_{21} & a'_{22} & a'_{23} \\ f_{31} & f_{32} & a''_{33} \end{matrix}$$

The matrix represents the LU decomposition of $[A]$.

$$[A] = [L][U] \quad \text{... 5.15}$$

where $[U] = \begin{bmatrix} a_{11} & a_{12} & a_{13} \\ f_{21} & a'_{22} & a'_{23} \\ f_{31} & f_{32} & a''_{33} \end{bmatrix}$ and $[L] = \begin{bmatrix} 1 & 0 & 0 \\ f_{21} & 1 & 0 \\ f_{31} & f_{32} & 1 \end{bmatrix}$

Example of LU decomposition

$$A = \begin{bmatrix} 0.6 & -0.2 & -0.4 \\ 0.2 & 14 & -0.6 \\ 0.6 & -0.4 & 20 \end{bmatrix}$$

After forward elimination the upper triangular matrix is obtained.

$$[U] = \begin{bmatrix} 6 & -0.2 & -0.4 \\ 0 & 14.0666 & -0.58666 \\ 0 & 0 & 20.024 \end{bmatrix}$$

$f_{21} = 0.2/6 = 0.0333$; $f_{31} = 0.6/6 = 0.1$ and element a'_{32} is eliminated by $f_{32} = -0.38/14.0666 = -0.02713$

The lower triangular matrix is:

$$[L] = \begin{bmatrix} 1 & 0 & 0 \\ f_{21} & 1 & 0 \\ f_{31} & f_{32} & 1 \end{bmatrix} = \begin{bmatrix} 1 & 0 & 0 \\ 0.0333 & 1 & 0 \\ 0.100 & -0.2713 & 1 \end{bmatrix}$$

$$[U] = \begin{bmatrix} 6 & -0.2 & -0.4 \\ 0 & 14.0666 & -0.58666 \\ 0 & 0 & 20.024 \end{bmatrix}$$

Giving [L] [U] = [A]

$$\begin{bmatrix} 6 & -0.2 & -0.4 \\ 0.2 & 14 & -0.6 \\ 0.6 & -0.4 & 20 \end{bmatrix}$$

Solving a Set of Equations by LU Decomposition Method

An example is; $\begin{bmatrix} 6 & -0.2 & -0.4 \\ 0.2 & 14 & -0.6 \\ 0.6 & -0.4 & 20 \end{bmatrix} \begin{pmatrix} x_1 \\ x_2 \\ x_3 \end{pmatrix} = \begin{pmatrix} 15.70 \\ -39.1234 \\ 140.1686 \end{pmatrix}$

The forward elimination of Gauss elimination is:

$$\begin{bmatrix} 6 & -0.2 & -0.4 \\ 0 & 14.00666 & -0.58666 \\ 0 & 0 & 20.024 \end{bmatrix} \begin{pmatrix} x_1 \\ x_2 \\ x_3 \end{pmatrix} = \begin{pmatrix} 15.70 \\ 39.1234 \\ 142.8 \end{pmatrix}$$

The forward substitution is:

$$\begin{pmatrix} 1 & 0 & 0 \\ 0.06666 & 1 & 0 \\ 0.20 & -0.05426 & 1 \end{pmatrix} \begin{pmatrix} d_1 \\ d_2 \\ d_3 \end{pmatrix} = \begin{pmatrix} 15.7 \\ -38.6 \\ 142.8 \end{pmatrix}$$

Multiplying left side ones:

$$d_1 = 15.70; \quad 0.0666\, d_1 + d_2 = -38.6; \quad 0.2\, d_1 - 0.05426\, d_2 + d_3 = 142.8$$

Giving $d_1 = 15.70$; $d_2 = -39.12$; $d_3 = 140.16$

So $$(D) = \begin{pmatrix} 15.70 \\ -39.12 \\ 140.16 \end{pmatrix}$$

The result can be substituted in $[U]\,(X) = (D)$ to give x_1, x_2 and x_3 values.

They are: $\begin{pmatrix} 6 \\ 5.0 \\ 14.0 \end{pmatrix}$

Inverse of a Matrix

For a square matrix: $[A]\,[A^{-1}] = [A^{-1}]\,[A] = [I]$

Numerical method of inverse using LU decomposition method:

Example: $[A] = \begin{pmatrix} 6 & -0.2 & -0.4 \\ 0.2 & 14 & -0.6 \\ 0.6 & -0.4 & 20 \end{pmatrix}$

$$[U] = \begin{pmatrix} 6 & -0.2 & -0.4 \\ 0 & 14.0666 & -0.5866 \\ 0 & 0 & 20.024 \end{pmatrix}; \; [L] = \begin{pmatrix} 1 & 0 & 0 \\ 0.067 & 1 & 0 \\ 0.20 & -0.054 & 261 \end{pmatrix}$$

***Solution*:** The lower triangular system can be:

$$\begin{pmatrix} 1 & 0 & 0 \\ 0.0666 & 1 & 0 \\ 0.200 & -0.5426 & 1 \end{pmatrix} \begin{pmatrix} d_1 \\ d_2 \\ d_3 \end{pmatrix} = \begin{pmatrix} 1 \\ 0 \\ 0 \end{pmatrix}$$

Solving for $(D)^T = (1, -0.067, -0.2018)$. This is used for right-hand side:

$$\begin{pmatrix} 3 & -0.1 & -0.2 \\ 0 & 14.0066 & -0.5866 \\ 0 & 0 & 20.024 \end{pmatrix} \begin{pmatrix} x_1 \\ x_2 \\ x_3 \end{pmatrix} = \begin{pmatrix} -1 \\ -0.0066 \\ -0.2018 \end{pmatrix}$$

giving (X^T) as: (0.009888; 0.2858; 0.00542) which is second column of matrix.

Finally, with right-hand side (for 3rd column) it is $(B)^T = (0, 0, 1)$.

Giving $(X^T) = (0.013596, 0.008366; 0.19976)$

which is final column of the matrix.

$$[A^{-1}] = \begin{pmatrix} 0.6698 & 0.009888 & 0.013596 \\ -0.01036 & 0.2858 & 0.008366 \\ -0.02016 & 0.00542 & 0.19976 \end{pmatrix} \text{ Ans.}$$

EXERCISES

1. (a) Obtain the Laplace transform of cos at where a is a constant.

   (b) Solve the differential equation by Laplace transform method.

$$\frac{d^2y}{dx^2} + \frac{dy}{dx} = \text{Constant. } y(0) = y'(0) = 0$$

2. Using Laplace transform method, calculate: $L(x^n)$, $L(e^{ax})$, $L(\sin ax)$ where L is Laplace transform operator.

3. Solve using Laplace transforms:

$$y'' + 3y' + 2y = 12\, e^{2x} \text{ where } y(0) = 1,\ y'(0) = 1.$$

4. Using Laplace transforms, calculate:

   (a) $y'' - 3y' + 2y = 4e^{2x};\ y(0) = -3;\ y'(0) = 5.$

   (b) $y'' + 2y' - 5y = e^{-x} \sin x$ where $y(0) = 0,\ y'(0) = -1.$

5. (a) Obtain Laplace transform of $t \sin \omega t$ where ω is a constant.

   (b) Solve using Laplace transforms: $\dfrac{d^3y}{dx^3} - 3\dfrac{d^2y}{dx^2} + 3\dfrac{dy}{dx} - y = t^2 e^t;$

$$y(0) = 1,\ y'(0) = 0,\ y''(0) = 2$$

6. Solve by Laplace transforms: $\dfrac{d^2y}{dt^2} + \dfrac{dy}{dt} = \text{Constant};\ y(0) = y'(0) = 0.$

7. Find L^{-1} of $\left(\dfrac{S+8}{S^2+4S+13}\right)$.

8. Solve the following equation by the method of Laplace transforms, $\dfrac{dy}{dt} - y = e^{at}$ with the boundary conditions $y = -1$ at $t = 0$. Give the solution for positive values of t.

9. Solve $20\dfrac{dx}{dt} + 5\dfrac{dy}{dt} + 15y = 0;\ 6\dfrac{dx}{dt} + 4x + 2\dfrac{dy}{dt} = 2;\ x = y = 0$ at $t = 0$.

10. Find $L f(t) = \cos\left(t - \dfrac{2\pi}{3}\right)$; for $t > \dfrac{2\pi}{3}$, $f(t) = 0$ and for $t < \dfrac{2\pi}{3}$.

❑❑❑

6

Optimization Methods

6.1 OPTIMIZATION

6.1.1 Introduction

Aims: 1. Minimum cost of production.

2. Maximum profit
3. Maximum yield
4. Number of days for regeneration of a catalyst
5. Optimum number of effects in a multiple effect evaporator
6. Monetary framework
7. Reduced maintenance costs
8. Less equipment wear
9. Less contaminants.

In a distillation column, for optimum reflux ratio: Operating cost is: Pumping + Heating + Condenser load. Initial cost: Number of plates less but column diameter is more.

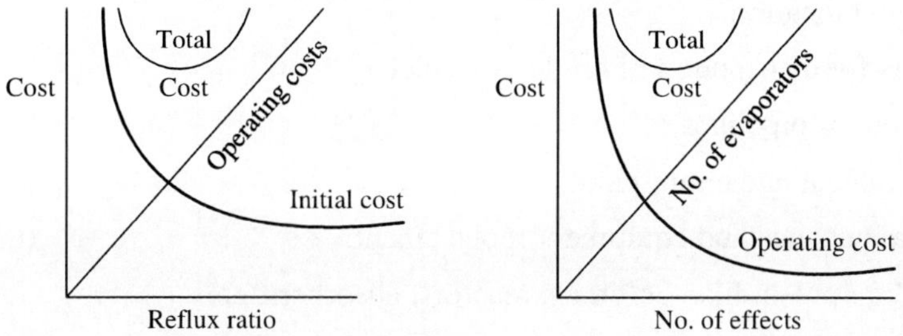

Fig. 6.1

In multiple effect evaporators: Total cost = Initial cost (number of evaporators) + Operating cost (steam).

6.2 SCOPE AND HIERARCHY OF OPTIMIZATION

Processor operating is represented by mathematical equations. Because of computers, fast and tedious calculations are done, to obtain optimum values.

Uses:

1. Production planning
2. Scheduling
3. Product distribution
4. Making optimum number of products and amounts, like by-products in a steel plant
5. Amounts of products in a petroleum distillation (uses linear programming).

All the above are based on cost or profit or some objective. In industry optimization is used in three areas or levels as follows:

1. **Management.** Covers allocation and scheduling R&D, location of new plants, product selection, etc.
2. **Process Design and Equipment Specifications.** Deals with design of reactors, selecting types of heat exchangers (batch or continuous), number of reactors, materials of construction.
3. **Plant Operations.** Covers shipping, temperature, pressure, flow rates, steam, consumption, cooling water.

Sources of data for optimization:

1. Profit and loss statements
2. Operating records of a plant

6.3 EXAMPLES OF OPTIMIZATION

1. Best site for plant location.
2. Routing tankers for distribution of crude or products.
3. Sizing and layout of pipelines.
4. Designing equipment and entire plant.
5. Scheduling maintenance and equipment replacement.
6. Operating the equipment like: columns, reactors, absorbers, etc.
7. Constructing a model of the process.
8. Minimising the inventory charges.
9. Allocating resources to several processes.
10. Planning and scheduling construction.

Example 6.1*a* Optimisation of insulation thickness.

The factors to be considered are:

(a) Adding insulation thickness saves money through reduced heat loss (operating cost).
(b) Cost of insulation increases with thickness (fixed cost).

Example 6.1*b* Operating conditions in boiler

The factors to be considered are:

(a) Thermal efficiency increases and then decreases with air/fuel ratio.
(b) Hydrocarbons emissions decrease with increase of air/fuel ratio.
(c) Nitrogen oxides emissions increase with increase of air/fuel ratio.

Example 6.1*c* Distribution of a single product made at different plants to customers at different distances from plants. Transport costs, production costs at each plant are to be considered.

6.4 OPTIMIZATION PROBLEM CONSISTS OF THREE ESSENTIAL CATEGORIES

1. Objective function (like profit, cost, etc.) economic model.
2. Equality constraints (equations): $x_1 + x_2 + S_1 = 10$, where x_1, x_2 are variables and S_1 is slack variable.
3. Inequality constraints: Some variables should not go below some minimum values or maximum values.

6.4.1 Terminology (Technical Terms Used)

1. **Objective function.** The criterion or index is known as the objective function. For example, minimum cost of production, maximum profit, payout time, rate of return, etc.
2. **Monotonic function.** A function $y(x)$ is increasing or decreasing continuously with change of independent variable is a monotonic function. Other one is a piecewise continuous function.

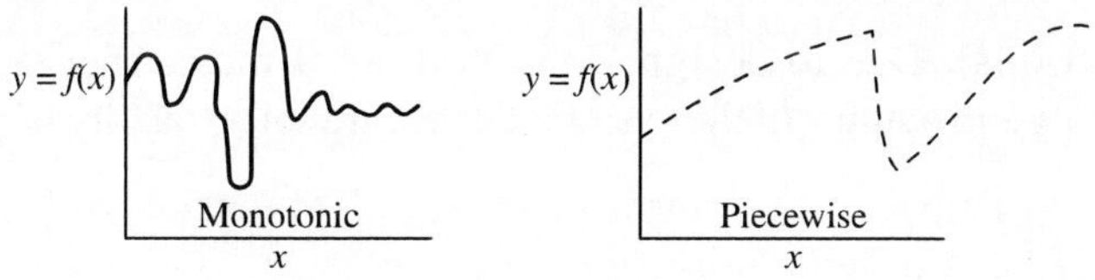

Fig. 6.2

3. **Unimodal Function.** Having only one maximum value or minimum value (only one peak).

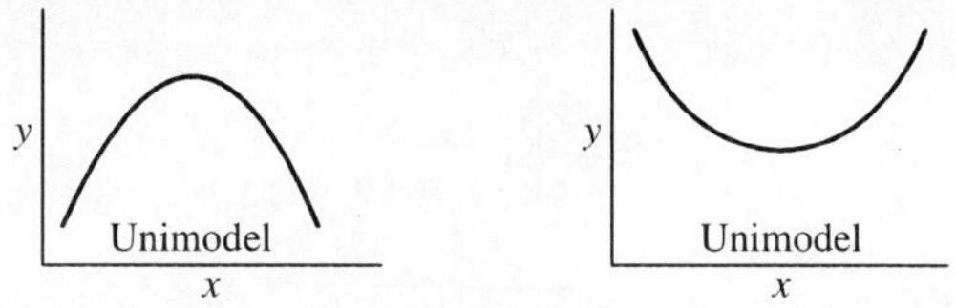

Fig. 6.3

Interval of Uncertainty. This is also called the region of interest. While searching for optimum, first x_1 ... x_2 is the interval of uncertainty, for optimum in the next search x_3 ...x_4 becomes the interval of uncertainity.

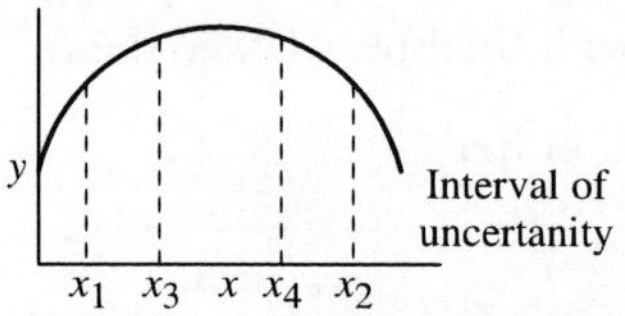

Fig. 6.4

Stochastic Process. Output is uncertain for a given input. For example, product from a chemical reactor of catalytic type becomes uncertain when the catalyst activity cannot be predicted accurately. Ambient humidity, temperature cannot be predicted correctly are other examples.

Deterministic Process. The output depends on the input entirely.

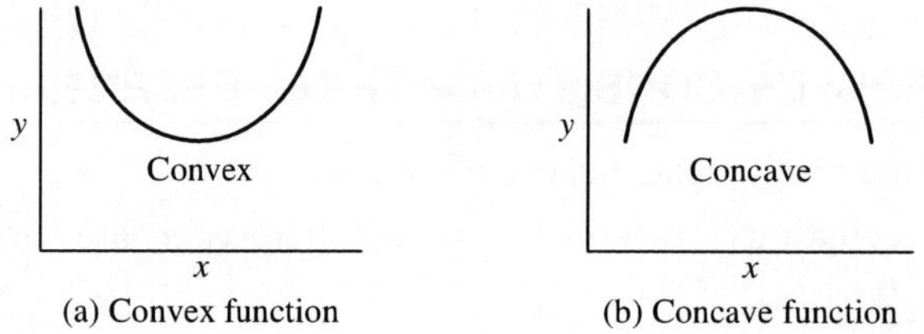

Fig. 6.5

Convex Set. *a*, *b* is a concave set, since *ba* line is outside the set.

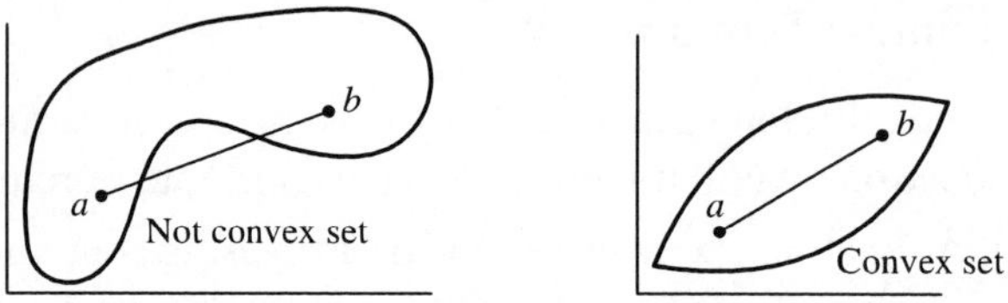

Fig. 6.6

Constraints (Restrictions). Due to design, conservation of mass, energy, momentum, rate equations, control the inputs and outputs of the system. Others are sales, quality of product acceptable by consumer, etc.

For example $T_{in} \leq T \leq T_{max}$; Concentration $\geq u$

Feasible Region. To search for an optimum of constrained optimisation problem. It must be in the interior region or along the boundaries fixed by the constraints. This region is called feasible region.

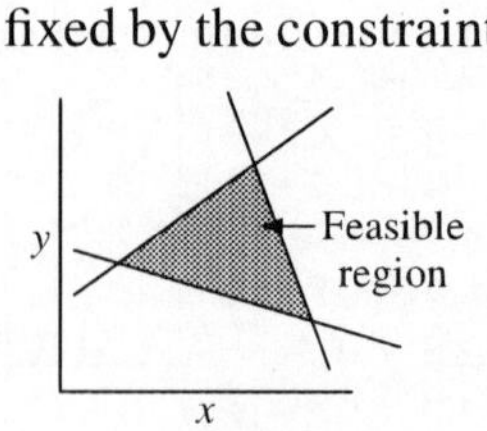

Fig. 6.7

Stationary Point. $\dfrac{dy}{dx_1} = 0;\ i = 1, 2, \ldots n$. Gradient or derivative vanishes.

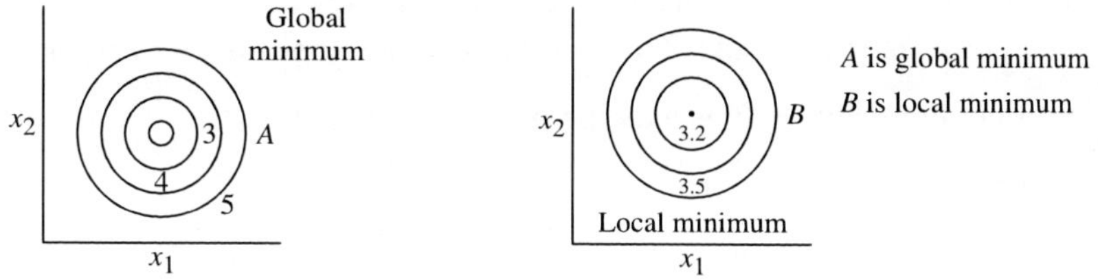

Fig. 6.8

Saddle Point. This is neither maximum nor minimum. Point A is a saddle point.

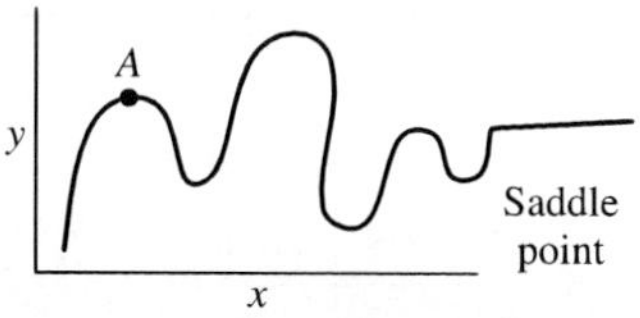

Fig. 6.9

Slack Variable. For converting the inequality to equality, for solving optimization problems, when constraints are present. These come in linear programming and dynamic programming problems.

Taylor's Series.
$$f(x + h) = f(x) + hf'(x) + \frac{h^2 f''}{2}(x)$$
$$f(x - h) = f(x) - hf'(x) - \frac{h^2}{2} f''(x)$$

Search Problem. Any set of instructions for placing the n experiments will be called search plan and any investigation seeking optimum value of unknown function will be called search problem. Search problems can be classified according to the number of independent variables and whether or not experimental error is present.

No experimental error – deterministic case.

Method of optimization can be used as a root finding procedures.

Deterministic Problem. Search problems having no unknown or random factors will be described as deterministic.

Stochastic Process. If experimental error cannot be neglected, the problem will be stochastic, a word meaning that known factors are involved.

Simultaneous Search Plan. Plans specifying the location of every experiment before any results are known, will be called simultaneous search plan.

Sequential Search. While a plan permitting the experiments to base the future experiments on past outcomes will be called sequential search.

When two responses are equal, the optimum must lie in between the two experiments.

6.5 GENERAL: INTEREST AND PAYMENT

Example 6.2 An investment of ₹ 23,000 in a process produces a net income of ₹ 10,000 a year for 5 years. There is no income after this time. To what interest rate does this correspond?

(a) Using straight line depreciation plus average interest.

(b) Using exact formula.

Solution: (a) $10{,}000 = Pi$ where P = present worth and i is interest.

$$P = 23{,}000$$

$$i = \frac{10{,}000}{23{,}000} = 43.5 \text{ per cent}$$

$i = 43.5$ per cent **Ans.**

(a)
$$R' = \frac{P}{n} + \frac{P \times i}{2} \frac{(n+1)}{n}$$

$$R' = 10{,}000,$$

$P = 23{,}000$, n (number of years) $= 5$

$$10{,}000 = \frac{23{,}000}{5} + \frac{23{,}000\, i}{2} \times \frac{6}{5}$$

$i = 39.2$ per cent **Ans.**

Example 6.3 An evaporator costs ₹ 25,000. At the end of 10 years it will be replaced and at that time, its salvage value will be ₹ 2500 with 6 per cent interest. Calculate:

(a) Annual fixed charges using exact method.

(b) Annual fixed charges by straight line depreciation plus average interest.

(c) The average interest rate.

Solution: (a)
$$R = (P - L)\left[\frac{i(1+i)^n}{(1+i)^n - 1}\right] + Li$$

L (salvage value) $= 2500$ $P = 25{,}000$,

$i = 0.06$, $n = 10$

$$R = (25000 - 2500)\,\frac{0.06(1.06^n)}{(1.06^n - 1)} + 2500 \times 0.06$$

$$R = 22500 \times \frac{0.06 \times 1.791}{0.791} + 150$$

$= 3206$ **Ans.**

(b) $$R = \left(\frac{P-L}{2}\right) i \left(\frac{n+1}{n}\right) + \left(\frac{P-L}{n}\right) + L \times i$$

$$= \frac{22500}{2} \times \frac{11}{10} \times 0.06 + \frac{22500}{10} + 2500 \times 0.06$$

$$= 742.5 + 2250 + 150 = ₹\ 3142.5$$

(c) $$i = \frac{i}{2}\left(\frac{n+1}{n}\right) = \frac{0.06}{2} \times \frac{11}{10}$$

$$= 3.3 \text{ per cent}$$ **Ans.**

Example 6.4 A pilot plant making samples for market development has a capacity of 100 litres/day. The uniform rate of sending samples is 20 litres/day. The optimum number of production runs per year is 10. Calculate the average inventory for optimum operation.

Solution: $$100 \times t \times n = 20 \times 365$$

$n = 10$, number of runs

t = days operated/cycle

$$t = \frac{73}{n}, \text{ for } n = 10,$$

$t = 7.3$ days/cycle

$$\text{Average inventory} = (100 - 20)\frac{t}{2}$$

$$= 40t = 40 \times 7.3$$

$= 292$ litres/cycle.

Example 6.5 With interest at 6 per cent, calculate the present worth of:

(a) ₹ 540 at the end of 10 years

(b) ₹ 37.30 per year for 10 years

(c) ₹ 20 per year for life.

Solution: (a) $S = P(1 + i)^n$, $S = 540$, $i = 0.08$

P = Present worth

$$P = \frac{540}{(1+0.08)^{10}} = ₹\ 250$$

(b) $$S = \frac{R[(1+i)^n - 1]}{[i(1+i)^n]}$$

Here, $R = 37.3$, $i = 0.08$

$$S = \frac{37.3[1.08^{10} - 1]}{[0.08(1 + 0.08)^{10}]} = ₹\,230$$

(c) n = Infinity

$$S = \frac{R}{i};\ R = 37.3;\ i = 0.08$$

$$S = ₹\,466$$

6.5.1 Single Variable Optimization

The first derivative is zero. Second derivative is negative for maximum and positive for minimum.

Example 6.6 Extraction: There is available 10,000 kg/hr of a solution containing 30 weight per cent acetic acid in water. An extractor is to be designed to extract most of the acetic acid from this solution, using a solvent (a mixture of 85 per cent ethyl acetate and 15 per cent benzene).

Make an economic analysis of the process and report:

(a) How many plates should be used in the extractor?

(b) How much acetic acid is recovered per hour and how much is lost in the raffinate?

(c) What is the cost of the installation?

Use a perforated plate extractor of 0.76 m diameter. The efficiency will be 50 per cent and the first cost of the extractor is ₹ 405 per plate installed.

Acetic acid is worth of ₹ 5 per kg.

Depreciation, maintenance, taxes (property) and profit should be calculated as 50 per cent of the investment.

Operation is 300 days/year. Costs of the operations are independent of the column design. The number of theoretical plates is calculated by the equation,

$$W = \frac{(E - 1)}{(E^{n+1} - 1)}$$

W = Solute lost as fraction of the solute in feed.

E = Extraction factor m B/A, where m is the distribution coefficient, B/A is ratio of solvent to raffinete, use E as 2.3. n is the number of theoretical stages.

Solution: (a)

$$W = \frac{\text{Acetic acid lost}}{\text{Acetic acid in feed}} = \frac{W_l}{3000}$$

$$W_l = 3000\ \text{W}$$

$$W_l = 3000 \times \frac{(E-1)}{(E^{n+1}-1)} = \frac{(3000 \times 1.3)}{(2.3^{n+1}-1)}$$

where $E = 2.3$

$$\text{Acetic acid recovered per hour} = 3000 - \frac{3000 \times 1.3}{(2.3^{n+1}-1)}$$

$$\text{Annual savings} = \left[3000 - \frac{3900}{(2.3^{n+1}-1)}\right] \times 300 \times 5 \times 24$$

$$= 1.08 \times 10^8 - \frac{1.404 \times 10^8}{(2.3^{n+1}-1)}$$

Annual costs are: $\frac{405}{0.5} \times 0.3 \times n = 243\,n$

$$\text{Total annual return } (T.R.) = \text{Annual costs} - \text{Annual savings}$$

$$= 243n - 1.08 \times 10^8 + \frac{1.404 \times 10^8}{(2.3^{n+1}-1)}$$

For optimum $\frac{d(T.R.)}{dn} = 0$

$$\frac{d(T.R.)}{dn} = 243 - \frac{2.3^{n+1} \ln 2.3 \times 1.404 \times 10^8}{(2.3^{n+1}-1)^2}$$

$$(2.3^{n+1} - 1)^2\, 243 = 1.169 \times 10^8 \times 2.3^{n+1}$$

By trial and error, $n = 16$

Actual plates = 2 × 16 = 32 **Ans.**

(b) $$W_l = 3000\, W = \frac{3000 \times 1.3}{2.3^{17}-1} = 0.0028$$

Acetic acid lost per hour = 0.0028 kg.

Acetic acid recovered = 3000 kg (neglecting fraction)

(c) Cost of installation: 2 × 16 × 405 = ₹ 12,960 **Ans.**

Example 6.7 Evaporator: Compute the optimum number of effects (N) for an evaporator system. Heat transfer rate is 4 million Kcal/hr. Useful life is 12 years. Fixed costs are 22 per cent per year. Days of operation per year are 200 (24 hr/day). Steam cost is ₹ 10/500 kg. One kg of steam will evaporate ($0.8\,N + 1.0$) kg of water. Latent heat of water is taken as 500 Kcal/kg. Each effect costs ₹ 25,000.

Solution: $q = 4 \times 10^6$ Kcal/hr

Fixed costs are = 25000 N × 0.22 per year = 5500 N

Variables costs:

$$\text{Steam required} = \frac{4 \times 10^6}{500(0.8N+1)} \text{ kg/hr}$$

$$\text{Steam cost} = \frac{4 \times 10^6 \times 24 \times 200 \times 10}{500(0.8N+1)\,500}$$

$$\frac{\text{Total cost}}{\text{Year}},\ \text{T.C.} = \frac{\text{Variable cost}}{\text{Year}} + \frac{\text{Fixed cost}}{\text{Year}}$$

$$\text{T.C.} = 5500\text{ N} + \frac{768{,}000}{(0.8N+1)}$$

$$\frac{dT.C.}{dn} = 0 = 5500 - \frac{0.8 \times 768000}{(0.8N+1)^2}$$

$N = 12$ effects **Ans.**

Example 6.8 Calculations for an agitated vessel gave these results.

N (rpm)	50	100	150	200	250
h	100	174	290	300	360
U	67	93	109	119	128
Annual power, cost	20	160	540	1280	2500

Cost of vessel plus agitator is ₹ 12,000 and annual fixed costs are 25 per cent of capital cost. Find the optimum speed from these figures.

Solution: Fixed cost = 12000 × ₹ 0.25 year = 3000

A table is made for *N*, F.C. (fixed cost). T.C. (total cost), *U* and power cost per year.

h = Heat transfer coefficient in vessel

U = Overall heat transfer coefficient

N	F.C.	P.C.	U	T.C.
50	3000	20	67	3020
100	3000	160	93	3160
150	3000	540	109	3540
200	3000	1280	119	4280
250	3000	2500	125	5500

Seeing total cost relative to U value, the optimum (economic) conditions are:

$$U = 109,$$

$$N = 150 \quad \textbf{Ans.}$$

Example 6.9 Calculate the optimum thickness, cost of insulation for the following conditions:

(a) Insulation cost factor:
Capital cost: ₹ 500/ cu.metre
Depreciation rate: 10 per cent
Interest rate: 10 per cent
Shape: Flat.

(b) Heat cost factor constant: ₹ 40 per million kcal.

(c) Thermal factors: Conductivity of insulation: 0.10 kcal/(hr.m°C)

Film coefficient on outside: 4 kcal/(hr.sqm °C)

Solution: Basis = 1 sq. metre

Fixed annual costs : 500 L (0.1 + 0.1) = 100 L

where L = Thickness of insulation, m

$$\text{Heat loss/hr/metre} = U\,\Delta\,T$$

$$\frac{1}{U} = \frac{1}{h} + \frac{L}{K} = \frac{1}{4} + \frac{L}{0.1} = (0.25 + 10\,L)$$

$$\text{Heat loss/hr} = 365 \times \frac{24\Delta T}{(0.25 + 10L)} \text{ kcal/hr}$$

$$\text{Value of heat loss} = \frac{8760 \times T \times 40}{10^6(0.25 + 10L)}$$

$$\text{Total annual cost (T.C.)} = \frac{100L + 0.35\Delta T}{(0.25 + 10L)}$$

$$\frac{d.T.C}{dL} = \frac{100 - 35\Delta T}{(0.25 + 10L)^2} = 0$$

$$100 = \frac{35\Delta T}{(0.25 + 10L)^2},$$

$$L = 0.1\sqrt{0.35\Delta T} - 0.025$$

Let $\Delta T = 100°\text{C}$

$$0.25 + 10L = 5.93,$$

$$L = 0.57 \text{ m. } \textbf{Ans.}$$

Example 6.10 Heat Exchanger: For a certain heat exchanger

$$h_S = 150\ (dP)^{0.3}$$

$$\text{Annual power cost} = 175\ dP$$

$$\text{Annual fixed cost} = 200\ dP^{-0.3}$$

where h_S = Heat transfer coefficient

dP = Pressure drop.

(a) Calculate the ratio of annual power cost/annual fixed cost at optimum dP.

(b) An increase in electric power rates raises the annual power cost by 20 per cent. For this new condition, find the ratio of annual power cost to annual fixed cost at the optimum dP.

Solution: (a)

$$\text{Total cost} = \text{Annual power cost} + \text{Annual fixed cost}$$

$$\text{T.C.} = 175\ dP + 200\ dP^{-0.3}$$

$$\frac{dT.C.}{d(dP)} = 175 - 60\ dP^{-1.3} = 0$$

$$60\ dP^{-1.3} = 175$$

$$dP^{-1.3} = 29.02$$

$$dP^{1.3} = 0.0345$$

$$dP = 0.445 \text{ at optimum. } \textbf{Ans.}$$

$$\frac{\text{Annual power cost}}{\text{Annual fixed cost}} = \frac{175 \times 0.445}{200 \times 0.445^{-0.3}}$$

$$= \frac{77.9}{255} = 0.306$$

(b)

$$\text{Total cost} = 175 \times 1.2\ dP + 200\ dP^{-0.3}$$

$$\frac{dT.C.}{d(dP)} = 0$$

$$dP^{-1.3} = 175 \times \frac{1.2}{60} = 3.5$$

$$dP^{1.3} = 0.286$$

$$dP = 0.382$$

$$\frac{\text{Annual power cost}}{\text{Annual fixed cost}} = \frac{1.2 \times 1.75 \times 0.382}{200 \times 0.382^{-0.3}} = 0.306$$

The ratio is same as in part (a).

6.6 MULTIVARIABLE OPTIMIZATION

The first derivatives are equated to zeroes and from these relations, optimum values are obtained.

Example 6.11 A chemical plant consists of a tubular reactor, a product separator, compressors and a mixer. The costs depend on two variables, the pressure (x_1) and the recycle ratio (x_2). The required production rate is 10 million kg/year. For this rate, the annual costs in rupees are:

$$\text{Compressing raw material} = 2000x_1$$

$$\text{Operating reactor} = \frac{5 \times 10^9}{(x_2\, x_1^{1.5})}$$

$$\text{Operating separator} = 2 \times 10^5 x_2$$

$$\text{Recirculation material} = 2 \times 10^5 x_2$$

(i) Calculate the optimum operating conditions.

(ii) Check to determine that the value of optimum obtained is an optimum, not an inflection, ridge or saddle.

Solution: (i) Cost equation (c) is

$$= 2000x_1 + \frac{5 \times 10^9}{x_1^{1.5} x_2} + 2 \times 10^5 x_2 + 2 \times 10^5 x_2$$

$$\frac{dc}{dx_1} = 2000 - \frac{5 \times 10^9 x_2 x_1^{0.5} \times 1.5}{x_1^3 x_2^2} = 0$$

$$\frac{dc}{dx_2} = 4 \times 10^5 - \frac{5 \times 10^9 x_1^{1.5}}{x_1^3 x_2^2} = 0$$

From these two relations, $x_1 = 382, \quad x_2 = 1.33.$

(ii) Let

$$A = \frac{\partial^2 c}{\partial x_1^2}$$

$$B = \frac{\partial^2 c}{\partial x_1 \partial x_2}$$

$$C = \frac{\partial^2 c}{\partial x_2^2}$$

If $B^2 - AC < 0$ and if $A < 0$ then relative maximum at (x_{10}, x_{20})

If $B^2 - AC < 0$ and if $A > 0$ then relative minimum at (x_{10}, x_{20})

If $B^2 - AC > 0$ then saddle point at (x_{10}, x_{20})

If $B^2 - AC = 0$ then nature of critical point is undetermined.

A saddle point is one which represents neither maximum nor minimum. It is a local minimum or maximum but not global minimum or maximum.

$$\frac{\partial^2 c}{\partial x_1^2} = \frac{7.5 \times 2.5 \times 10^9}{(x_1^{3.5} x_2)}$$

$$\frac{\partial^2 c}{\partial x_2^2} = \frac{10^{10}}{x_1^{1.5} x_2^3}$$

$$\frac{\partial^2 c}{\partial x_1 \partial x_2} = \frac{7.5 \times 10^9}{x_1^{2.5} x_2^2}$$

At $x_{10} = 382;\ x_{20} = 1.33,$

$$A = 12.7;\ B = 1485;\ C = 576{,}000$$

In this problem, $B^2 - AC$ is less than zero and A is greater than zero. So, relative minimum condition is satisfied.

For two or more variables 'geometric programming' is used to obtain the optimum values. For geometric programming, the objective function/equation should be of form:

$$Y = \sum_{j=1}^{m} C_j P_j(X)$$

i.e.,
$$Y = C_1 P_1 + C_2 P_2 + C_3 P_3 + \ldots$$

$$P_j = \prod_{i=1}^{m} x_i^{a_{ij}},$$

i.e.,
$$P_1 = x_1^{a_{11}} x_2^{a_{21}} x_3^{a_{31}} \ldots$$

$$P_2 = x_1^{a_{12}} x_2^{a_{22}} x_3^{a_{32}} \ldots$$

At minimum or maximum (optimum), the first derivatives must vanish giving,

$$\sum_j a_{kj} w_j = 0,\ k = 1, 2, \ldots, n$$

and
$$\sum w_j = 1.0$$

which on expanding are

$$a_{11}w_1 + a_{12}w_2 + a_{13}w_3 + \ldots = 0$$

$$a_{21}w_1 + a_{22}w_2 + a_{23}w_3 + \ldots = 0$$

$$a_{31}w_1 + a_{32}w_2 + a_{33}w_3 + \ldots = 0$$

$$Y^* = \pi_j \left(\frac{C_j}{w_j}\right)^{w_j} \quad \text{(optimum value).}$$

Example 6.12 Catalyst Replacement: The cost of replacing the catalyst in a catalytic reactor is ₹ 90,000. Since the catalyst will last less than a year, this should be considered as an expense, not a capital investment. Regeneration cost is ₹ 1000. The time needed for regeneration or replacement is neglected in the cost analysis.

During operation, the decrease in catalyst activity causes the product value to decrease at the rate of ₹ 2000/day. Regeneration reduces this loss to 20 per cent of the value at the beginning of the regeneration cycle. Make all the regeneration cycles of same length. Find the optimum cycle of regeneration and replacement.

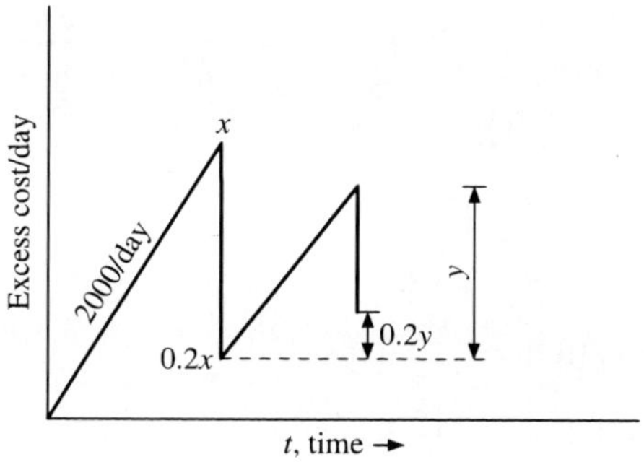

Fig. 6.10

Solution:

$$\text{Total excess cost} = \int_0^t E_e dt$$

$$n \text{ triangles} = n\frac{2000}{2}m^2 = 1000\, m^2 n$$

where n = Number of regenerations.

m = Days/regeneration

Number of days/replacement = mn

n	Rectangles
1	0
2	1
3	3
4	6

$$\text{No. of Rectangles} = \frac{n(n-1)}{2}$$

$$\text{Rectangles area} = \frac{n(n-1)m \times 400\, m}{2} = 200\, m^2(n^2 - n)$$

$$\frac{\text{Total cost}}{\text{Replacement cycle}} = 1000\, m^2 n + 200\, m^2(n^2 - n) + 1000(n - 1) + 90{,}000$$

$$\frac{\text{(T.C.) Total cost}}{\text{Day}} = \frac{1000 m^2 n}{mn} + \frac{200 m^2 (n^2 - n)}{mn} + \frac{1000(n-1)}{mn} + 90{,}000/mn$$

$$\frac{dT.C}{dn} = 200\, m - \frac{89{,}000}{mn^2} = 0$$

$$\frac{dT.C}{dm} = 800 + 200n - \frac{1000}{m^2} - \frac{89{,}000}{mn} = 0$$

Solving both equations simultaneously,

$$n = 18.85$$

$$m = 1.12.$$

$$\text{Cycle of regeneration} = 1.12 \text{ days}$$

$$\text{Days/replacement} = 1.12 \times 18.85$$

$$= 21.1 \text{ days.}$$

Example 6.13 Using geometric programming, calculate the optimum for

$$y = 2x^{1.5} + \frac{3}{x^2}$$

Solution:

$$C_1 = 2;\ C_2 = 3$$

$$a_{11} = 1.5;\ a_{12} = -2$$

$$a_{11} w_1 + a_{12} w_2 = 0$$

$$w_1 + w_2 = 1$$

$$1.5 w_1 - 2 w_2 = 0$$

$$w_1 + w_2 = 1.0$$

$$w_1 = \frac{2}{3.5}$$

$$w_2 = \frac{1.5}{3.5}$$

$$Y^*(\text{Y optimum}) = \pi_j \left(\frac{C_j}{w_j}\right)^{w_j} = \left(\frac{2 \times 7}{4}\right)^{4/7} \left(\frac{3 \times 7}{3}\right)^{3/7} = 4$$

First term: $\frac{4}{7} \times 4.7 = 2x^{1.5}$, $x = 1.22$ **Ans.**

Example 6.14 A fluid is pumped through a heater and from the heater through a long pipeline. Costs are shown below. Calculate the optimum values of diameter, D and temperature T.

$$\text{Annual cost of pumping} = ₹\frac{1200}{D^2}\ /\text{year}$$

$$n \text{ is the viscosity} = \frac{10000}{(T+20)}$$

$$\text{Pipe cost} = ₹\,10{,}000\,D^{1.2}$$

$$\text{Pump cost} = ₹\,\frac{25n}{D^2}$$

$$\text{Heater cost} = ₹\,500\,T$$

$$\text{Annual cost of heat} = ₹\,150\,T/\text{year.}$$

Annual fixed costs are 20 per cent of first cost.

Write equations as function of D and T.

(1) Total capital cost, (2) Annual fixed cost, (3) Annual variable cost, (4) Total annual cost, (5) dT_C/dD, (6) dT_C/dT.

Solution: (1) $10{,}000D^{1.2} + \dfrac{25n}{D^2} + 500T$

(2) $0.2\left[10{,}000D^{1.2} + \dfrac{25n}{D^2} + 500\,T\right]$

(3) Annual variable cost + $\dfrac{1200}{D^2}\dfrac{10{,}000}{(T+20)} + 150\,T$

(4) $2000D^{1.2} + \dfrac{5n}{D^2} + 100\,T + \dfrac{1200}{D^2}\left(\dfrac{10{,}000}{T+20}\right) + 150\,T$

(5) $\dfrac{dT_C}{dD} = 2.4\times10^3 D^{3.2} - \dfrac{10^5}{(T-20)} - \dfrac{24\times10^6}{(T+20)} = 0$

(6) $\dfrac{dT_C}{dT} = 250 - \dfrac{5\times10^4}{D^2(T+20)^2} - \dfrac{12\times10^6}{D^2(T+20)^2} = 0$

Solving 5 and 6 simultaneously,

$$D = 2.02,\ T = 15.25\ \textbf{Ans.}$$

Example 6.15 It is required to evaporate 50000 kg/hr of water in an evaporator. The latent heat is 1000, $U_0 = 300$, $U = 150$, (time) = 50 hr (total time).

$$\frac{1}{U^2} = \frac{1}{U_0^2} + \beta\theta$$

(a) Calculate the cycle to give maximum production.

Solution:

$$\frac{1}{U^2} = \frac{1}{150^2} + 50\beta\,,\ U = 150$$

$$\beta = 6.67 \times 10^{-7}$$

$$\frac{\lambda\, dw}{d\theta} = UA\ dt,$$

$$\lambda = 1{,}000$$

$$\frac{dw}{d\theta} = \frac{166.7}{\left[\left(\frac{1}{300^2}\right) + (6.67 \times 10^{-7}\theta)\right]^{1/2}}$$

$$w = \int_0^{50} \frac{166.7}{\left(\frac{1}{300^2} + 6.67 \times 10^{-7}\theta\right)^{1/2}} = 1.67 \times 10^6$$

$$\text{Average rate} = 1.67 \times \frac{10^6}{50} = 33{,}400$$

$$\frac{\text{Average rate}}{\text{Cycle}} = 50(\text{evaporation}) + 2(\text{starting}) + 8(\text{cleaning}) = 60$$

$$= 1.67 \times \frac{10^6}{60} = 27830 \text{ kg/hr.}$$

Maximum production in a cycle:

$$\frac{w}{(\theta + 10)} = 50.6 \times 10^7 \left[\frac{(11.1 \times 10^{-6} + 6.67 \times 10^{-7}\theta)^{0.5} - (11.1 \times 10^{-6})^{0.5}}{\theta + 10}\right]$$

Let $\theta_1 = \theta + 10$

Equating $\frac{dw}{d\theta_1} = 0$ gives 35.8 hours **Ans.**

Cycle time = 10 + 35.8 = 45.8 hours.

Example 6.16 $2000x_1 + \frac{5 \times 10^9}{x_1^{1.5} x_2} + 4 \times 10^5 x_2$ using geometric programming.

Solution:

$$TC = 2000x_1 + \frac{5 \times 10^9}{x_1^{1.5} x_2} + 4 \times 10^5 x_2$$

$$Y = \sum_{j=1}^{m} C_j P_j$$

$$Y = C_1P_1 + C_2P_2 + C_3P_3$$

$$P_j = \prod_{i=1}^{n} x_i^{a_{ij}}$$

In this problem, $C_1 = 2000$, $C_2 = 5 \times 10^9$, $C_3 = 4 \times 10^5$,

$$P_1 = x_1,\ P_2 = x_1^{-1.5} \times x_2^{-1},\ P_3 = x_2$$

$$\sum_j a_{kj}\, w_j = 0,\ k = 1, 2, \ldots, n$$

$$\sum w_j = 1.0$$

where w_j are weights

$$w_1 - 1.5\, w_2 = 0$$

$$-w_2 + w_3 = 0$$

$$w_1 + w_2 + w_3 = 1.0$$

Solving simultaneously,

$$w_1 = \frac{1.5}{3.5}$$

$$w_2 = \frac{1}{3.5}$$

$$w_3 = \frac{1}{3.5}$$

$$\text{Minimum cost } (Y^*) = \prod_j \left(\frac{C_j}{w_j}\right)^{w_j}$$

$$Y^* = \left(\frac{2000 \times 3.5}{1.5}\right)^{1.5/3.5} \left(\frac{5 \times 10^9 \times 3.5}{1}\right)^{1/3.5} \left(\frac{4 \times 10^5 \times 3.5}{1}\right)^{1/3.5}$$

$$= 1.66 \times 10^6$$

$$\text{Compressor cost} = 2000x_1 = Y^*\omega_1$$

$$x_1 = 7.1 \times \frac{10^5}{2000} = 355 \textbf{ Ans.}$$

For x_2 : Operating separator + Recalculating material

$$= Y^* w_3 = 1.66 \times \frac{10^6}{3.5} = 4 \times 10^5\, x_2$$

$$x_2 = 1.2.$$

Example 6.17 In the design of a chemical plant, it is found that the profit depends on two variables:

Temperature T and purity x.

The equation relating profit P to these variables is

$$P = AT^{0.5} + \frac{100,000}{Tx^{0.5}} + 40,000x^2$$

The value of A is not known, but it is known that the optimum P is 100,000. Find A using geometric programming.

Solution: Using geometric programming:

$$Y^* = 100,000$$

$$a_{11} = 0.5, a_{12} = -1, a_{13} = 0$$

$$a_{21} = 0, a_{22} = -0.5, a_{23} = 2$$

$$a_{11}w_1 + a_{12}w_2 + a_{13}w_3 = 0 = 0.5w_1 - w_2$$

$$a_{21}w_1 + a_{22}w_2 + a_{23}w_3 = 0 = -0.5w_1 - 2w_2$$

$$w_1 + w_2 + w_3 = 1$$

Solving simultaneously,

$$w_1 = \frac{8}{13}$$

$$w_2 = \frac{4}{13}$$

$$w_3 = \frac{1}{13}$$

$$Y^* = \left(\frac{C_1}{w_1}\right)^{w_1}\left(\frac{C_2}{w_2}\right)^{w_2}\left(\frac{C_3}{w_3}\right)^{w_3}$$

$$= \left(\frac{A}{8} \times 13\right)^{8/13}\left(\frac{100,000 \times 13}{4}\right)^{4/3}\left(\frac{40,000 \times 13}{1}\right)^{1/3}$$

$$= 191A^{0.6162} = 100,000$$

$A = 26,200$ **Ans.**

6.7 LINEAR PROGRAMMING

The general linear programming problem is to find a vector (a set of values of $x_1, x_2, \ldots, x_n$) which minimises the linear objective function.

$$C_1 x_1 + C_2 x_2 + C_3 x_3 + \ldots.. + C_n x_n$$

There are linear constraints.

$$x_j \geq 0, j = 1, 2, \ldots, n$$

and

$$a_{11}x_1 + a_{12}x_2 + \ldots + a_{1n}x_n = b_1$$

$$a_{21}x_1 + a_{22}x_2 + \ldots + a_{2n}x_n = b_2$$

$$\ldots\ldots\ldots\ldots\ldots\ldots\ldots\ldots$$

$$\ldots\ldots\ldots\ldots\ldots\ldots\ldots\ldots$$

$$a_{m1}x_1 + a_{m2}x_2 + \ldots + a_{mn}x_n = b_m$$

This basic theorem of linear programming is: In the optimal solution of any linear programming problem, the total number non-zero variables is exactly equal to the number of constraints (this does not count the requirements that all x values are non-negative).

Any solution in which the number of non-zero variables is equal to the number of constraints is called a basic solution. The basic solution is not necessarily optimal.

Constraints are often called inequalities. These are converted to equalities by adding a 'slack variable'. For example, the equation $x_1 + x_2 \leq 10$ would be written as

$$x_1 + x_2 + S_1 = 10$$

Example 6.18 Maximum $P = 10x_1 + 4x_2 + x_3$ with the constraints

$$x_1 + x_2 \leq 5$$

$$2x_1 + x_2 + x_3 \leq 20$$

Solution: **Simplex Method**

Procedure:

Step 1: Convert in equalities to equalities by introducing slack variables. Convert equations to

$$\text{Minimise } P = -10x_1 - 4x_2 - 1x_3$$

$$1x_1 + 1x_2 + 1S_1 = 5$$

$$2x_1 + x_2 + x_3 + S_2 = 20$$

Step 2: Set up a table here:

(a) A column for each variable, including slack variables

(b) Coefficients C

(c) Constraint coefficients a

(d) Choose basis variables, equal in number to the number of constraints. These should have a coefficient a of 1.0 and 0 for all other constraint coefficients.

			x_1	x_2	x_3	S_1	S_2
Basis	c	b	–10	–4	–1	0	0
S_1	0	5	1	1	0	1	0
S_2	0	10	2	1	1	0	1

(e) List c coefficients for these basis variables in column labelled c.

(f) List b values for each constraint equation in column b. These are also the values of the basis variables, when all the other variables are zero.

Step 3: Calculate for each variable,

$$Z_j = \Sigma c_i a_i, \; i = 1, 2, \ldots, m$$

$$S_j = Z_j - C_j$$

List S_j in column j at the bottom. For example, in the tableau below,

$$Z_1 = 0 \times 1 + 0 \times 2 = 0$$

$$S_1 = 0 - (-10) = 10$$

			x_1	x_2	x_3	S_1	S_2
Basis	c	b	–10	–4	–1	0	0
S_1	0	5	①	1	0	1	0
S_2	0	20	2	1	1	0	1

Step 4: By inspection, locate the largest non-negative value of S_j. If there are no values greater than 0, the basis is optimal. The column with the largest non-negative S_j is called column K. In this column find root for which b/a_k is minimum but greater than zero. Circle this pivot element and call this row: row i, variable k replaces variable i as the basis.

Step 5: In this row for which the basis is now x_k, divide all the coefficients by a constant to reduce, a_k to 1.0. In all other rows, eliminate the coefficients in column k and then multiply by a constant to get the value of $a = 1.0$ in the column for which that row is the basis. In operating on the a values, the same operations are performed on column b.

For example, to get from step 1 above to step 2 below, $j = 1$, $S_j = 10$ pivot element is a_{11}. x_1 replaces S_1 as basis. Row 1 remaining the same since the value is already 1.0. Eliminate the coefficient in row 2 under x_1 by multiplying row 1 by 2, and subtracting from row 2. For column 1, this gives $2 - 2(1) = 0$, column 2, this gives $1 - 2(1) = -1$, etc., for column b, $20 - 2(5) = 10$.

			x_1	x_2	x_3	S_1	S_2
Basis	c	b	–10	–4	–1	0	0
x_1	–10	5	1	1	0	1	0
S_2	0	10	0	–1	①	–2	1
		–50	0	–6	1	–10	0

Step 6: This is a new tableau. Go back to 3 and repeat until there are no positive values of S_j.

			x_1	x_2	x_3	S_1	S_2
Basis	c	b	–10	–4	–1	0	0
x_1	–10	5	1	1	0	1	0
x_3	–1	10	0	–1	1	–2	1
			0	–5	0	–8	–1

Since all the S values are zero or negative, the procedure is finished. The answer is

$$x_1 = 5$$

$$x_3 = 10$$

This gives $P_{max} = 60$ **Ans.**

6.8 DYNAMIC PROGRAMMING

A feed containing A kg solvent and solute at a concentration of x_F kg solute/kg. A is to be extracted in a two-stage cross current extractor with solvent B. The solvent contains y_s kg solute/kg B.

Example 6.19 The distribution coefficient is constant and A and B are insoluble. If the cost of delivering 1 kg of solvent B at this concentration is same as the value of L kg solute. Determine the optimum amount of solvent to be used and its distribution to each stage. It is assumed that the cost of the stage remain fixed.

(Minimise the cost of extracted solute less cost of solvent)

$$x_n = (x_{n-1} - y_s/m)\left(\frac{1}{\varepsilon - 1}\right) + \frac{y_s}{m}$$

$$\varepsilon = \frac{mB}{A}$$

where, x = Raffinate composition

y = Extract composition

m = Slope of equilibrium curve

B/A = Ratio of solvent to raffinate

Solution: (a) **Step 1.** (Stage 1): Find optimum x_2 and B_2 as a function of x_1 (procedure starts from last stage to first)

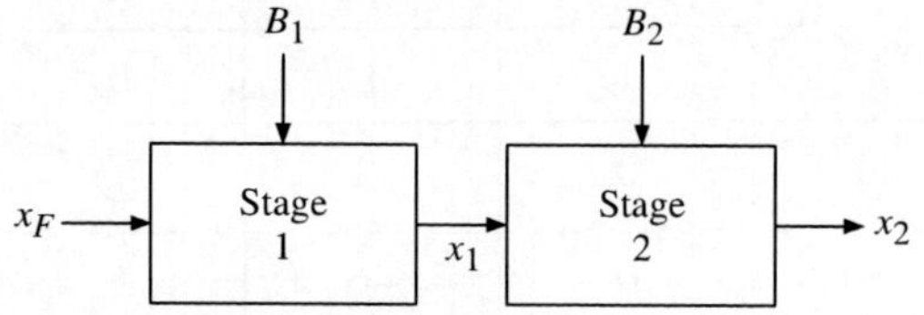

Fig. 6.11

$$\text{Income } I_2 = A(x_1 - x_2) - LB_2$$

To optimize,
$$\frac{dI_2}{dB_2} = 0 = \frac{-A\,dx_2}{dB_2} - L$$

$$B_2 = \left(\frac{x_1 - y/m}{x_2 - y/m} - 1\right)\frac{A}{m} \quad \text{(This is obtained by rearranging)}$$

$$x_2 = \left(x_1 - \frac{y}{m}\right)\left(\frac{1}{\varepsilon + 1}\right) + \frac{y}{m}$$

$$\varepsilon = \frac{mB_2}{A}$$

$$dB_2 = \left(\frac{A}{m}\right)\left(\frac{x_1 - y/m}{(x_2 - y/m)^2}\right)dx_2$$

$$\frac{dx_2}{dB_2} = \frac{(x_2 - y/m)^2}{(x_1 - y/m)}\frac{m}{A}$$

$$L = \frac{m(x_2 - y/m)^2}{(x_1 - y/m)} \text{ from } \frac{dI_2}{dB_2} = 0 \text{ relation}$$

$$\text{Optimum } x_2 = \sqrt{\frac{L}{m}\left(x_1 - \frac{y}{m}\right)} + \frac{y}{m}$$

$$\text{Optimum } B_2 = \left(\frac{(x_1 - y_1/m)}{\left(\sqrt{\frac{L}{m}\left(x_1 - \frac{y}{m}\right)} + \frac{y}{m}\right) - \frac{y}{m}} - 1\right)\frac{A}{m}$$

$$= \left(\sqrt{\frac{L}{m}\left(x_1 - \frac{y}{m}\right)} - 1\right)\frac{A}{m}$$

Step 2. (Stage 2): For stage 1, relation between B_1 and x_1 is

$$B_1 = \frac{A}{m}\left(\frac{x_F - y/m}{x_1 - y/m} - 1\right)$$

Step 3. Let total income in terms of x_1 using optimum stage 2 values:

$$I = A(x_F - x_2) - L(B_1 + B_2)$$

$$= A\left(x_F - \sqrt{\frac{L}{m}(x_1 - y/m)} - \frac{y}{m}\right) - \frac{LA}{m}$$

$$= \left[\frac{x_F - y/m}{x_1 - y/m} - 1 + \sqrt{\frac{m}{L}(x_1 - y/m)} - 1\right]$$

Step 4. Optimised I by differentiating and setting derivative = 0

$$\frac{1}{A}\frac{dI}{dx_1} = 0 = -\frac{1}{2}\sqrt{\frac{L}{m(x_1 - y/m)}} - \frac{1}{2}\sqrt{\frac{m}{L(x_1 - y/m)}} + \frac{L}{m}\frac{x_F - y/m}{(x_1 - y/m)^2}$$

$$\sqrt{\frac{L}{m(x_1 - y/m)}} = \frac{L}{m}\frac{(x_F - y/m)}{(x_1 - y/m)^2}$$

$$B_1 = \frac{A}{m}\left(\frac{x_F - y/m}{x_1 - y/m} - 1\right) = \frac{A}{m}\left(\frac{x_F - y/m}{3\sqrt{\frac{L}{m}(x_F - y/m)^2}} - 1\right)$$

$$B_1 = \frac{A}{m}\left(\sqrt[3]{\frac{m x_F - y}{L}} - 1\right)$$

$$B_2 = \left(\sqrt{\frac{m}{L}(x_1 - y/m)} - 1\right)\frac{A}{m} = \left(\frac{m}{L}\sqrt[3]{\frac{L}{m}(x_F - y/m)^2} - 1\right)\frac{A}{m}$$

$$= \frac{A}{m}\left(\sqrt[3]{\frac{m x_F - y}{L}} - 1\right)$$

So, $$B_1 = B_2$$

(b) Another way of solving this is to make tables.

Let $L_1 = 0.01, A = 1000, x_F = 0.40, y_S = 0, m = 0.02$

$$x_1 = \frac{0.4 \times 1000}{(0.2B_1 + 1000)} = \frac{400}{(1000 + 0.2B_1)} \text{ rearranging,}$$

or $$B_1 = \left(\frac{2000}{x_1} - 5000\right)$$

$$x_2 = x_1\frac{A}{(mB_2 + A)} = \frac{1000x_1}{0.2B_2 + 1000}$$

Rearranging, $$B_2 = 5000\left(\frac{x_1}{x_2}\right) - 5000$$

$$\text{Profit for stage 2} = P_2 = A(x_1 - x_2) - LB_2$$

At $x_1 = 0.1$, $$B_2 = \left(\frac{500}{x_2} - 5000\right)$$

$$P_2 = 1000(x_1 - x_2) - 0.01\, B_2$$

$$\text{Profit for stage 1} = P_1 = A(x_F - x_1) - LB_1$$

$$= 1000(0.4 - x_1) - 0.01\, B_1$$

Tables are made for P_2, P_1 for different x_1 and x_2 values.

For $x_F = 0.4$

x_1	x_2	B_2	P_2	B_1	P_1	$P_1 + P_2$
0.10	0.10	0	0	15000	150	150
	0.08	1250	7.5	15000	150	157.5
	0.06	3335	6.65	15000	150	156.65
	0.4	7500	–15.0	15000	150	135.0
0.2	0.2	0	0	5000	150	150
	0.15	1665	33.35	5000	150	183.35
	0.10	5000	50	5000	150	200.0
	0.05	15000	0	5000	150	150.0
0.3	0.3	0	0	1700	83	83
	0.25	1000	40	1700	83	123
	0.20	2500	75	1700	83	158
	0.15	5000	100	1700	83	183
	0.10	10000	100	1700	83	183
	0.05	25000	0	1700	83	0

x_1	x_2	B_1	B_2	$B_1 + B_2$	$P_1 + P_2$
0.10	0.08	15000	1250	6250	157.5
0.2	0.10	5000	5000	10000	**200**
0.3	0.15	1700	5000	6750	183

The optimum value of x_2 is marked for each x_1 value. Then a table of x_1 with optimum values of x_2, B_1, B_2 and total profit is made. From this table, the optimum values of x_1 and x_2 are read by seeing the maximum profit ($P_1 + P_2$). So, optimum values for 2-stage extraction set up are $x_1 = 0.2$, $x_2 = 0.1$, $B_1 = 5000$, $B_2 = 5000$ and $P_1 + P_2 = 200$.

Nomenclature

a_{ij} = Constant

B = Solvent

$C_1, C_2, \ldots, C_j$ = Constant coefficients

D = Diameter

dP = Pressure drop

h = Individual heat transfer coefficient

i = Interest rate

k = Thermal conductivity of metal wall

L = Salvage ₹ value

N = Number of effects in evaporator

n = Number of years

P_1, P_2 = Profit

P = Present ₹ worth

q = Heat transfer rate

U = Overall heat transfer coefficient

R = Annual fixed charges or ₹ deposit

R' = Income per year ₹

S_1, S_2 = Slack variables

S = Worth ₹ after some years

T = Temperature

w_j = Weights

$x_1, x_2, \ldots,$ = Variables or raffinate compositions

x_F = Feed composition

y = Objective function or extract compositions

y^* = Optimum y

ΔT = Temperature difference

β = Constant

θ = Time

6.9 ONE-VARIABLE OPTIMIZATION

If f is a function of one variable (x) only then the condition for minimum or maximum (optimal) is

$$\frac{df(x)}{dx} = 0$$

For maximum value, $\frac{d^2 f(x)}{dx^2}$ should be negative.

For minimum value, $\frac{d^2 f(x)}{dx^2}$ should be positive.

6.10 TWO-VARIABLE OPTIMIZATION

For checking the value obtained is optimum or not.

$$A = \frac{d^2 y}{dx_1^2}(x_1, x_2) = f_{xx}$$

$$B = \frac{d^2 y}{dx_1\, dx_2} = f_{xy}$$

$$C = \frac{d^2 y}{dx_2^2} = f_{yy}$$

$B^2 - AC < 0$ if $A < 0$ relative maximum at x_0, y_0

$B^2 - AC < 0$ if $A > 0$ relative minimum at x_0, y_0

$B^2 - AC > 0$ saddle point at (x_0, y_0)

$B^2 - AC = 0$ nature of critical point undetermined.

A saddle point is one which represents neither maximum nor minimum.

Optimisation values are obtained using the differentiation of $f(x, y)$ as

$$\frac{df(x, y)}{dx} = 0$$

$$\frac{df(x, y)}{dy} = 0$$

From these relations, optimum values of x, y are obtained sometimes by solving simultaneously.

6.11 DICHOTOMOUS SEARCH (ONE VARIABLE)

x_1 and x_2 are placed symmetrically at the centre of interval of uncertainity I_0 with gap of E.

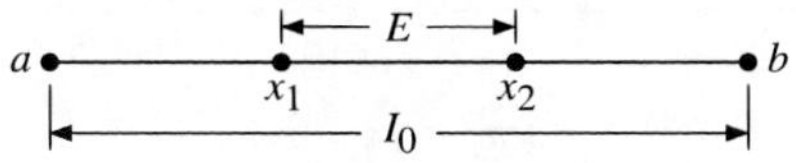

Fig. 6.12

It is assumed that the function is unimodal. Depending on $f(x_1)$ is greater than $f(x_2)$ or $f(x_1)$ is less less than $f(x_2)$, the new interval of uncertainty is ax_2 or bx_1.

$$I_1 = \frac{I_0}{2} + \frac{E}{2} \quad \text{...(1)}$$

Using second trial

$$I_2 = \frac{I_1}{2} + \frac{E}{2}$$

or

$$I_2 = \frac{I_0}{2^2} + \left(1 - \frac{1}{2^2}\right)E \quad \text{...(2)}$$

Number of function evaluations = $n = 2k$

where k is the number of experiments.

For k times,

$$I_k = \frac{I_0}{2^k} + \left(1 - \frac{1}{2^k}\right)E \quad \text{...(3)}$$

If E is very small

$$I_k = \frac{I_0}{2^k}$$

$$\frac{I_k}{I_0} = \left(\frac{1}{2}\right)^k$$

Example 6.20 Find the minimum value of y for the function $y = x^2 - 0.5x + 1.0$ between 0 and 1 using two experiments with $E = 0.05$. Define the region of uncertainty after two experiments and compare with equation value.

Solution:

$$x_1 = 0.5 - \frac{0.05}{2} = 0.475$$

$$x_2 = 0.5 + \frac{0.05}{2} = 0.525$$

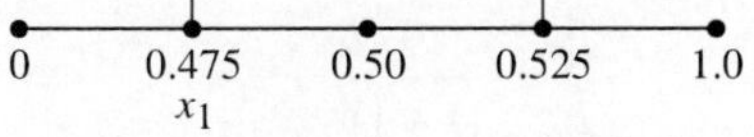

Fig. 6.13

$$y(x_1) = y(0.475) = 0.9881 = y_1$$

$$y(x_2) = 1.0132 = y_2(0.525)$$

Since $y_2 > y_1$, so leave right of $x_2 \cdot E = 0.05$

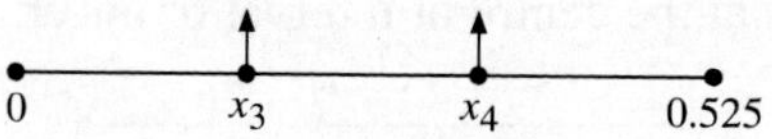

Fig. 6.14

$$x_3 = \frac{0.525}{2} - \frac{E}{2} = 0.2375$$

$$x_4 = \frac{0.525}{2} + \frac{E}{2} = 0.2875$$

$$y_3 = 0.93764$$

$$y_4 = 0.92704$$

Since y_3 is greater than y_4, so leave left of x_3. The interval of uncertainty after two experiments is

$$0.525 - 0.2375 = 0.2875$$

The interval of uncertainty from theoretical equation after two experiments is

$$I_k = \frac{I_0}{2^k} + \left(1 - \frac{1}{2^k}\right) E$$

For $k = 2$

$$I_2 = \frac{I_0}{2^2} + \left(1 - \frac{1}{2^2}\right) 0.05$$

$$= 0.2875.$$

6.11.1 Fibonacci – Search (One Variable)

To make the search more efficient or reduction fast, this search uses Fibonacci numbers.

The numbers are given by the relation

$$F_n = \frac{1}{5^{1/2}}\left[\left(\frac{1+5^{1/2}}{2}\right)^{n+1} - \left(\frac{1-5^{1/2}}{2}\right)^{n+1}\right]$$

$$F_0 = 1; F_1 = 1; F_2 = 2; F_3 = 3; F_4 = 5; F_5 = 8 \text{ etc.}$$

Now trial points are

$$x_3^i = \frac{F_n - (i+1)}{F_n - (i-1)}(x_2^{i-1} + x_1^{i-1}) + x_1^{i-1} \quad \ldots(1)$$

$$x_4^i = \frac{F_n - (i+1)}{F_n - (i-1)}(x_2^{i-1} - x_1^{i-1}) + x_1^{i-1} \quad \ldots(2)$$

$$\text{Reduction } (\alpha) = \frac{I_{n-1}}{I_0} = \frac{1}{F_n}$$

where n is the number of function equations.

Example 6.21 $y = x^2 - 0.5x + 1$, find the location of minimum value of y between 0 and 1. Define the region of uncertainty after four function evaluations and compare with equation values.

Solution: $n = 4$

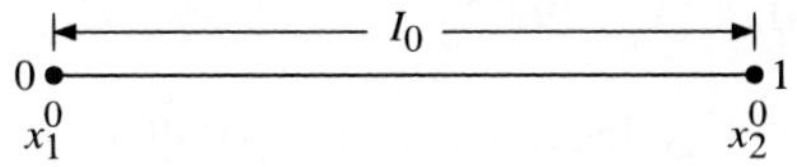

Fig. 6.15

$$x_3^1 = \frac{F_{n-2}}{F_n}(x_2^0 - x_1^0) + x_1^0$$

$$x_4^1 = \frac{F_{n-1}}{F_n}(x_2^0 - x_1^0) + x_1^0$$

$$x_3^1 = \frac{F_2}{F_4} \times 1 + 0 = \left(\frac{2}{5}\right) \times 1 = 0.40$$

$$x_4^1 = \frac{F_3}{F_4} \times 1 + 0 = \left(\frac{3}{5}\right) \times 1 = 0.60$$

$$y(x_3^1) = 0.9600$$

$$y(x_4^1) = 1.0600$$

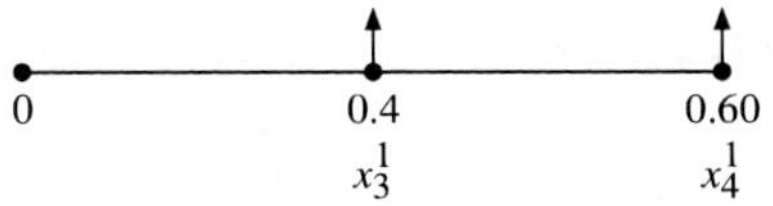

Fig. 6.16

Since $y(x_3^1)$ is less than $y(x_4^1)$ so leave right of x_4^1.

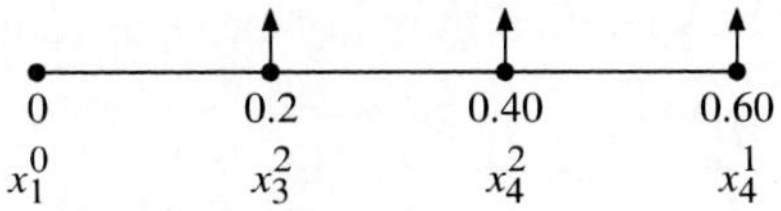

Fig. 6.17

$$x_3^2 = \frac{F_1}{F_3} \times I_1 + 0 = \left(\frac{1}{3}\right) \times 0.60 + 0 = 0.20$$

$$x_4^2 = \frac{F_2}{F_3} \times I_1 + 0 = \left(\frac{2}{3}\right) \times 0.60 = 0.40$$

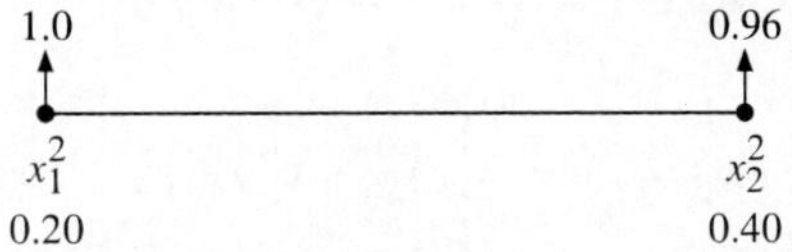

Fig. 6.18

$$y(0.20) = 0.9400$$

$$y(0.40) = 0.9600$$

$$y(0.60) = 1.0600$$

Since $y(x_4^2)$ is greater than $y(x_3^2)$ so leave right of $x_4^2(0.40)$

$$x_3^3 = \frac{F_0}{F_2}(x_2^2 - x_1^2) + x_1^2 = \left(\frac{1}{2}\right) \times 0.40 + 0 = 0.20$$

$$x_4^3 = \frac{F_1}{F_2}(\cdot \ddot{u} \; \ddot{u} - \quad) + \quad = \left(\frac{1}{2}\right) \times 0.4 + 0 = 0.20$$

Since both points x_3^3 and x_4^3 are coming same, so take $E = 0.05$

$$x_3^3 = 0.150$$

$$x_4^3 = 0.250$$

$$y(x_3^3) = 0.9475$$

$$y(x_4^3) = 0.9375$$

Since $y(x_4^3)$ is less than $y(x_3^3)$, so leave left of x_3^2.

So observed interval of uncertainty is from 0.150 to 0.400 = 0.250.

Theoretical or equation interval of uncertainty = $I_3 = \left(\frac{F_1}{F_4}\right) \times I_0 = \left(\frac{1}{5}\right) \times 1 = 0.20$ **Ans.**

6.12 GOLDEN SECTION SEARCH (ONE VARIABLE)

The number of experiments cannot be specified in many cases as required in Fibonacci search.

Using
$$\frac{I_0}{I_1} = r = \frac{I_1}{I_2} = \ldots = \frac{I_{n-1}}{I_n}$$

$r = 1.618$:
$$\frac{I_{n-1}}{I_0} = \frac{1}{r^{n-1}}$$

$$x_3^i = \frac{(r-1)}{r}(x_2^{i-1} - x_1^{i-1}) + x_1^{i-1}$$

$$x_4^i = \left(\frac{1}{r}\right)(x_2^{i-1} - x_1^{i-1}) + x_1^{i-1}$$

$$x'_3 = \left(\frac{r-1}{r}\right)(x_2^0 - x_1^0) + x_1^0$$

$$x'_4 = \left(\frac{1}{r}\right)(x_2^0 - x_1^0) + x_1^0$$

Example 6.22 Using Golden section, find the interval of uncertainty between 0 and 1 for the function $y = x^2 - 0.5x + 1.0$ with $n = 4$ function evaluations.

Solution:

Fig. 6.19

$$x'_3 = \frac{r-1}{r}(x_2^0 - x_1^0) + x_1^0 = \frac{0.618}{1.618} \times 1 + 0 = 0.383$$

$$x'_4 = \left(\frac{1}{r}\right)(x_2^0 - x_1^0) + x_1^0 = \left(\frac{1}{1.618}\right) \times 1 + 0 = 0.617$$

$$y'_3 = 1.0744$$

$$y'_4 = 0.9544$$

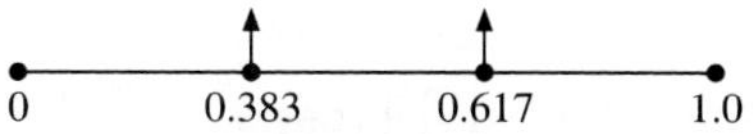

Fig. 6.20

Since $y(x'_4) = y'_4$ is less than y'_3 so leave right of y'_3. Functions done are two. Left interval of uncertainty is 0.617.

$$x_3^2 = \left(\frac{0.618}{1.618}\right) \times 0.617 = 0.234$$

$$x_4^2 = \left(\frac{1}{1.618}\right) \times 0.617 = 0.383$$

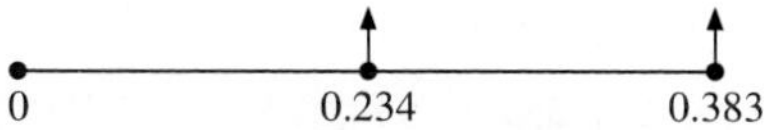

Fig. 6.21

$$y(0.234) = 0.9379$$

$$y(0.383) = 0.9544$$

Since $y(x_3^2)$ is greater than $y(x_4^2)$, then leave right of 0.383 for y_{min}. Functions done up to now are three (One is repeated).

$$x_3^3 = \frac{0.383 \times 0.618}{1.618} = 0.149$$

$$x_4^3 = \frac{0.383}{1.618} = 0.234$$

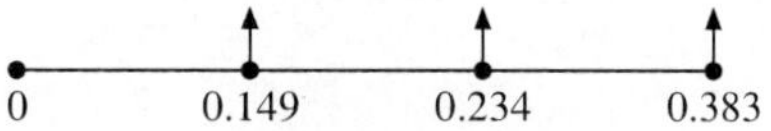

Fig. 6.22

$$y(0.149) = 0.9479$$

$$y(0.234) = 0.9379$$

Since $y(0.149)$ is greater than $y(0.234)$, so leave left of 0.149 for y_{min}. The number of function evaluations done are four. Now the interval of uncertainty is

$$0.383 - 0.149 = 0.234.$$

Equation value of interval of uncertainty $= \dfrac{1}{r^{n-1}} = \dfrac{1}{1.618^{4-1}} = 0.234$. So both are agreeing.

6.13 OPTIMIZATION METHODS (TWO OR MORE VARIABLES)

Example 6.23 The following equation shows the effect of the variable x and y on the total cost of an operation. Determine the values of x and y which will give the least cost using geometric programming.

$$C = 2.33x + \frac{11900}{xy} + 1.86y + 10$$

Solution:

$$C - 10 = 2.33x + \frac{11900}{xy} + 1.86y$$

Using geometric programming,

$$y = C_1 P_1 + C_2 P_2 + C_3 P_3$$

$$C_1 = 2.33$$

$$C_2 = 11900$$

$$C_3 = 1.86$$

$$P_1 = x^1$$

$$P_2 = x^{-1} y^{-1}$$

$$P_3 = y^1$$

$$P_1 = \sum_{i=1}^{n} x_i^{a_{i1}} = x_1^{a_{11}} x_2^{a_{21}} x_3^{a_{31}} = x$$

Here, $x_2 = y$; no x_3 present; So $a_{11} = 1$

$$P_2 = \sum_{i=1}^{n} x_i^{a_{i2}} = x_1^{a_{12}} x_2^{a_{22}} x_3^{a_{32}} = x^{-1} y^{-1}$$

So, $a_{12} = -1$; $a_{22} = -1$

$$P_3 = \sum_{i=1}^{n} x_i^{a_{i3}} = x_1^{a_{13}} x_2^{a_{23}} x_3^{a_{33}} = y^1. \text{ So } a_{23} = 1$$

$a_{kj}\,\omega_j = 0$ for $k = 1, 2, \ldots, n$; $\sum \omega_j = 1.0$

$$\sum_j a_{1j}\,\omega_j = a_{11}\,\omega_1 + a_{12}\,\omega_2 + a_{13}\,\omega_3 = 0$$

$$\sum_j a_{2j}\,\omega_j = a_{21}\,\omega_1 + a_{22}\,\omega_2 + a_{23}\,\omega_3 = 0$$

$$\sum_j a_{3j}\,\omega_j = a_{31}\,\omega_1 + a_{32}\,\omega_2 + a_{33}\,\omega_3 = 0$$

Substituting for a_{11}, a_{12}

$$\omega_1 - \omega_2 = 0$$

$$-\omega_2 + \omega_3 = 0$$

$$\omega_1 + \omega_2 + \omega_3 = 0$$

Solving simultaneously,

$$\omega_1 = \omega_2;$$

$$\omega_2 = \omega_3$$

So, $$\omega_1 = \omega_2 = \omega_3 = \frac{1}{3}$$

$$(C - 10)^* = \prod_j \left(\frac{C_j}{\omega_j}\right)^{\omega_j}$$

$$= \left(\frac{2.33}{(1/3)}\right)^{1/3} \left(\frac{11900}{1/3}\right)^{1/3} \left(\frac{1.86}{1/3}\right)^{1/3} = 111.66$$

$$2.33x^* = \omega_1^*(C-10)^* = 111.66 \times \frac{1}{3}$$

$x^* = 15.974$ **Ans.**

$$1.86\,y^* = (C - 10) \times \omega_3$$

$y^* = 20.11$ **Ans.**

$C^* = 121.66$ **Ans.**

Example 6.24 Using geometric programming find the optimal values for the given equation (y, x_1, x_2)

$$y = 1000\,x_1 + \frac{4 \times 10^9}{x_1 x_2} + 2.5 \times 10^5 x_2$$

Solution:

$$y = \sum_{j=1}^{n} C_j\,P_j(x)$$

$$y = C_1 P_1 + C_2 P_2 + C_3 P_3$$

$$C_1 = 1000$$

$$C_2 = 4 \times 10^9$$

$$C_3 = 2.5 \times 10^5$$

$$P_j = \prod_{i=1}^{n} x_i^{a_{ij}}$$

$$P_1 = \prod_{i=1}^{n} x_i^{a_{ij}} = \prod_{i=1}^{n} x_i^{a_{i1}} = x_1^{a_{11}} x_2^{a_{21}} x_3^{a_{31}} = x_1$$

$$P_2 = \prod_{i=1}^{n} x_i^{a_{i2}} = x_1^{a_{12}} x_2^{a_{22}} x_3^{a_{32}} = x_1^{-1} x_2^{-1}$$

$$P_3 = \prod_{i=1}^{n} x_i^{a_{i3}} = x_1^{a_{13}} x_2^{a_{23}} x_3^{a_{33}} = x_2$$

At optimum $\sum_j a_{kj}\, \omega_j = 0; K = 1, 2, \ldots, n.$

$$\sum \omega_j = 1.0; \text{ where } \omega_j \text{ are weights.}$$

Expanding $\sum_j a_{1j}\, \omega_j = a_{11}\omega_1 + a_{12}\omega_2 + a_{13}\omega_3 = 0$

$$\sum_j a_{2j}\, \omega_j = a_{21}\omega_1 + a_{22}\omega_2 + a_{23}\omega_3 = 0$$

$$\sum_j a_{3j}\, \omega_j = a_{31}\omega_1 + a_{32}\omega_2 + a_{33}\omega_3 = 0$$

$$a_{11} = 1;\ a_{21} = 0;\ a_{31} = 0;\ a_{12} = -1,\ a_{22} = -1;\ a_{32} = 0$$

$$a_{13} = 0,\ a_{23} = 1;\ a_{33} = 0$$

$$\omega_1 - \omega_2 = 0;\ -\omega_2 + \omega_3 = 0;\ \omega_1 + \omega_2 + \omega_3 = 1.0$$

Solving gives $\omega_1 = \omega_2 = \omega_3 = \frac{1}{3}$

$$y^* = \left(\frac{1000}{1/3}\right)^{1/3} \left(\frac{4 \times 10^9}{1/3}\right)^{1/3} \left(\frac{2.5 \times 10^5}{1/3}\right)^{1/3}$$

$$= 3.0 \times 10^6$$

$$1000 x_1^* = 3 \times 10^6 \times \frac{1}{3}$$

$$x_1^* = 1000$$

$$2.5 \times 10^5 x_2^* = 3 \times 10^6 \times \frac{1}{3}$$

$$x_2^* = 4$$

$$y^* = 3 \times 10^6 \textbf{ Ans.}$$

Example 6.25 (a) P is a function of x and y as given by

$$P = Ax + B\left(\frac{xy - a}{y}\right) + \frac{C}{(xy)^{1/2}} \text{ (two variables)}$$

where A, B, C and a are constants. Obtain the expression for the values of x and y in terms of the constants corresponding to the maximum or minimum value of P.

(b) In the design of certain piece of equipment it is found that the two costs involved may each be expressed as a power function of a design variable x.

Total cost = $A + B$ where $A = ax^n$ and $B = b/x^n$

Showing for the optimum design, i.e., the value of x corresponding to the minimum total cost, the ratio of individual costs A and B is m/n.

Solution: (a)

$$P = Ax + Bx - \frac{Ba}{y} + Cx^{-1/2}y^{-1/2}$$

$$\frac{dP}{dx} = A + B - 0.5\,Cx^{-3/2}y^{-1/2} = 0$$

$$A + B = 0.5Cx^{-3/2}y^{-1/2} \qquad \ldots(1)$$

$$\frac{dP}{dy} = \frac{Ba}{y^2} - 0.5Cx^{-1/2}y^{-3/2} = 0$$

$$Ba = 0.5y^{-1/2}y^{1/2}C \qquad \ldots(2)$$

From Eq. (2),

$$y^{+1/2} = \frac{Bax^{1/2}}{0.5C}$$

Substituting in Eq. (1),

$$A + B = \frac{0.5C \times 0.5C}{Bax^{1/2}}x^{-3/2}$$

Substituting in Eq. (1),

$$A + B = \frac{0.25C^2x^{-2}}{Ba}$$

or

$$x^2 = \frac{0.25C^2}{Ba(A+B)};$$

$$x = \frac{0.5C}{((A+B)Ba)^{1/2}}$$

(b) Total cost: $ax^n + \dfrac{b}{x^m}$; $\dfrac{dT}{dx} = nax^{n-1} - mbx - m - 1 = 0$

$$x^{m+n} = \frac{bm}{(na)}$$

$$x = \left(\frac{bm}{na}\right)^{1/(m+n)}$$

$$A = ax^n = a\left(\frac{bm}{na}\right)^{n/(m+n)}$$

$$B = \frac{b}{x^m} = \frac{b}{\left(\dfrac{bm}{na}\right)^{m/m+n}}$$

$$\frac{A}{B} = \frac{m}{n} \quad \textbf{Ans.}$$

Example 6.26 A hypothetical chemical plant consists of a tubular reactor, a product separator, compressor and a mixer. The costs depend on two variables: the pressure: x_1 and the recycle ratio: x_2. The required reduction rate is 10 million kg/year. For this flow rate the costs are in rupees.

Compressing raw material = 2000 x_1

Operating the reactor = $\dfrac{5 \times 10^9}{(x_1^{1.5} x_2)}$

Operating the separator = $2 \times 10^5 x_2$

Recirculating the material = $2 \times 10^5 x_2$

(a) Calculate the optimum conditions by partial differentiation.

(b) Check to determine that the value of optimum obtained is an optimum, not an inflection, ridge or saddle.

Solution: Cost Eq. (2) is

$$C = 2000x_1 + \frac{5 \times 10^9}{x_1^{1.5} x_2} + 2 \times 10^5 x_2 + 2 \times 10^5 x_2$$

$$\frac{dC}{dx_1} = 2000 - \frac{5 \times 10^9 x_2 x_1^{0.5} \times 1.5}{x_1^3 x_2^2} = 0 \qquad \ldots(1)$$

$$\frac{dC}{dx_2} = 4 \times 10^5 - \frac{5 \times 10^9 x_1^{1.5}}{x_1^3 x_2^2} = 0 \qquad \ldots(2)$$

Solving Eqs. (1) and (2) simultaneously, gives

$$x_1^* = 382$$

$$x_2^* = 1.33$$

$$\text{Total cost} = 2000x_1^* + \frac{5 \times 10^9}{x_1^{*1.5} x_2^*} + 4 \times 10^5 x_2^*$$

(b) Checking the value obtained is an optimum or not:

$$A = \frac{d^2y}{dx_1^2}(x_1, x_2) = f_{xx}$$

$$B = \frac{d^2y}{dx_1 dx_2} = f_{xy}$$

$$C = \frac{d^2y}{dx_2^2} = f_{yy}$$

$B^2 - AC < 0$, if $A > 0$ gives relative maximum at x_0, y_0.

$B^2 - AC < 0$, if $A > 0$ gives relative minimum at x_0, y_0.

$B^2 - AC > 0$ gives saddle point at x_0, y_0.

$B^2 - AC = 0$ nature of critical point undetermined.

$$\frac{d^2y}{dx_1^2} = \frac{7.5 \times 2.5 \times 10^9}{(x_1^{3.5} x_2)} = A$$

$$\frac{d^2y}{dx_2^2} = \frac{10^{10}}{(x_1^{1.5} x_2)} = C$$

$$\frac{d^2y}{dx_1\, dx_2} = \frac{7.5 \times 10^9}{(x_1^{2.5} x_2^2)} = B$$

At $x_1 = 382$; $x_2 = 1.33$

$$A = 12.7$$

$$B = 1.485$$

$$C = 5.76 \times 10^5$$

$$B^2 - AC = (1.485 \times 10^3)^2 - 12.7 \times 5.76 \times 10^5$$

This is less than zero and A is greater than zero. This implies the obtained conditions are for relative minimum.

Example 6.27 Maximum: $Z = 7x_1 + 5x_2$

Subject to:

$$x_1 + 2x_2 \leq 6$$
$$4x_1 + 3x_2 \leq 12$$
$$x_1, x_2 \geq 0$$

Solution: $Z - 7x_1 - 5x_2 = 0$

Subject to: $x_1 + 2x_2 + S_1 = 6$

$$4x_1 + 3x_2 \leq 12, x_1, x_2 \geq 0$$

	x_1	x_2	S_1	S_2	RHS
Z	–7	–5	0	0	0
S_1	1	2	1	0	6
S_2	$\boxed{4}$	3	0	1	12

	x_1	x_2	S_1	S_2	RHS
Z	0	1/4	0	7/4	21
S_1	0	5/4	1	–1/4	3
X_1	1	3/4	0	1/4	3

Giving $x_1 = 3$; $x_2 = 0$; $Z = 7 \times 3 + 5 \times 0 = 21$ **Ans.**

Example 6.28 Minimise: $Z = 3x_1 + 2x_2$

Subject to:

$$3x_1 + x_2 > 8$$
$$4x_1 + 3x_2 > 6$$
$$x_1 + x_2 \leq 3$$
$$x_1, x_2 > 0$$

Here, $Z - 3x_1 - 2x_2 = 0$

Subject to:

$$-3x_1 - x_2 \leq -3$$
$$-4x_1 - 3x_2 \leq -6$$
$$x_1 + x_2 \geq 0$$
$$x_1, x_2 \leq 0.$$

i.e., equating inequalities to equalities:

$$-3x_1 - x_2 + S_1 = -3$$
$$-4x_1 - 3x_2 + S_2 = -6$$
$$x_1 + x_2 + S_3 = 3$$
$$x_1 + x_2 \geq 0$$

Solution: Dual-Simplex Method

	x_1	x_2	S_1	S_2	S_3	RHS
Z	–3	–2	0	0	0	0
S_1	–3	–1	1	0	0	–3
S_2	–4	–3	0	1	0	–6
S_3	1	1	0	0	1	3

	x_1	x_2	S_1	S_2	S_3	RHS
Z	–1/3	0	0	–2/3	0	4
S_1	–5/3	0	1	–1/3	0	–1
x_2	4/3	1	0	1/3	0	2
S_3	–1/3	0	0	1/3	1	1

	x_1	x_2	S_1	S_2	S_3	RHS
Z	0	0	–1/5	–3/5	0	21/5
x_1	1	0	–3/5	1/5	0	3/5
x_2	0	1	4/5	1/15	0	6/5
S_3	0	0	–1/5	2/5	1	6/5

Optimal values are: $x_1 = \frac{3}{5}$

$$x_2 = \frac{6}{5}$$

$$S_3 = \frac{6}{5}$$

$$Z = \frac{21}{5}$$ **Ans.**

Example 6.29 Simplex method:

$$Z = 3x_e + 2x_i;\ e\text{–external};\ i\text{–internal}$$

Subject to: $x_e + 2x_i \le 6$

$$2x_e + x_i \le 8$$

$$-x_e + x_i \le 1$$

$$x_i < -2$$

Rewriting:

$$Z - 3x_e - 2x_i = 0$$
$$x_e + 2x_i + S_1 = 6$$
$$2x_e + x_i + S_2 = 8$$
$$-x_e + x_i + S_3 = 1$$
$$x_i + S_4 = 2$$

	x_e	x_i	S_1	S_2	S_3	S_4	RHS
Z	–3	–2	0	0	0	0	0
S_1	1	2	1	0	0	0	6
S_2	2	1	0	1	0	0	8
S_3	–1	+1	0	0	10	0	1
S_4	0	1	0	0	0	1	2

	x_e	x_i	S_1	S_2	S_3	S_4	RHS
Z	0	–1/2	0	3/2	0	0	12
S_1	0	3/2	1	–1/2	0	0	2
x_e	1	1/2	0	1/2	0	0	4
S_3	0	3/2	0	1/2	1	0	5
S_4	0	1	0	0	0	1	2

	x_e	x_i	S_1	S_2	S_3	S_4	RHS
Z	0	0	1/3	4/3	0	0	38/3
x_i	0	1	2/3	–1/3	0	0	4/3
x_e	1	0	0	0	0	0	10/3
S_3	0	0	0	0	1	0	3
S_4	0	0	0	0	0	1	2/3

The solution yields

$$x_e = \frac{10}{3}$$

$$x_i = \frac{4}{3}$$

$$Z = \frac{38}{3} \quad \textbf{Ans.}$$

Example 6.30 Linear programming by graphical method.

Maximise Z = $5x_1 + 7x_2$ for integer values of x_1 and x_2.

Subject to:

$$x_1 + x_2 \le 4$$

$$5x_1 + 8x_2 \le 24$$

$$10x_1 + 7x_2 \le 35$$

and x_1, x_2 greater than zero.

Solution:

$$Z = 5x_1 + 7x_2 \quad \text{...(1)}$$

Draw lines of

$$x_1 + x_2 \le 4 \quad \text{...(2)}$$

$$5x_1 + 8x_2 \le 24 \quad \text{...(3)}$$

and

$$10x_1 + 7x_2 \le 35. \quad \text{...(4)}$$

On a graph sheet with a suitable scale (Fig. 6.23).

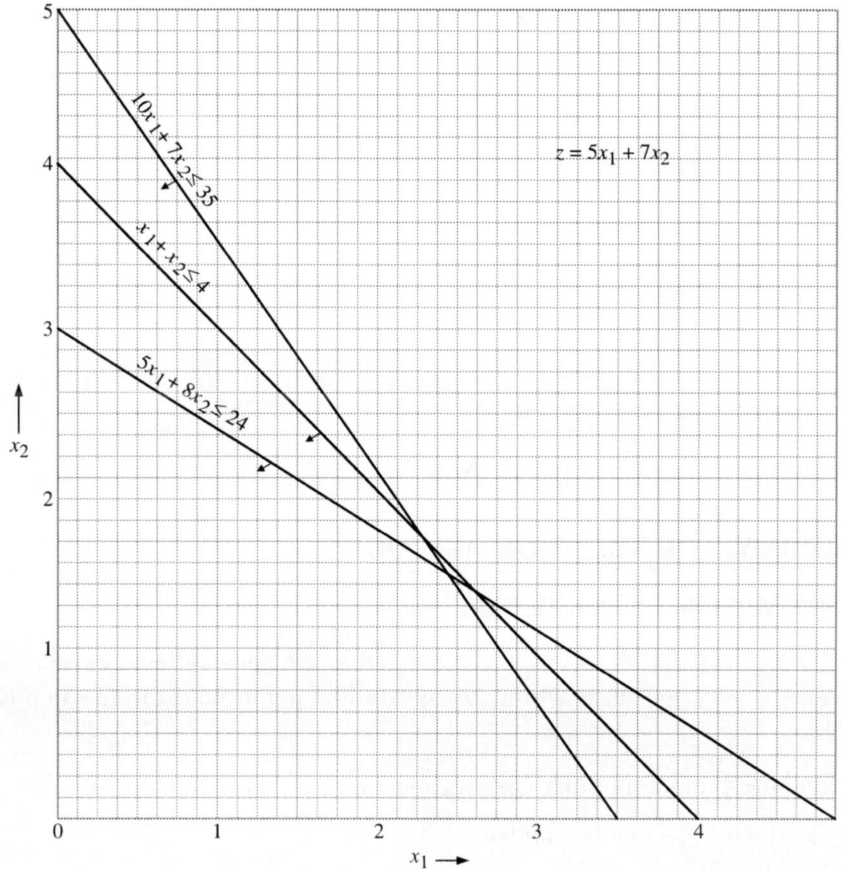

Fig. 6.23

x_1	0	1	2	3	4
x_2	4	3	2	1	0
x_1	0	1	2	3	4
x_2	3	2.4	1.75	1.1	0.50
x_1	0	1	2	3	3.5
x_2	5	3.6	2.1	0.70	0

$$\text{Eq. (2): } x_1 + x_2 = 4$$

$$\text{Eq. (3): } 5x_1 + 8x_2 = 24$$

$$\text{Eq. (4): } 10x_1 + 7x_2 = 35$$

Taking nearby integers:

Extremes are: $x_1 = 0$

$x_2 = 3$

$Z = 21$

$x_1 = 3$

$x_2 = 1$

$Z = 22$

$x_1 = 2$

$x_2 = 2$

$Z = 26$

So, maximum is at $Z = 26$ **Ans.**

6.14 MULTIVARIABLE OPTIMIZATION METHODS

6.14.1 Alternate Variable Search Method

One variable is varied, keeping the others as constants to locate apparent extremism. Each of the variable is changed in turn starting from the previously loci point, until the function cannot be improved further.

Example 6.31 Find the minimum of the objective function $y = 7x_1^2 + 10x_2^2$ starting at $x_1 = 3$ and $x_2 = -9$; and verify with the differential approach.

Solution: By differential approach:

$$\frac{dy}{dx_1} = 14x_1 = 0$$

gives $x_1 = 0$
$$\frac{dy}{dx_2} = 20x_2$$

$$x_2 = 0$$

By alternate variable search method: Starting $x_1 = 3$; $x_2 = -9$

$$y_1 = 7(3^2) + 10(-9)^2 = 873$$

Keep x_2 constant, varying x_1 until y becomes minimum is $x_1 = 0$

$$y_2 = 7(0)^2 + 10(-9)^2 = 810 \text{ (since } y_2 \text{ is } < y_1\text{, so improvement)}$$

Next change x_2 to decrease y to the lowest value, which at $x_2 = 0$.

$$y_3 = 0$$

So, $y_{min} = 0$ at $x_1 = 0$; $x_2 = 0$ **Ans.**

6.14.2 Exploratory and Pattern Move Method

This method is by Hookes and Jeeves (1961). B_0 is initial base point. B_1 is obtained by exploratory move, from B_0 by changing each variable by a small amount and testing for improvement in the objective function. Now from B_1 a pattern move is made by extrapolating to point $P_1 = B_1 + (B_1 - B_0)$. P_1 is not executed/evaluated. Another base point B_2 is obtained by changing x_1 and x_2. If B_2 shows improvement then the direction from B_1 to B_2 is considered, worth of further investigation, by making a pattern move to P_2.

If the function value at B_2 is worse than at B_1 then the direction is a failure. The search then retreats to B_1 and where new moves are tried from B_1 or with a decreased step size.

Example 6.32 $y = 100(1.0 - (-1.2)^3)^2 + (1 + 1.2)^2 = 749.04$ has a minimum of $y = 0$ at $x_1 = x_2 = 1.0$. Starting search from $x_1 = -1.2$ and $x_2 = 1.0$.

Solution: Starting search from $x_1 = -1.2$ and $x_2 = 1.0$

$$B_0 = 100(1.0 + 1.2^3)^2 + (1 + 1.2)^2 = 749.04$$

Let $dx_1 = 0.20$; $dx_2 = -0.20$

$$B_1 = x_1 = -1.2 + 0.20 = -1.0$$

$$x_2 = 1.0 - 0.20 = 0.80$$

$$B_1 = (-1, 0.8)$$

$$y = 328$$

Since B_1 is $< B_0$, so improvement, so continue in the same way.

Pattern, $P_1 = B_1 + (B_1 + B_0) = -1 + (-1 + 1.2) = -0.8(x_1)$

$$x_2 = 0.8 + 0.8 - 1.0 = 0.60$$

Do not evaluate the function (y) at P_1 (–0.8, 0.60)

$$B_2 = P_1 + dx = x_1 + dx_1;\ x_2 + dx_2$$

$$B_2 = (0, -0.20)$$

$y(B_2) = 5$, so improvement over B_1 (40.5).

So, the process is continued with same dx_1, dx_2 (steps).

$$P_2 = B_2 + (B_2 - B_1) = 0 + 0 - (-0.60)$$

$$= 0.60(x_1) = -0.2(-0.20 - 0.40) = -0.80$$

$$B_3 = -0.80, -1.0$$

$y = 228.6$; no improvement, over $y(B_2) = 5$.

So go back to B_2 and change x_1 and x_2 in different way or in small steps. Now B_2 is base point.

$$B_2 = (0, -0.20)$$

Change dx_1 and dx_2 and continue. Final point is $x_1 = 1.0$; $x_2 = 1.0$ giving $y = 0$

6.14.3 Steepest Ascent/Descent Method and Contour Tangent Method

Example 6.33 $y = 100(x_2 - x_1^2)^2 + (1 - x_1)^2;$

$$x_0 = (2, 2)$$

$$y(2, 2) = 401$$

$$\frac{dy}{dx_1} = -400(x_1 x_2 - x_1^3) - 2(1 - x_1)$$

$$\frac{dy}{dx_2} = 200(x_2 - x_1^2)$$

$$\frac{dy}{dx_1} \text{ at 2, 2 is} = m_1 = 1602$$

$$\frac{dy}{dx_2} \text{ at 2,2 is} = m_2 = -400$$

Contour tangent equation: $dy = m_1 dx_1 + m_2 dx_2 = 1602dx_1 - 400dx_2 = 0$

y (is constant (or not changing).

$$1602dx_1 = 400dx_2$$

$$\frac{dx_2}{dx_1} = \frac{1602}{400} = 4.005$$

Steepest ascent or descent line ⊥ to contour tangent line. This is done graphically.

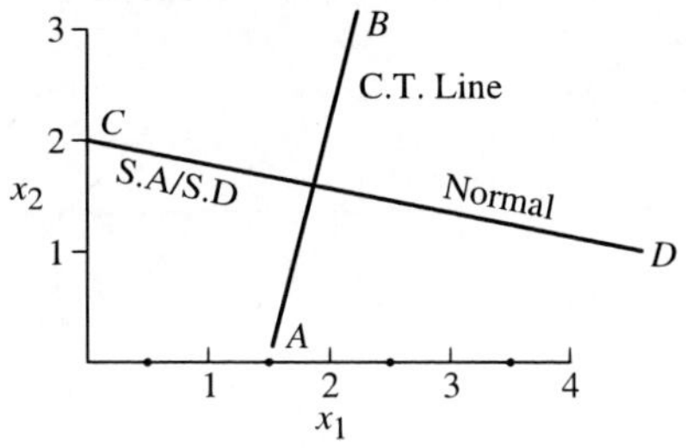

Fig. 6.24

AB is contour tangent line with slope 4.005. *CD* is normal line to *AB*. Now, we have to try on *CD* lines.

$$y\,(2, 2) = 401$$

$$y\,(1.0, 2.25) = 169 \text{ (improvement over 2, 2)}$$

$$y\,(1.5, 2.15) = 1.25 \text{ (improvement over 1.0, 2.25)}$$

$$y\,(0, 2.5) = 677 \text{ (no improvement)}$$

$$y\,(2.5, 1.85) = 1938.25 \text{ (no improvement)}$$

So, the points on steepest ascent or steepest descent line are tested.

Best is 1.5, 2.15. Now at 1.5, 2.15 another contour tangent line is drawn with a new slope. Then a perpendicular line for steepest ascent/descent. Final optimum (min. *y*) is 1.0, 1.0, is obtained.

6.14.4 Partan Search (Parallel Tangents)

Contour tangent equation is $dy = m_1 dx_1 + m_2 dx_2 = 0$

$$m_1 = \frac{dy}{dx_1}$$

$$m_2 = \frac{dy}{dx_2}$$

e.g.,

$$y = 7{x_1}^2 + 10{x_2}^2$$

$$x_0(10, 3);\ y(x_0) = 790$$

$$m_1 = \frac{dy}{dx_1} \text{ at } 10, 3 = 14x_1 = 140$$

$$\frac{dy}{dx_2} \text{ at } 10, 3 \text{ is } 20x_2 = 60 = m_2$$

$$140\,dx_1 + 60dx_2 = dy;\ dy = 0$$

$$\frac{dx_2}{dx_1} = \frac{-140}{60} = -2.33$$

Contour tangent equation is $y - 790 = 140(x_1 - 10) + 60(x_2 - 3)$

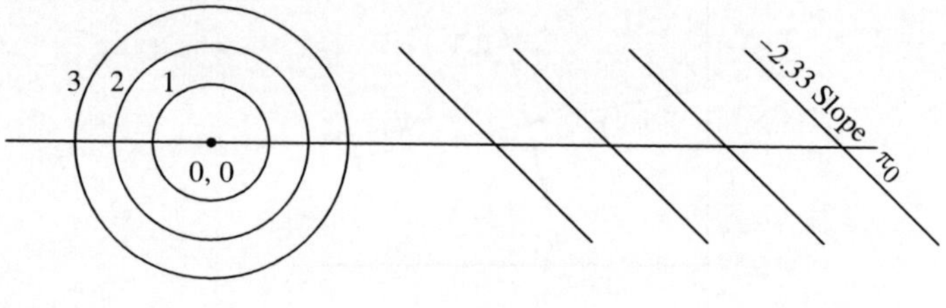

Fig. 6.25

The next point is selected at any place on the line parallel to contour Π_0.

6.14.5 Second Order Methods

Use second order derivatives.

Newton–Raphson method, for solving algebraic equation. $y(x) = 0$

$y(x)$ can be expanded as: $y(x^0) + (x - x^0)\ \dot{y}\,(x^0)$

$$y(x^i) + y\,(x^{i+1} - x^0)\ \dot{y}\,(x^i) = 0$$

$$x^{i+1} = x^i - \frac{y(x_i)}{\dot{y}(x_i)}$$

If $y(x)$ is replaced by $\dot{y}(x)$ then the equation takes the form:

$$x^{i+1} = x^i - \frac{\dot{y}(x_i)}{\ddot{y}(x_i)}$$

For several variables, $x^{i+1} = x^i - H^{-1}\frac{dy}{dx}(x)$ where H is given by

$$H = \begin{bmatrix} \frac{d^2y}{dx_1^2} & \frac{d^2y}{dx_1 dx_2} & \cdots & \frac{d^2y}{dx_1 dx_n} \\ \frac{d^2y}{dx_2 dx_1} & \frac{d^2y}{dx_2^2} & \cdots & \frac{d^2y}{dx_2 dx_n} \\ \vdots & \vdots & \cdots & \vdots \\ \frac{d^2y}{dx_n dx_1} & \frac{d^2y}{dx_n dx_2} & \cdots & \frac{d^2y}{dx_n^2} \end{bmatrix}$$

Example 6.34 Let

$$y = 3x^2 - 2x;$$

$$x(0) = 1$$

$$y(x_0) = 3 - 2 = 1$$

$$\dot{y}(x) = 6x - 2$$

$$\dot{y}x_0(x^0) = 6 - 2 = 4$$

$$x^1 = x^0 - \frac{y(x^0)}{\dot{y}(x^0)} = 1 - \frac{1}{4} = 0.75$$

$$x^2 = x^1 - \frac{y(x^1)}{\dot{y}(x^1)}$$

$$y(x^1) = \frac{3 \times 3^2}{4^2} - \frac{2 \times 3}{4} = \frac{3}{16}$$

$$\dot{y}(x^1) = \frac{6 \times 3}{4} - \frac{2 \times 3}{4} = \frac{4 \times 3}{4} = 3$$

$$x^2 = 0.75 - \frac{3}{16 \times 3} = \frac{11}{16} = 0.70 \text{ (decreasing)}$$

Analytically, $\dot{y} = 6x - 2 = 0$

$$x = \frac{2}{6} = 0.333$$

For y_{min} $3x^2 - 2x = y = 3(0.333^2) - 2 \times 0.333$

$$= -0.37 \textbf{ (Ans.)}$$

6.14.6 Convergence Method (Root Finding Methods)

The various methods are:

1. Newton method
2. Direct substitution
3. Wegstein method
4. Successive bisection method
5. Secant method.

6.14.6.1 Newton's Method

This is Newton–Raphson method, shown earlier as:

$$x_{n+1} = x_n - \frac{f(x_n)}{f'(x_n)}$$

6.14.6.2 Regula–Falsi method (Method of False Position)

Approximating curve by chord. The chord intersects at the point.

$$x = \frac{af(b) - bf(a)}{f(b) - f(a)}$$

This is an approximation for the root x_0 of equation $f(x) = 0$.

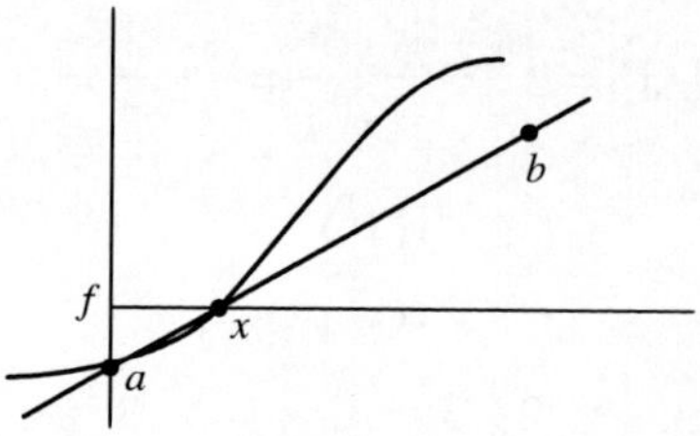

Fig. 6.26

Example 6.35 $$f(x) = x^3 + x - 1 = 0$$

Near $x = 1$; $$f(1) = 1; f(0.5) = -0.375$$

Choosing $a = 0.5$; $b = 1$;

By above equation for x:

$$x_1 = \frac{0.5 \times 1 - 1 \times -0.375}{1 - (-0.375)} = 0.64$$

$f(0.64) = -0.0979$ less than zero.

In the next step, choose $a = 0.64$; $b = 1$ and by equation of x, obtain accurate value of $x(x_2)$ as 0.672 and so on.

6.14.6.3 Iteration Method

$$f(x) = 0; x = g(x);$$

$$x_2 = g(x_1);$$

$$x_3 = g(x_2);$$

$$x_{n+1} = \frac{1}{\left(1 + x_n^2\right)}$$

e.g., $$f(x) = x^3 + x - 1 = 0$$

$$x(x^2 + 1) = 1;$$

$$x = \frac{1}{1 + x^2}; \; x_{n+1} = \frac{1}{(1 + x_n^2)}$$

Taking $x_1 = 1$; Right Hand Side (RHS) is

$$\frac{1}{(1 + 1)} = 0.50 = x_2$$

$$x_3 = \frac{1}{(1 + 0.5^2)} = 0.80$$

$$x_4 = \frac{1}{(1 + 0.8^2)} = 0.61$$

$$x_5 = \frac{1}{(1 + (0.61)^2)} = 0.6$$

Until x does not change

$$\text{LHS} = \text{RHS}$$

6.14.6.4 Bisection Method

Example 6.36 Find the real root of the equation

$$x^3 - 2x - 5 = 0$$

Let $f(x) = x^3 - 2x - 5 = 0$

$$f(2) = -1$$

$$f(3) = 16$$

Root lies between 2 and 3.

$$x_0 = \frac{(2 + 3)}{2} = 2.50$$

$$f(2.5) = 5.625$$

So choose 2 and 2.5.

n	a	b	$x = (a + b)/2$	$f(x)$
1	2	3	2.5	5.625
2	2	2.5	2.25	1.891
3	2	2.25	2.125	0.3457
4	2	2.125	2.0625	–0.3513
5	2.0625	2.125	2.09375	–0.0089
6	2.09375	2.125	2.10938	0.1668
7	2.09375	2.10938	2.10156	0.07856
8	2.09375	2.10156	2.09766	0.05471
9	⋮	⋮	⋮	⋮
10	⋮	⋮	⋮	⋮
11	2.09375	2.09473	2.09424	$-0.0035(\approx 0)$

$x = 2.09424$ **Ans.**

6.14.6.5 *Wegstein Method of Convergence*

Similar to projection method:

$$x_3 = \frac{x_1 \times x_{c2} - x_{c1} \times x_2}{x_1 - x_2 + x_{c2} - x_{c1}}$$

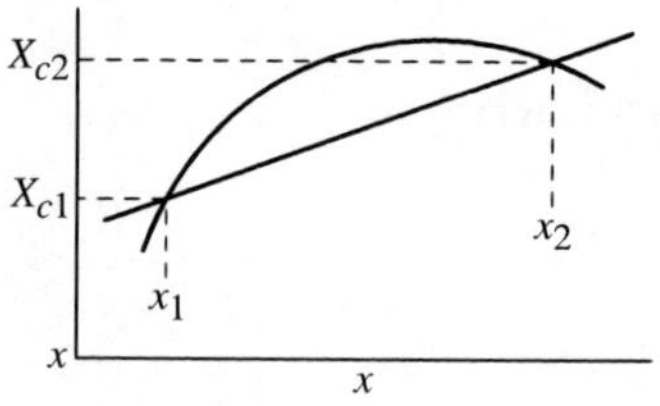

Fig. 6.27

$f(x_3)$ is calculated. Based on $f(x_3)$ is positive or negative, x_2, x_3 and x_{c2}, x_{c3} are obtained. Next calculate x_4. This procedure is repeated until $f(x_n)$ is nearly zero.

6.15 COMPARISON BETWEEN DIRECT AND GRADIENT METHODS, MERITS AND DEMERITS

While gradient methods are generally more effective than direct search methods, there are situations, for example, when the function gradient is not analytically defined in which case the direct search methods are more approximate. Those methods vary from the elementary univariant form to the sophisticated Powell and Resenbrock techniques. This univariant search method should rarely be used due to its tendency to stick on ridges and its slow progress especially when the ridges are not parallel to the coordinate axes.

The Rosenbrock approach overcomes all the disadvantages of the univariant method, and its coordinate rotation feature allows it to follow ridges closely which it has no satisfactory convergence criterion and is sometimes slow to get started, particularly when one of the steps is too large and has to be reduced several times. It is effective and has found widespread applications. The methods exhibiting quadratic convergence are said to be superior to others. Powell's development is more economical and possesses good convergence, in most applications. Each of the methods has some basic strengths and some weaknesses. A universal search with or without gradient knowledge has to be proposed.

Direct search is more efficient in terms of computation time. Gradient methods converge more rapidly than the direct search methods, provided the number of experiments required to establish the gradient at a point is not excessive. Direct methods are not feasible for constraint problems. It can be said that variable metric method is a powerful technique for unconstrained multivariable function optimization. This method is superior to direct search method of Powell. The method of conjugate is more economical, in which merely location of optimization is desired.

6.16 TRANSPORTATION PROBLEM

(Reference: Operations Research, S.D. Sharma, Kedar Nath, Ram Nath and Co., Meerut, 2007.)

Example 6.37 Find the initial basic feasible solution of the following transportation problem:

Warehouse/ Factory	w_1	w_2	w_3	w_4	Factory capacity
F_1	19	30	50	10	7
F_2	70	30	40	60	9
F_3	40	8	70	20	18
Warehouse requirement	5	8	7	14	14

Solution: North-west corner method or steeping stone method.

1. First Method: Construct an empty 3 × 4 matrix table with complete row and column requirements. See the following Table.

	w_1	w_2	w_3	w_4	Available
F_1					7
F_2					9
F_3					18
Requirements	5	8	7	14	

Insert a set of allocations in the cells in such a way that the total in each row and column is the same as shown against the respective rows and columns. Starting with the cell (1, 1) at north-west corner (upper left hand corner) allocate as much as possible in 1, 1 element.

Maximum we can put in 1, 1 is 5 (shown below in Table). So $x_{1,1} = 5$.

This allocation of 5 in $x_{1,1}$ leaves the surplus of 2(7 – 5) for row 1.

So allocating 2 for $x_{1,2}$. Now allocation of first row and first column is complete. Now there is a deficiency of 6 units in column 2. These 6 are put in 2, 2($x_{2,\,2}$).

	w_1	w_2	w_3	w_4	Available
F_1	5(19)	2(30)			7
F_2		6(50)	3(40)		9
F_3			4(70)	14(20)	18
	5	8	7	14	

Column 1 and column 2 requirements are met.

Now left amount for row 2 is (9 – 6) is 3. So, allocate $x_{2,3}$ as 3. Now column 3 requires another $7 - 3 = 4$ units. Now keeping these 4 units in $x_{3,3}$. Now there is a deficiency of $(18 - 4 = 14)$ in row 3. Now keeping 14 units in $x_{3,4}$ completes row 3. Column 4 has 14 in $x_{3,4}$ in south-east corner. Now the cost is $5 \times 19 + 2 \times 30 + 6 \times 50 + 3 \times 40 + 4 \times 70 + 14 \times 20 =$ ₹ 1015.

Second Method: Row minima method; Third method: Column minima method; Fourth method: Lowest cost entry method and Fifth method: Vogel's approximation method.

6.16.1 Row Minimum Method

Table 1

	w_1	w_2	w_3	w_4
F_1	19	30	50	10
F_2	70	30	40	60
F_3	40	8	70	20

Step 1: In the first row, minimum of 19, 30, 50, 10 is 10. The first allocation is made in the cell (1, 4), the amount of allocation is given by x_{14} = minimum of (7, 14) = 7. This exhausts the availability for factory F_1 and thus we cross-out the first row from transportation Table (Table 2).

Table 2

	w_1	w_2	w_3	w_4	
F_1	19	30	50	10	X
F_2	70	30	40	60	9
F_3	40	80	70	20	18
	5	8	7	7	

Step 2: In the second row minimum of 70, 30, 40, 60 is 30. The second allocation is made in cell 2, 2, the amount of allocation is $x_{2,2}$ = minimum of (9, 8) = 8. This satisfies the requirement of warehouse w_2 and thus we cross out the second column from transportation Table, obtaining (Table 3).

Table 3

	w_1	w_2	w_3	w_4	
F_1					X
F_2	70	30	40	60	9
F_3	40		70	20	18
	5	8	7	7	

Step 3: In Table 3, minimum of 70, 40, 60, is 40. The third allocation is made in cell 2, 3, the amount being $x_{2,\,2}$ = minimum of (1, 7) = 1. This exhausts the availability from factory F_1 and thus crossing out the second row from Table 4 and Table 5.

Table 4

	w_1	w_2	w_3	w_4	
F_1					X
F_2	20	30	40	60	X
F_3	40		70	20	18
	5	X	6	7	

Table 5

	w_1	w_2	w_3	w_4	
F_1					X
F_2	70	30	40	60	X
F_3	40		70	20	18

Step 4: The next allocation is made in the cell (3, 4) since minimum of 40, 70, 20 is 20. The amount of allocation being $x_{3,4}$ = minimum of (7, 18) = 7. This exhausts the requirement of warehouse w_4 and thus we cross out the fourth column from Table 6.

Step 5: The next allocation is made in cell 3, 1. Since minimum is 40, the amount of allocation being $x_{3,\,1}$ = min (5, 11) = 5. This seems the requirement of warehouse w_1 and so we cross out the first column, to the new Table (Table 7).

Table 6

	w_1	w_2	w_3	w_4	
F_1					X
F_2					X
F_3			70		11
	5	X	6	X	

Table 7

	w_1	w_2	w_3	w_4	
F_1				7	X
F_2		8	1		X
F_3	5		6 (70)	7	X
	X	X	X	X	

Step 6: The last allocation of amount $x_{3,3} = 6$ is obviously made in the cell (3, 3). This exhausts the availability of factory F_3 and requirement of warehouse w_3 simultaneously. So, we cross out third row and third column to get final table (Table 8). The solution is in Table 8.

Table 8

	w_1	w_2	w_3	w_4	
F_1				7	X
F_2		8	1		X
F_3	5		6	7	X

The transportation cost is given by

$$Z = 7 \times 10 + 8 \times 30 + 1 \times 40 + 5 \times 40 + 6 \times 70 + 7 \times 20$$

$$= ₹\ 1110 \textbf{ Ans.}$$

ASSIGNMENT PROBLEM

Example 6.38 An H.O.D. has four assistants and four tasks are to be performed. Time each assistant taken to perform each task is given in the effectiveness matrix. How the tasks should be allocated to each person so as to minimize the total man hours?

Table 1

		I	II	III	1V
T	*A*	8	26	17	11
A	*B*	13	28	4	26
S	*C*	18	19	18	15
K *S*	*D*	19	26	24	10

Solution: **Step 1:** Subtracting the smallest element from each element in that row, we get the reduced matrix, Table 2.

Table 2

	I	II	III	1V
A	0	18	9	3
B	9	24	0	22
C	23	4	3	0
D	9	16	14	0

Step 2: Next subtract the smallest element in each column from every element of that column, get the second reduced matrix (Table 3).

Table 3

	I	II	III	IV
A	0	14	9	3
B	9	20	0	22
C	23	0	3	0
D	9	12	14	0

Step 3: Test whether it is possible to make an assignment using only zeroes. Zero assignment is possible in Table 3 as follows.

(a) Starting with row I of matrix (Table 4), examine rows one by one until a row containing exactly a single zero element is found. Then an experimental assignment, indicated by ☐ is marked to that cell. Now cross all other zeroes in the column in which the assignment has been already made. This eliminates the possibility of making further assignments in that column. The procedure is in Table 4.

Table 4

	I	II	III	IV
A	[0]	14	9	3
B	9	20	[0]	22
C	13	[0]	3	0
D	9	12	14	[0]

(b) When a set of rows has been completely examined, an identical procedure is applied successively to columns. Starting with column 1, examine all columns until a column containing exactly one zero is found. Then make an experimental assignment in that position and cross other zeros in the row in when the minimum total man hours as follows.

Optimal	A–I	B-III	C–II	D–I
Assignment main-hours	8	4	19	10

= 41 Hrs. **Ans.**

(Reference: Operations Research: Theory, Methods and Applications (2010) S.D. Sharma; Kedar Nath Ram Nath & Co., Meerut, U.P.)

TRAVELLING SALESMAN

Example 6.39 Given, $C_{11} = \infty$; $C_{12} = 20$; $C_{13} = 4$; $C_{23} = 5$; $C_{34} = 6$; $C_{25} = 10$; $C_{35} = 6$; $C_{45} = 20$; where $C_{ij} = C_{ji}$; and there is no route between cities i and j if the value for C_{ij} is not shown.

Solution: **Step 1:** First make table as shown below (Table 1).

Table 1

	1	2	3	4	5
1	∞	20	4	10	∞
2	20	∞	5	∞	10
3	4	5	∞	6	6
4	10	∞	6	∞	20
5	∞	10	6	20	∞

Step 2: Subtract the smallest element in each row from each element of that row. Then subtract the smallest element in each column from each element of the column (Tables 2 and 3).

Table 2 (Rows)

	1	2	3	4	5
1	∞	16	0	6	∞
2	15	∞	0	∞	5
3	0	1	∞	2	2
4	4	∞	2	∞	16
5	∞	4	0	14	∞

Table 3 (Columns)

	1	2	3	4	5	
1	∞	15	[0]	4	∞	
2	15	∞	~~0~~	∞	3	
3	[0]	~~0~~	∞	~~0~~	~~0~~	L_2
4	4	∞	~~0~~	∞	12	
5	∞	3	~~0~~	12	∞	

Step 3: Mark zero assignment (shown by ☐ and cross out other zeroes in that row or column.

Step 4: 1. Draw minimum number of lines (L_1 and L_2) to cover all zeroes (Table 2).

2. Pick up minimum element among all uncovered elements (here it is 3).

3. Subtract 3 from all uncovered elements and add at the intersection of two lines (Table 4).

Table 4

	1	2	3	4	5	
1	∞	12	0	1	∞	
2	12	∞	∞	∞	9	
3	0	$\not{0}$	0	0	0	L_1
4	1	∞	0	∞	9	
5	∞	0	0	9	∞	
		L_4	L_1		L_3	

Step 5: Again make zero assignments in Table 4. Repeat step 4.

Step 6: Minimum number of lines to cover all zeroes is four here, and minimum element among all uncovered elements is 1. So subtract 1 from all uncovered elements and add 1 at the intersection of the lines. Thus, Table 5 is obtained.

Table 5

	1	2	3	4	5
1	∞	12	$\boxed{0}$	⊠	∞
2	11	∞	⊠	∞	$\boxed{0}$
3	$\boxed{X}$	$\boxed{X}$	∞	$\boxed{0}$	1
4	$\boxed{0}$	∞	$\boxed{X}$	∞	0
5	∞	$\boxed{X}$	$\boxed{X}$	8	∞

Step 7: Make zero assignments in Table 5. Here the solution of the assignment problem is (1, 3); (3, 4); (4, 1); (5, 2); (2, 5). This is not the solution of the travelling salesman problem as the frequency obtained is not in the cyclic order.

Step 8: It is seen that the next lowest number other than zero is 1. Therefore, instead of making zero assignment in the cell (5, 2) make assignment in the cell (3, 2) having element 1. Consequently, make assignment in cell (5, 4) having element 8, instead of zero assignment in the cell (3, 2). Thus, following assignment table is obtained having shortest route for the travelling salesman, i.e., $1 \to 3 \to 2 \to 5 \to 4 \to 1$. Shown in Table 6.

Table 6

	1	2	3	4	5
1	∞	12	[0]	#	∞
2	11	∞	[#]	∞	[0]
3	[#]	[1]	∞	#	1
4	[0]	*m*	#	∞	8
5	∞	[#]	#	[0]	∞

EXERCISES

1. Determine the maximum and minimum values of the following function and saddle point if any,

$$f(x) = 12x^5 - 45x^4 + 40x^3 + 5$$

2. (a) Explain the principle of geometric programming.

   (b) The cost of production of a plant is given by

$$y = \left[1000x_1\right] + \left[\frac{4\times 10^9}{(x_1 x_2)}\right] + \left[2.5\times 10^5 x_2\right]$$

   Using geometric programming, find the values of y minimum, x_1^* and x_2^*.

3. Calculate for an agitated vessel, the optimum speed.

N (RPM)	50	100	150	200	250
h	100	174	240	300	360
U	67	93	109	119	128
Annual power cost	20	160	540	1280	2500

   Cost of vessel plus agitator is ₹ 12,000 and annual fixed costs are 25% of capital costs.

   (**Ans.** $N = 150$)

4. For a certain heat exchanger, annual power costs = 175 dP and annual fixed cost is 2000 $dP^{-0.3}$. Calculate the ratio of annual power cost to fixed cost at optimum pressure drop. (**Ans.** 0.31)

5. A fluid is pumped through a heater and from there through a long pipeline. Costs are shown below. Calculate the optimal value of diameter D and temperature T.

$$\text{Annual cost of pumping} = \frac{1200n}{D^2} \text{ ₹/year}$$

$$n \text{ is the viscosity} = \frac{10{,}000}{(T + 20)}$$

$$\text{Pipe cost} = 10{,}000\ D^{1.2} \text{ ₹}$$

$$\text{Pumping cost} = \text{₹ } \frac{25n}{D^2}$$

$$\text{Annual cost of heat} = 150\ T \text{ ₹/year}$$

Annual fixed costs are 20% of the first cost.

$$\text{Heater cost} = \text{₹ } 500\ T$$

$$(\textit{Hint:} \text{ Total cost} = C + 10{,}000\ D^{1.2} + \frac{25n}{D^2} + 500T\ )$$

(F) Annual fixed costs = 0.20 × Total cost.

$$\text{Annual variable cost} = V = \frac{1200}{D^2}\left(\frac{1000}{T+20}\right) + 150T$$

$$\text{Total annual costs} = V + F$$

Then differentiate for optimum values.

6. Calculate the maximum profit of a company, manufacturing two machines x and y by using graphical method. The objective function is given by: $f(x, y) = 50x + 100y$ and the constraints are

$$10x + 5y \leq 2500$$

$$4x + 10y \leq 2000$$

$$x + 1.5y \leq 450$$

$$x, y \geq 0$$

7. An open cylindrical vessel is to be constructed to transport 80 cu.m of grain from a warehouse to a factory. The sheet metal used for the bottom and sides cost: ₹ 300 and ₹ 400 per sq. metre, respectively. If it costs ₹ 5 for each round trip of the box, find the dimensions of the vessel for minimizing transportation cost. Assume that the vessel has no loss upon completion of the operation.

8. A butane recovery plant recovers butane from natural gas by absorption in absorber oil. The recovered butane has a market value of ₹ 3/litre. Cost of processing butane after absorption is: ₹ 1 per litre. After operating the plant for a time, it is found that the plant operating data can be correlated by equation. $A = 0.004 \times R^{1.05}$, where A = absorption oil rate in million litres/hr. R = recovery of butane in litres/hr. Calculate the optimum butane recovery and oil circulation rate?

9. Lourisch gives this equation for calculating the heat transfer coefficient in a heat exchanger,

$$h_S = q_S\left(\frac{16}{L}dP\right)^{0.3}$$

The term q_S is the heat transfer quality number. Values for this quantity is/are given in a table by the author.

$$\frac{1}{U} = \frac{1}{h} + C$$

$$A = \frac{q}{U}dt$$

Capital cost AS; Annual fixed charges are 20 per cent of capital cost.

Power required per year = $VH\ dp/69{,}000$

Calculate the optimum pressure drop for the following:

q = 100,000; V = 282; q_S = 171; dt = 1455; L = 12; S = 800; J = 100; N = 6000; q = heat/hr.; V = volume/hr.; dt = temperature difference: L = length of heat exchanger; S = cost per unit area of heat exchanger material: J = cost of power, kW/hr.; H = number of houres/year; dp = pressure and A = area of heat exchanger. (**Ans.** dp = 4.6)

10. The kinds of crude are available for purchase of an oil company.

100,000 barrels/weak for crude 1

200,000 barrels /weak for crude 2

Crude 2 can be processed in two ways: to make fuel oil to make lube oil. The yields, demands and profits are shown in Table below.

Product yields, fraction	Crude-1	Crude-2		Demand/week
Gasoline	0.5	0.5	0.4	110,000
Fuel oil	0.4	0.3	0.2	80,000
Lube oil	0.1	0.1	0.0	40,000
Profit, ₹ per thousand	1000	150	2500	

Calculate the amounts of each crude to be used for maximum profit, without exceeding the demand of products.

x_1 = barrels of crude-1

x_2 = barrels of crude-2; used to produce fuel oil.

x_3 = barrels of crude-2 used to produce lube oil.

❑❑❑

A

Appendix

Conversion of Units and Constants (SI)

Length	Meter	Mass	Kgs.
1 foot	0.3048	1 lb	0.4535
1 inch	0.0254	1 grain	6.48×10^{-5}
1 mile	1609	**Velocity**	**m/sec**
1 Å	10^{-10}	1 ft/hour	8.467×10^{-5}
1 micron	10^{-6}	1 ft/sec	0.3048

Pressure	**Bar**
1 N/m^2	10^{-5}
1 atm.	1.0133 = 101,325 pascals
1 kg/cm^2	0.9807
1 mm Hg	1.333×10^{-3}
1 m H_2O	0.0981
1 Psi	0.069
1 inch Hg	0.0339
1 bar = 10^5 pascals.	
Force	**N**
1 dyne	10^{-5}
1 kg	9.8067

1 lb	4.4482
Energy	**kJ**
1 NM	10^{-3}
1 kcal	4.1868
1 kWh	3600
1 kgf. m	9.81×10^{-2}
1 BTU	1.0551
1 ft. lb	1.356×10^{-3}
1 erg.	10^{-10}
1 therm	1.055×10^{5}
1 hp–hr.	2.648×10^{3}
Area	**m^2**
1 ft^2	0.0929
1 $inch^2$	6.452×10^{-4}
Volume	**m^3**
1 ft^3	2.8317×10^{-2}
1 $inch^3$	1.6387×10^{-5}
Volume flow rate	**m^3/sec.**
1 ft^3/hr.	7.8658×10^{-6}
1 U.S. gal/hr.	1.0515×10^{-6}
1 barrel/day	1.84×10^{-6}
Density	**kg/m^3**
1 lb/ft^3	16.018
1 gm/c.c.	1000
Viscosity	**NS/m^3**
1 C.P.	$10^{-3} = 10^{-3}$ pascals. sec. (PaS)
1 kg/m hr.	0.2778
1 lb/ft. hr.	4.134×10^{-7}
1 lb/ft. sec.	1.4882

1 stoke (1 cm^2/sec.)	$10^{-4}m^2$/sec.
1 m^2/hr.	2.778×10^{-4} m^2/sec.
1 ft^2/hr.	2.581×10^{-5} m^2/sec.
Surface tension	**N/m**
1 dyne/cm	10^{-3}
1 kg f/m	9.814
1 lb/ft.	1.46
Enthalpy	**kJ/kg.**
1 kcal/kg.	4.1868
1 BTU/lb	2.326
Specific heat	**kJ/(kg.K)**
1 kcal/(kg°C)	4.1868
1 BTU/(lb. °F)	4.1868
Power	**kJ/sec.**
1 kW	1
1 hp	0.7355
1 BTU/h.	0.293×10^{-3}
1 kcal/h.	1.163×10^{-3}
1 erg./sec.	10^{-10}
Thermal conductivity	**W/(m.K)**
1 kcal/(hr. m. °C)	1.163
1 BTU/(h. ft. °F)	1.7308
Coefficient of expansion	**kg/(m^3. K)**
1 lb/(ft^3. °C)	16.018
1 lb/(ft^3. °F)	28.833
Heat transfer coefficient	**W/(m^2. K)**
1 kcal/(h. m^2. °C)	1.163
1 BTU/(h. ft^2. °F)	5.6784
Heat flux	**kW/m^2**

1 kcal/hr. m^2	1.163×10^{-3}
1 BTU/h. ft^2	3.153×10^{-3}
Mass flux	**kg/(m^2. sec)**
1 lb/hr. ft^2	1.356×10^{-3}

Mass transfer coefficient

1 lb/hr.ft^2 = 4.8823 kg/h. m^2

1 kg/hr. m^2. atm. = 0.98687 kg/hr. m^2. bar

1 kg/S.N. = 3.6×10^8 kg/hr. m^2. bar

1 lb/hr. ft^2. atm. = 4.8182 kg/hr. m^2. bar

PHYSICAL CONSTANTS

Gas constant, R = 8.314 kJ/kg. mol. K

= 1.987 kcals/kg. mol. K

= 0.08206 m^3 atm./(kg mole, K)

= 1.545 ft-lb/(lb. mole °R)

Avogadro's number, N = 6.025×10^{23} atoms/gm. mole

Standard gas conditions: At NTP (273 K, 1.033 bar = 1 atm)

= 22.42 m^3/kg. mole = 359.01 ft^3/lb. mole

Stefan Boltzmann constant = 1.381×10^{-23} J/K = 5.65×10^{-24} ft. lb f/R.

Planck's constant = 6.625×10^{-34} JS

Gravitational constant = 9.81 kgm/kgf. S^2

= 32.17 ft. lb/lb sec^2

Viscosity of water = 0.001 kg/(m.S)

Viscosity of air = 1.8×10^{-5} kg/(m.S)

Accelaration due to gravity (g) = 9.81 m/S^2

Atmospheric Composition

Volume % – Nitrogen = 78.05, Oxygen = 21.0, Argon = 0.93, Carbon dioxide = 0.02 – 0.03

Weight % – Nitrogen = 75.5, Oxygen = 23.2, Argon = 1.25, Carbon dioxide = 0.50 – 0.40.

References

1. J.O. Osburn and Karl Kammermeyer. *Money and the Chemical Engineer*, Prentice Hall Inc. New Jersey, 1958.
2. Asghar Hussain and Kota Gangaiah. *Optimization Techniques for Chemical Engineers*, ELBS.
3. Robert S. Aries and Robert D. Newton. *Chemcial Engineering Cost Estimation*, McGraw Hill Co., New York, 1955.
4. Chaplin Tyler and C.H. Wrinter, Jr. *Chemcial Engineering Economics*, McGraw Hill Co., New York, 1959.
5. Beveridge, G.S.G. and Schechter, R.S. *Optimization Theory and Practice*, McGraw Hill Co., New York, 1970.
6. Wilde, D.J. *Optimum Seeking Methods*, Prentice Hall Inc., Englewood Cliffs, New Jersey, 1964.
7. Harold S. Mickley. *Applied Mathematics in Chemical Engineering*, McGraw Hill Co., New York, 1957.
8. V.G. Jenson and G.V. Jeffreys. *Mathematical Methods in Chemical Engineering*, Academic Press, New York, 1963.
9. Grewal, B.S. *Higher Engineering Mathematics*, Khanna Publishers, New Delhi, 2001.
10. Statistical Methods books.
11. Heat Transfer books (For P.D.Es)
12. Chemical Engineering Journals (for Applications).
13. Numerical Methods books.
14. Rao, S.S. *Engineering Optimization: Theory and Applications*, New Age International, New Delhi, 1996.

❑❑❑

Index

□□□